MOLECULAR BIOLOGY
INTELLIGENCE
UNIT

Branching Morphogenesis

Jamie A. Davies, M.A.(Cantab.), Ph.D.

Centre for Integrative Physiology
Edinburgh University College of Medicine
Edinburgh, Scotland, U.K.

LANDES BIOSCIENCE / EUREKAH.COM
GEORGETOWN, TEXAS
U.S.A.

SPRINGER SCIENCE+BUSINESS MEDIA
NEW YORK, NEW YORK
U.S.A.

BRANCHING MORPHOGENESIS

Molecular Biology Intelligence Unit

Landes Bioscience / Eurekah.com
Springer Science+Business Media, Inc.

ISBN: 0-387-25615-6   Printed on acid-free paper.

Springer Science+Business Media, Inc., 233 Spring Street, New York, New York 10013, U.S.A.
http://www.springeronline.com

Please address all inquiries to the Publishers:
Landes Bioscience / Eurekah.com, 810 South Church Street, Georgetown, Texas 78626, U.S.A.
Phone: 512/ 863 7762; FAX: 512/ 863 0081
http://www.eurekah.com
http://www.landesbioscience.com

Printed in the United States of America.

9 8 7 6 5 4 3 2 1

Library of Congress Cataloging-in-Publication Data

Branching morphogenesis / [edited by] Jamie A. Davies.
     p. ; cm. -- (Molecular biology intelligence unit)
  Includes bibliographical references and index.
  ISBN 0-387-25615-6 (alk. paper)
  1. Morphogenesis. 2. Branching processes. I. Davies, Jamie A. II. Series: Molecular biology intelligence unit (Unnumbered)
  [DNLM: 1. Morphogenesis. 2. Developmental Biology. 3. Physiology, Comparative. 4. Tissue Engineering. QH 491 B816 2005]
  QH491.B72 2005
  571'.833--dc22
                              2005018283

To all members of our laboratories,
past and present, in gratitude for their inspiration,
their dedication, and their good company.

CONTENTS

EDITOR

Jamie A. Davies
Centre for Integrative Physiology
Edinburgh University College of Medicine
Edinburgh, Scotland, U.K.
Chapters 1, 8, Afterword

CONTRIBUTORS

Mina J. Bissell
London Regional Cancer Center
Departments of Biochemistry/Oncology
London, Ontario, Canada
Chapter 7

Daniel Bouyer
Botanical Institute III
University of Köln
Köln, Germany
Chapter 3

Karen D. Carpenter
Center for Animal Biotechnology
  and Genomics
Department of Animal Science
Texas A&M University
College Station, Texas, U.S.A.
Chapter 11

Roger W. Davenport
Department of Biology
University of Maryland College Park
College Park, Maryland, U.S.A.
Chapter 2

Marcus Dejmek
Department of Pediatric Surgery
University of Southern California
Los Angeles, California, U.S.A.
Chapter 12

Vincent Fleury
Laboratoire de Physique de la Matière
  Condesée Ecole Polytechnique
Palaiseau Cedex, France
Chapter 12

Kanako Hayashi
Center for Animal Biotechnology
  and Genomics
Department of Animal Science
Texas A&M University
College Station, Texas, U.S.A.
Chapter 11

Jianbo Hu
Center for Animal Biotechnology
  and Genomics
Department of Animal Science
Texas A&M University
College Station, Texas, U.S.A.
Chapter 11

Martin Hülskamp
Botanical Institute III
University of Köln
Köln, Germany
Chapter 3

Tina Jaskoll
Laboratory for Developmental Genetics
University of Southern California
Los Angeles, California, U.S.A.
Chapter 9

Katherine M. Kollins
Department of Biology
University of Maryland College Park
College Park, Maryland, U.S.A.
Chapter 2

Igor A. Kosevich
Department Invertebrate Zoology
Biological Faculty
Moscow State University
Moscow, Russia
Chapter 5

Anke Lindner
Laboratoire des Milieux Désordonnés
et Hétérogènes
Université Paris 6
Paris Cedex, France
Chapter 12

P.C. Marker
Department of Genetics, Cell Biology,
and Development
University of Minnesota Cancer Center
Minneapolis, Minnesota, U.S.A.
Chapter 10

Liam J. McNulty
School of Biological Sciences
The University of Manchester
Manchester, U.K.
Chapter 4

Michael Melnick
Laboratory for Developmental Genetics
University of Southern California
Los Angeles, California, U.S.A.
Chapter 9

Audrius Meskauskas
Gediminas Technical University
Saulétekio al. 11
Vilnius, Lithuania
Chapter 4

David Moore
School of Biological Sciences
The University of Manchester
Manchester, U.K.
Chapter 4

Minh Binh Nguyen
Department of Pediatric Surgery
University of Southern California
Los Angeles, California, U.S.A.
Chapter 12

Thi-Hanh Nguyen
Laboratoire de Physique de la Matière
Condesée Ecole Polytechnique
Palaiseau Cedex, France
Chapter 12

Sybill Patan
Department of Anatomy
and Cell Biology
SUNY Downstate Medical Center
Brooklyn, New York, U.S.A.
Chapter 6

Laurent Schwartz
Laboratoire LPICM
Ecole Polytechnique
Palaiseau Cedex, France
Chapter 12

Thomas E. Spencer
Center for Animal Biotechnology
and Genomics
Department of Animal Science
Texas A&M University
College Station, Texas, U.S.A.
Chapter 11

A.A. Thomson
MRC Human Reproductive
Sciences Unit
Centre for Reproductive Biology
University of Edinburgh
Edinburgh, Scotland, U.K.
Chapter 10

Eva A. Turley
Life Sciences Division
Lawrence Berkeley National Laboratory
Berkeley, California, U.S.A.
Chapter 7

Mathieu Unbekandt
Laboratoire de Physique de la Matière
  Condesée Ecole Polytechnique
Palaiseau Cedex, France
Chapter 12

David Warburton
Department of Pediatric Surgery
University of Southern California
Los Angeles, California, U.S.A.
Chapter 12

Tomoko Watanabe
Laboratoire de Physique de la Matière
  Condesée Ecole Polytechnique
Palaiseau Cedex, France
Chapter 12

CHAPTER 1

Why a Book on Branching, and Why Now?

Jamie A. Davies

In a world overloaded with information, in which university library shelves bend under the weight of worthy tomes and the number of journals has been doubling every fifteen years,[1,2] a prospective reader is fully entitled to eye any new text suspiciously and to ask whether there is really a need for yet another book. The question is always a valid one, and was perhaps summed up most clearly by the reviewer who remarked of a manuscript under his scrutiny that *'this paper fills a much needed gap in the literature'*.* It is therefore a duty of any author or editor to begin the introduction of a new work with a justification for its existence.

For this particular volume, providing such a justification is easy. The subject matter is of critical importance to our understanding of the normal development of animals and plants and is a necessary component in the emerging technology of tissue engineering. Study of branching is changing quickly and is expanding through new links between cell biology and mathematical modelling. Most critically of all, its subject material has traditionally been scattered through the texts and journals of many different disciplines and has not been brought together in one place before. The recent emergence of general principles behind branching morphogenesis, and the observation that apparently disparate systems seem to share deep biological similarities,[4] is a strong and timely reason for considering them together, now, in a single volume in which each chapter is contributed by a world expert in a particular field.

Branching Morphogenesis Is Important and Pervasive

The development of repeatedly-branched structures is an important mechanism of morphogenesis across a wide range of phyla and scales. In some organisms, such as trees, branching shapes the complete body plan and is their most obvious morphological attribute. Most plants and multicellular fungi share this property, although in the case of fungi the branched structures are very fine and, to the naked eye, are not as obvious as unbranched reproductive structures such as mushrooms. Some animals also have a branched body plan but, in most phyla, branching is hidden away in the internal anatomy and is not obvious from external form. We are examples of such creatures, having unbranched exteriors but having insides riddled with interlinked networks of branched endothelial and epithelial tubes.

Branching usually arises where there is a reason to maximise the total area of contact between a structure and the environment that surrounds it, particularly where there is also a reason to pack this area of contact in a small volume (that is, an organism gains some functional selective advantage by doing this). For plant shoot systems, the 'aim' is to maximise the area for light capture and gas exchange: while large areas could be produced by the growth of a single enormous leaf, mechanical constraints (gravity, wind damage etc) limit this strategy to

---

* A history of this acerbic phrase has been reviewed elsewhere,[3] but so cliché d has it become that some reviewers now seem to miss its precise meaning, and use it even in very positive reviews: type the phrase into a web search engine to find many examples.

Branching Morphogenesis, edited by Jamie A. Davies.
©2005 Eurekah.com and Springer Science+Business Media.

very small plants or those supported by, for example, floating on water; self-supporting large plants are forced to use branched structures. Similarly, plant root systems need to achieve large areas of contact between themselves and the soil, an important source of water and minerals, and the production of fine branches has the added advantage that they can penetrate between particles of soil and thus expand into an almost 'solid' environment.

In animal tissues, branching is normally used to pack a large surface area for exchange between the external environment and internal tissues, or between two internal 'compartments', into a small volume. The branched structures of mammalian lungs are an example and here, as in most other systems, the branching tubes are not themselves the main surfaces over which substance exchange takes place. Rather, gas exchange takes place in specialised air sacs, alveoli, that appear at the ends of the finest branches; the branching system itself is simply a means of connecting many alveoli to the outside world while minimising the total distance from each to the final exit from the body (an alternative design, connecting all of the alveoli to a single long tube, would suffer the disadvantage that the most distant alveoli would be able to exchange gases with the outside only very inefficiently). A range of other branched epithelia act as 'drains' for substances (saliva, urine, seminal fluid, tears, milk etc) produced in specialised terminal structures, sometimes also called alveoli.

The blood and lymphatic systems, based on endothelia, are specialized for exchange of substances between body compartments (the 'tissues' and 'the circulation', each of which really constitutes several different functional compartments in itself). The aim is to ensure that no part of the tissue is more than a short distance from a blood vessel, and arterial flows ramify ever more finely in tissues to achieve this. In the case of vertebrate blood systems, the finest branches then connect with fine branches of a venous system, collecting post-exchange blood and draining it to successively larger-bore vessels to return it to the heart (in some organs, such as kidney and gut, blood is collected from the arterial system by an intermediate branching system that takes it to a second set of capillaries before it returns to the venous system proper, but such complications are beyond the scope of this introductory chapter).

The means for physical substance exchange is not the only system that has to spread throughout the body; there is also the need for distribution of 'command and control' information. Some of this is achieved using exchange of signalling molecules (hormones) between tissues and the general circulation but much of it is achieved by connecting specific tissue elements via nerves. In simple body plans such as those of cnidaria (sea anemonies, jellyfish etc), this is achieved by a distributed nerve net. In more complex animals it is done by connection of tissues to a central information processing unit—a ganglion or a brain. Nerves, which are bundles of neuronal cell processes (axons), run from central nervous system out to the tissues where they divide increasingly finely and eventually single axons may branch to make connections with multiple targets, for example, muscle fibres. Within the central nervous system, highly branched systems of neuronal processes are used to collect and integrate signals from multiple input neurons. In a clear reference to their shape, these are called dendritic arbours (from Greek *dendros,* = branch, and Latin *arbor,* = tree: US English retains the Latin spelling).

Branching morphogenesis produces structures on scales ranging from micrometres (the processes of a single cell) through centimeters (branching epithelia in mammalian organs) to tens of metres (trees). Indeed, the current record holder for the world's largest organism is a woodland fungus, *Armillaria ostoyaei,* which is over 2000 years old and spans about 10 square kilometers of forest floor;[5] it is composed almost entirely of a huge network of branched hyphae.

Branching morphogenesis is therefore an important and pervasive mechanism of development.

Patterns of Branching

All branched biological structures are generated by variations on just a few general mechanisms. One very common mechanism is branching of an elongated structure, such as a plant stem, an epithelial tubule or a cell process, by division of its growing tip into two or more new

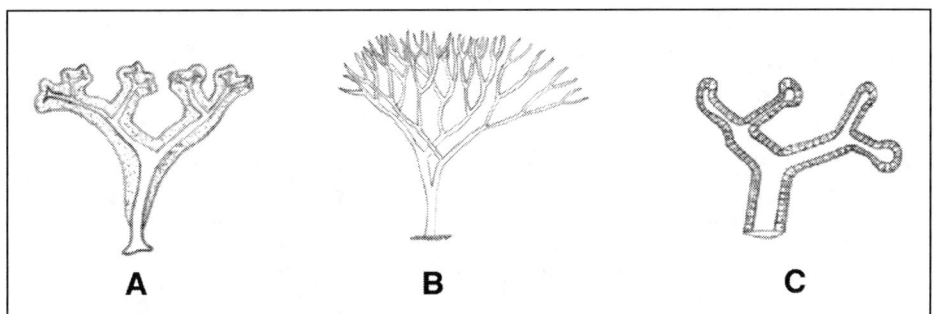

Figure 1. Examples of dipodial branching, in A) a green alga of the *Fucus* genus, B) a red alga of the *Polyides* genus and C) in the developing airway epithelium of a mouse embryo.

tips (Fig. 1). Variations on this theme certainly exist, both in terms of the numbers of tips formed and the method by which the tips divide, but the general process accounts for a great deal of branching morphogenesis over a huge range of scales. In its simplest form—dipodial branching—the daughters of each branching event are 'equal' and no one branch dominates the structure. A common variation is monopodial branching (Fig. 2), in which secondary branches form from one dominant stalk. This pattern is obvious in many trees, but is also found in animal tissues such as the mammary gland. In most examples of monopodial branching, the dominant stalk develops first and the side branches appear as later additions.

A very different mechanism for generating branched tubes is to divide up a large tube into many smaller ones by the introduction of longitudinal barriers (Fig. 3). This process—intussusceptive branching—creates a much larger surface area over which exchange can take place between the fluid in the tubes and their surroundings. For this reason, and also because it is well-adapted for tubes that form part of a circulation system rather than having closed ends,

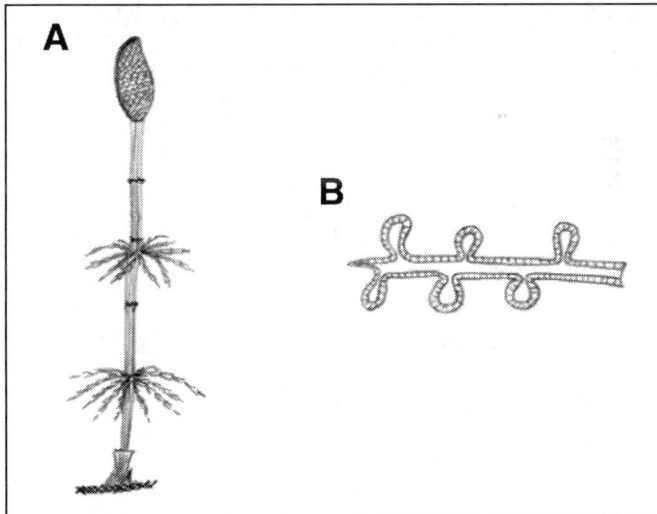

Figure 2. Examples of monopodial branching, in A) the horsetail *Equisetum* and B) mouse mammary epithelium budding alveoli during pregnancy.

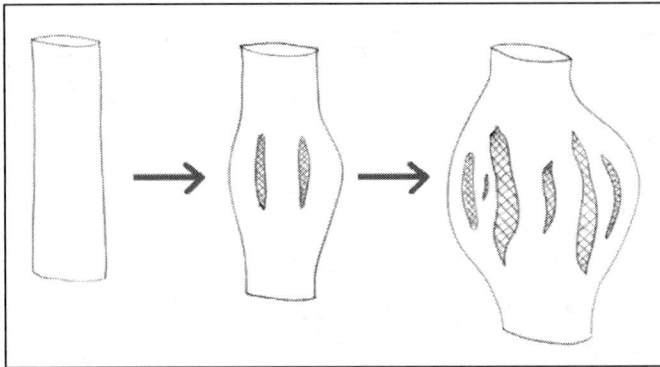

Figure 3. Intussusceptive branching, by the invasion of a vessel space by other tissue (following infolding of the vessel wall). The longitudinal dimension of this diagram has been compressed for clarity.

intussusceptive branching is common in the development of blood vessels. It is also how most river deltas form.

The above mechanisms fall squarely into the category of 'branching morphogenesis' because they operate by division of one thing into many. There are other ways of making branched structures which are not normally considered *bona fide* examples of branching morphogenesis but which are worth mentioning in this introductory chapter, if only because they are not considered elsewhere in this book. One is related to intussusceptive branching, and consists of producing a 'branched' gross structure by deletion of cell populations: an example is the 'branched' structure of chicken feet, which arises, in part from apoptosis of the cells that would otherwise form a continuous web between the toes[6] (Fig. 4). The other fairly common mechanism for creating a branched biological structure is the fusion of elements that originate separately and then converge. An example of branching by convergence is seen in the aggregation of myxamoebae of *Dictyostelium discoideum*, in which migratory cells form streams that join together and converge on one point[7] (Fig. 5). Another is seen in the mesonephros (temporary kidney) of mammalian embryos, in which tubules form independently but converge on to a common duct.[8]

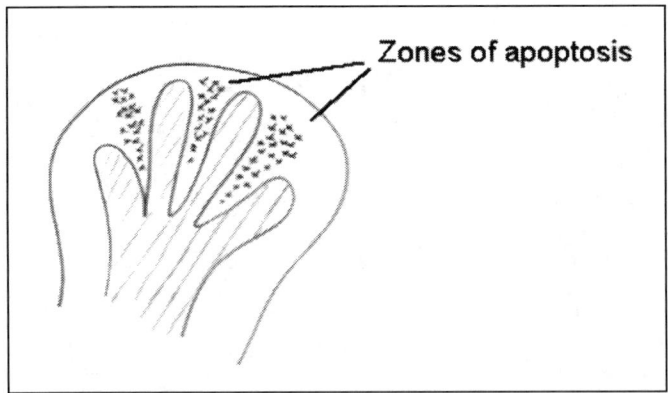

Figure 4. The role of apoptosis in separating the "branches" of the foot (the toes).

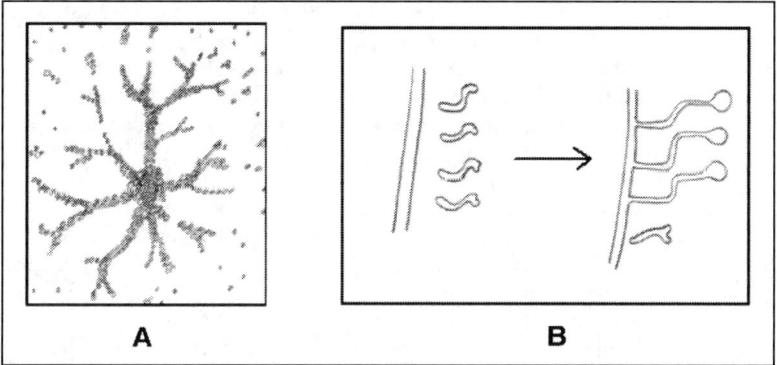

Figure 5. Formation of branched structures by convergence, in A) aggregating *Dictyostelium* myxamoebae, and B) mensonephric tubules joining a common nephric duct in mammalian embryogenesis.

Research into Branching

Branching morphogenesis can be studied at many different levels from molecular genetics to mathematical modelling. Some of the earliest work on branched systems was mathematical; Leornardo da Vinci studied the dimensions of trees that develop by dipodial branching, and showed that the ratio of the diameter of a branch of generation *n* to the diameter of one of generation *n+1* was constant for all *n* (Fig. 6). The constancy of this number, now called da Vinci's number,[9] implied that such structures are scale-free (one cannot deduce, from the ratio of branch sizes alone, whether the branches in question are the very largest or the very finest) and da Vinci's work on branching was one of the earliest examples of what would now be considered the mathematics of fractals. More recently, similar analyses have been performed on branched structures in animals, for example canine airway epithelium,[10] with the result that da Vinci's rule holds for a large range of *n*. It is clear, though, that the very first branching events

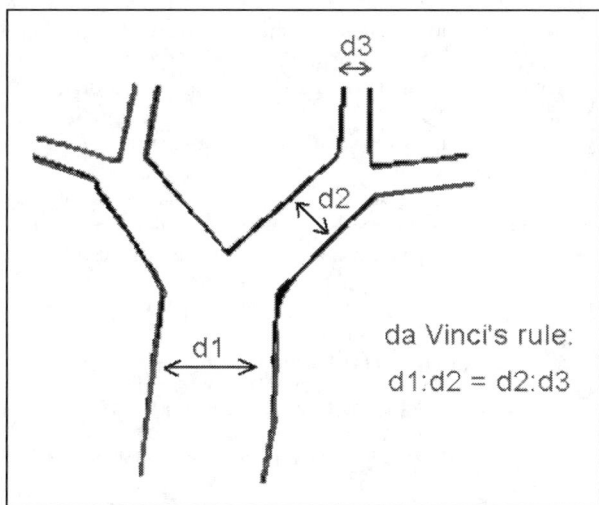

Figure 6. Leonardo da Vinci's rule.

and the very last depart from the rule in most systems, and fractal geometry is only a useful approximation for the middle stages of growth. Even then, additional systems must operate to set up the characteristic shape of each organ and to avoid collisions.[11]

Fractal studies have been joined by mathematical approaches based on the characteristics of purely physical models, in an attempt to understand the extent to which biological branching morphogenesis might rely on gross physical properties of their components, such as viscosity, pressure and mechanical stress, rather than on any especially 'biological' characteristics such as cytoskeletal remodelling. Some of the most interesting of these models have used 'viscous fingering'—the branching phenomenon that takes place when a liquid of low viscosity is forced into one of higher viscosity—to model epithelial branching in animals.[12] Other mathematical approaches involve a rule-based approach, in which 'rules' (that represent molecular systems such as those that connect a receptor to the changing transcription of a gene to changing cell behaviour) in a computer program are used to interpret simulated physicochemical parameters such as morphogenetic fields and which, over a broad range of parameters, reproduce biologically-plausible branching patterns.[13]

As mathematical study of branching has grown, so has study at the cell-biological level. Light-microscopic studies of the Victorian era indicated the basic arrangements of cells in tissues undergoing branching and identified key cellular components such as neuronal growth cones, which can control branching at a single-cell level.[14] In the last century, ultrastructural studies, made possible by the electron microscope, indicated that cells driving branching could show various specialisations such as altered extracellular matrix etc.[15] Over the last fifty years or so, these observations have been joined by biochemical analyses, by culture techniques and by experimental interventions that have allowed specific biochemical constituents to be correlated with particular aspects of morphogenesis.[16] Most recently, genetic manipulation has enabled experimenters to make exact and known changes to the genome and correlate these with both normal development and also with congenital disease.

Historically, most researchers into branching have made their strongest connections with others working on different aspects of their chosen organism or tissue, rather than with those studying branching in other systems. This pattern has begun to change, with the realization that the same families of molecules and the same patterns of cell behaviour seem to turn up in system after system. It is still not clear to what extent mechanisms of branching morphogenesis are truly conserved across organs and organisms,[17] but a number of conferences devoted to aspects of branching have shown the value of experimentalists immersing themselves in the biology of each other's systems and, most particularly, of an improved dialogue between biological data and mathematical modelling.

It is for this reason that a set of people involved in studying many different aspects of branching have come together to produce this book. It is impossible, in a volume of reasonable size, to cover everything and, recognizing this, we have tried to pick the topics in which understanding is growing at its fastest and which seem to relate naturally to each other. As Editor, I have very much enjoyed reading the contributions of all of the other authors of this book; I hope that you do too.

References

1. Pendlebury D. Science's go-go growth: Has it started to slow? The Scientist 1989; 3:14.
2. de Solla Price DJ. Little Science, Big Science. Columbia University Press, 1963.
3. Jackson A. Chinese acrobatics, an old-time brewery, and the "much needed gap": The life of mathematical reviews. Notices of the AMS 1997; 44:330-337.
4. Davies JA. Do different branching epithelia use a conserved developmental mechanism? Bioessays 2002; 24:937-948.
5. Ferguson BA, Dreisbach TA, Parks CG et al Coarse-scale population structure of pathogenic Armillaria species in a mixed-conifer forest in the blue mountains of northeast oregon. Can J For Res 2004; 33:612-623.

6. Merino R, Rodriguez-Leon J, Macias D et al. The BMP antagonist gremlin regulates outgrowth, chondrogenesis and programmed cell death in the developing limb. Development 1999; 126:5515-5522.
7. Palsson E. Othmer HG. A model for individual and collective cell movement in dictyostelium discoideum. Proc Natl Acad Sci USA 2000; 97:10448-10453.
8. Sainio K. Development of the mesonephric kidney. In: Vize PD, Woolf AS, Bard JBL, eds. The Kidney: From Normal Development to Congenital Disease. Academic Press, 2003:75-84.
9. Long CA. Leonardo da Vinci's rule and fractal complexity in dichotomous trees. J Theor Biol 2004; 167:107-113.
10. Nelson TR, West BJ, Goldberger AL. The fractal lung: Universal and species-related scaling patterns. Experientia 1990; 46:251-254.
11. Metzger RJ, Krasnow MA. Genetic control of branching morphogenesis. Science 1999; 284:1635-1639.
12. Fleury V, Watanabe T. Morphogenesis of fingers and branched organs: How collagen and fibroblasts break the symmetry of growing biological tissue. C R Biol 2002; 325: 571-583.
13. Moore D, McNulty LJ, Meskauskas A. Branching in fungal hyphae and fungal tissues: Growing mycelia in a desktop computer. In: Davies JA, ed. Branching Morphogenesis. Georgetown: Landes Biosciences, 2005: 70-85.
14. Ramon y, Cajal S. Sur l'origene et las ramifications des fibres nerveuses de la moelle embryonnaire. Anat Anz 1890; 5:111-119.
15. Williams JM, Daniel CW. Mammary ductal elongation: Differentiation of myoepithelium and basal lamina during branching morphogenesis. Dev Biol 1983; 97:274-290.
16. Grobstein C, Cohen J. Collagenase: Effect on the morphogenesis of embryonic salivary epithelium in vitro. Science 1965; 150:626-628.
17. Davies JA. Do different branching epithelia use a conserved developmental mechanism? Bioessays 2002; 24:937-948.

CHAPTER 2

Branching Morphogenesis in Vertebrate Neurons

Katherine M. Kollins and Roger W. Davenport

Abstract

Within the developing vertebrate nervous system, strict control of branching morphogenesis is essential for establishing appropriate circuitry, since the geometry of neuronal arbors critically influences their functional properties. Thus, identification of the specific molecules and mechanisms involved in regulating neuronal branching morphogenesis has been the focus of intense study within recent decades, producing tremendous advances in the understanding of neuronal differentiation. Intrinsic regulation of branching morphogenesis arises through a combination of background gene expression, structural constraints imposed by cellular dimensions, biophysical properties of intracellular cytoskeletal elements, and cell-autonomous control of arbor topology and branching probability. Epigenetic influences on the pattern of branching morphogenesis instead arise from temporally or spatially constrained microenvironmental cues including homotypic and heterotypic cell-cell interactions, substrate-bound and diffusible chemoattractants and chemorepellents, hormones and growth factors, and patterns and levels of electrical activity. Ultimately, such signaling must converge at the level of the cytoskeleton, with the structural changes characteristic of neuronal branching arising through dynamic regulation of the actin cytomatrix, microtubules, and a variety of microtubule-associated proteins. This review provides a comprehensive summary of the current understanding of branching morphogenesis in developing vertebrate neurons, emphasizing recent findings describing key cellular mechanisms and molecular signaling pathways underlying branch formation and stabilization.

Introduction to Branching Morphogenesis in Vertebrate Neurons

Since the pioneering neuroanatomical studies of Santiago Ramón y Cajal beginning in the late 1800s, morphology has emerged as one of the main criteria used for identifying and characterizing distinct populations of neurons.[1-13] As the vertebrate nervous system develops, individual neurons undergo significant morphological changes through a sequence of neurite outgrowth, arborization, and synaptogenesis, leading to the maturation of an astoundingly diverse array of phenotypes (Fig. 1).[3,6-8,14-19] Due to the generation of particular structural attributes through this sequence of morphogenesis, critical functional properties begin to emerge, allowing the establishment of appropriate activity patterns within the developing neuronal network.[8,14,18,20-33] For example, since the majority of CNS synapses are localized to neuronal arbors, changes in arbor surface area influence the reception, integration, and transmission of electrical activity.[2,8,24,25,34-38] Consequently, dynamic regulation of neuronal branching morphogenesis is especially critical for both generating and maintaining the functional organization of the nervous system. While many of the mechanisms underlying branching morphogenesis are likely to be conserved among neuronal populations, the tremendous diversity in neuronal

---

Branching Morphogenesis, edited by Jamie A. Davies. ©2005 Eurekah.com and Springer Science+Business Media.

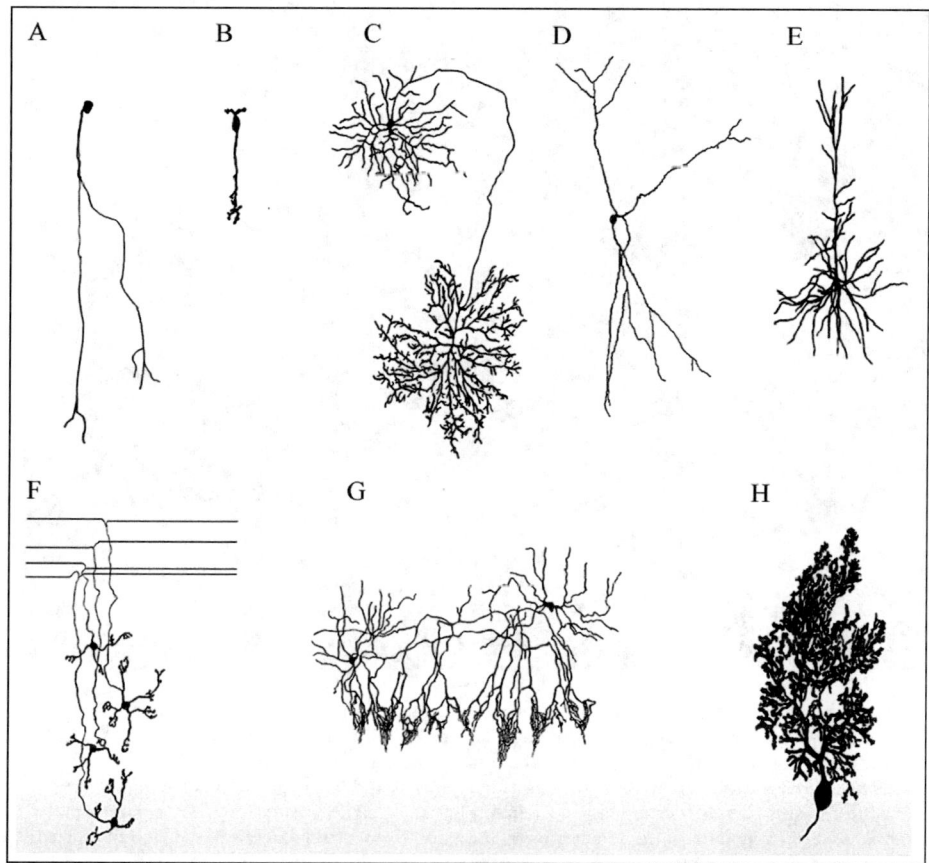

Figure 1. Variations in vertebrate neuronal morphology. A) Sensory neurons of the dorsal root ganglion, categorized as pseudo-unipolar neurons, produce a single elongated and fused axonal process that bifurcates into two functionally distinct branches, but no dendrites. In contrast, B) retinal and olfactory bipolar neurons develop both a single axonal process and a single arborizing dendrite. C) Visual system amacrine and horizontal cells lack typical axons, although specialized presynaptic and postsynaptic regions of dendritic processes exist. D) Neurons of the lateral geniculate nucleus, providing the link between retinal input and the primary visual cortex, are characterized by robust axonal arborization, but limited dendritic elaboration. Conversely, E) hippocampal pyramidal neurons possess distinct functional subpopulations of apical and basal dendrites, and an elongated axon which gives rise to multiple collateral branches. Within the cerebellum, F) granule neurons develop a signature "T-shaped" axon and several unbranched dendrites ending in claw-like termini. A significantly elaborated axonal structure is achieved instead by basket cells (G), which produce moderately arborized dendrites but numerous axon collaterals that form basket-like cages around Purkinje cell somata. In contrast, (H) Purkinje cells develop a planar highly-arborized dendritic tree studded with actin-enriched dendritic spines, but a relatively simple axon. Figure constructed with permission from data presented in references 1, 3, 14, 17, 40, 79, 440, 454 and 635.

cytoarchitecture suggests that some cell-type-specific differences in the regulation of arborization also must arise.[3,6-8,14-19,21-33] Through decades of intensive research, significant progress has been made toward identifying and characterizing the numerous extracellular molecules and cell-cell interactions that regulate aspects of branching morphogenesis. However, the subcellular molecular regulation of branching remains poorly understood. In fact, the degree to which variation in branching architecture among neuronal populations reflects fundamental

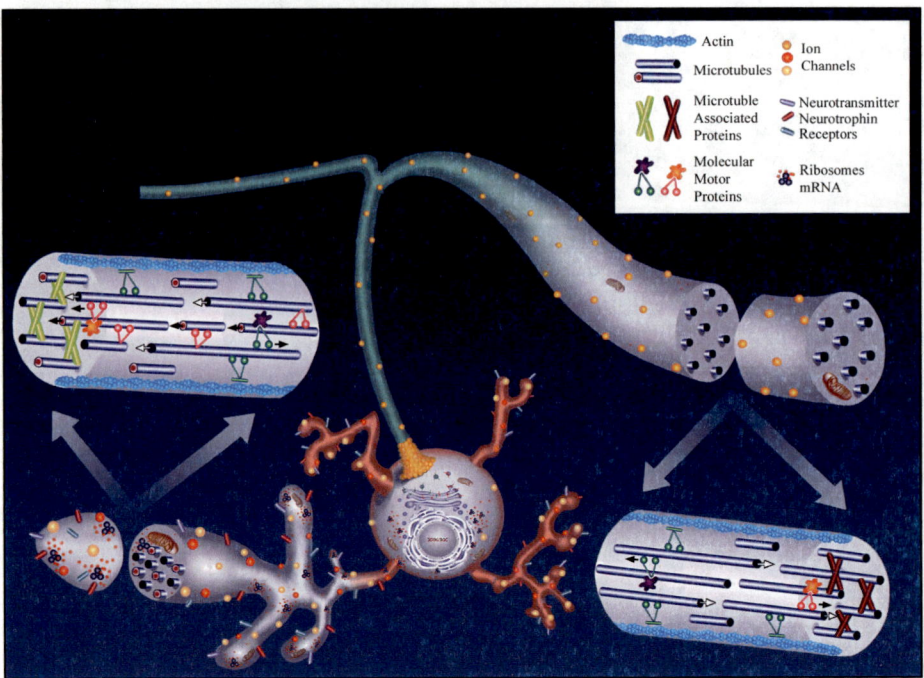

Figure 2. Schematic illustration of the distribution of cellular constituents in a well-polarized neuron. Neuronal morphogenesis typically culminates in the extension of an elongated primary axon and multiple short primary dendrites. Mechanisms that produce local subcellular differences in molecular and cytoskeletal constituents are necessary both for generating neuronal polarity and for triggering branching morphogenesis. At the ultrastructural level, the dendritic domain contains Golgi elements, rough endoplasmic reticulum, mRNA, tRNA, polyribosomes, transcription initiation factors, and a mixed population of plus-end and minus-end distal oriented microtubules. In contrast, the axonal domain excludes biosynthetic machinery but retains specific synapse-related proteins including Na+ and K+ channel isoforms, and a characteristic uniform population of plus-end distal oriented microtubules. Fully differentiated neurons thus display an asymmetrical distribution of cytoskeletal elements, cytoskeletal-associated stabilizing proteins, molecular motor proteins, organelles and vesicles, cytoplasmic and cell-surface proteins, and plasma-membrane components. Figure reprinted with permission from K.M. Kollins, constructed from data presented in references 8, 9, 36, 39, 40, 44, 47, 49 and 53.

differences in the control of arborization, or instead represents cell context-dependent modulation of common regulatory pathways, remains largely unclear. Therefore, one significant aim of current research is to identify regulatory pathways that underlie branching morphogenesis for all neuronal populations, and the degree to which branching may be modified through cell context-specific cues.[2,8,26-29,38,49,50,54,85-87,156-163]

The Importance of Neuronal Polarity during Branching Morphogenesis

Neurons are specialized secretory cells characterized by the heterogeneous compartmentalization of various cellular constituents into discrete and physiologically significant domains, the axon and dendrite (Fig. 2).[8,39-44] In turn, this asymmetric organization of cellular constituents allows the partitioning of cellular responses, such as electrical impulse reception, integration, propagation, and release of signaling molecules. Most significantly, the cellular specializations that support appropriate functional polarity also generate inherently different influences on the process of branching morphogenesis within the axonal and dendritic domains.[8,39-44]

Since all neuronal populations arise from neuroepithelial precursors, it has been postulated that developing neurons and epithelial cells may employ similar constraints on structural and functional polarity.[8,39,44] For epithelial cells, the luminal plasma membrane, or apical domain, is structurally separated from the parenchymal plasma membrane, or basolateral domain, by a circumferential band of tight junctions that produce a physiological fence. In addition, subjacent attachment regions of desmosomes and adherens junctions provide cytoskeletal-based structural support and some degree of chemical coupling between the plasma membrane surfaces of closely apposed epithelial cells.[8,9,16,39,41,44,45,77] For developing neurons, which do not possess tight junctions, the axonal domain instead appears to be functionally separated from the somatodendritic domain by a diffusion barrier intrinsic to the initial segment hillock, with the adherens junctions of epithelial cells thought to be an evolutionary antecedent of neuronal synapses.[8,39-43] Indeed, careful analysis of both the morphology and ontogeny of epithelial and neuronal cells has prompted the suggestion that the entire neuronal surface may in fact be equivalent to the epithelial basolateral domain, with distinct axonal and dendritic compartments arising through unique subpartitioning mechanisms evolved by neurons. If this interpretation is correct, axons may represent a highly specialized protein-sorting region, analogous to an elongated Golgi apparatus directly linked to the cell surface.[8,9,39-41,44,45]

Typically, neuronal morphogenesis culminates in the extension of a single elongated primary axon displaying periodic collateral branches, or distal branching within synaptic target regions, and multiple short primary dendrites that may be robustly arborized.[8,9-41,44] At the ultrastructural level, while the dendritic domain is continuous with the soma to a certain degree, containing Golgi elements, rough endoplasmic reticulum, mRNA, tRNA, polyribosomes, and transcription initiation factors, by contrast the axonal domain primarily excludes biosynthetic machinery but retains specific synapse-related proteins including Na^+ and K^+ channel isoforms.[8,39,40,44] Arguably, the most functionally important subcellular difference arising during neuronal development is the uniform plus-end distal microtubule orientation established within axons, and the mixed plus-end and minus-end distal microtubule orientation evident within dendrites.[2,8,39,40,44-49] As a consequence of this asymmetric cytoskeletal arrangement, distinctly different biomechanical forces are generated within the axon and dendrites, contributing to the directed transport of organelles and cytoskeletal polymers.[16,18,39,45,50-59]

As a rule, organelles and vesicles that selectively translocate toward the minus-ends of microtubules are expected to be mechanically excluded from the axons, while those that instead translocate toward microtubule plus-ends can be conveyed into both the axon and dendrites, due to the mixed microtubule polarity orientation characteristic of dendrites.[8,44,50,60-64] Additionally, it has been suggested that certain biosynthetic components are restricted from entering the mature axon due to the tight bundling of uniformly oriented microtubule polymers, achieved through axon-specific microtubule-associated proteins.[50,53,54,65,66] Consequently, localization of the majority of axonal proteins must be achieved through targeted delivery of proteins initially synthesized within the soma.[8,40,44,50,54,65,66] For dendrites, however, mRNAs and polyribosomes readily accumulate at postsynaptic sites underlying dendritic branches, suggesting a role for local protein synthesis in response to physiological cues.[39,40,44,66-75] Due to the microtubule orientation differences that arise between developing axonal and dendritic domains, fully differentiated neurons display an asymmetrical distribution of cytoskeletal elements, cytoskeletal-associated stabilizing proteins, molecular motor proteins, organelles and vesicles, mRNA, cytoplasmic and cell-surface proteins, and plasma-membrane components.[8,40,44] As a consequence, mature neurons are equipped to produce local responses to microenvironmental cues that are essential for generating appropriate neuronal branching morphology and for modulating synaptic plasticity.[8,39,40,44]

Mechanisms that produce local differences in molecular and cytoskeletal constituents are necessary both for generating neuronal polarity and for triggering branching morphogenesis, processes that are similarly initiated through symmetry breaking.[8,16,18,76,77] First applied to biological systems by Alan Turing[78] as a theoretical construct to explain embryonic morphogenesis, spontaneous symmetry breaking describes the emergence of asymmetry from initially symmetric but unstable conditions, through internal dynamic processes. Accordingly, when a dynamic system, such as a neuron, reaches an internal state of biochemical or structural instability, the small irregularities that are produced through stochastic fluctuations, such as protein synthesis, turnover, or localization, become increasingly amplified. As a consequence, subcellular symmetry is typically abolished, leading to a new and stable state that, in the case of neuronal morphogenesis, typically produces neurite extension or branching.[8,9,18,39,76-78] Research conducted to determine what changes in cellular or molecular constituents are both necessary and sufficient to trigger neuronal symmetry breaking have uncovered critical roles for mitochondria and Ca^{2+} homeostasis, actin depolymerization, microtubule dynamics and motor protein activity, selective organelle localization, and directed membrane insertion, all of which can be affected by changes in the microenvironment of a developing neuron.[8,9,16,18,44]

Mechanisms of Axonal Branching Morphogenesis

Immediately after the commencement of neuronal morphogenesis triggered by symmetry breaking, significant cytoskeletal changes begin to sculpt the developing neuronal form. The earliest of these modifications involve rapid reorganization of the actin matrix that establishes dynamic membranous specializations termed growth cones.[2,79-94] Significant progress has been made toward describing axonal and dendritic growth cones at the ultrastructural level, and in characterizing physiological functions mediated by axonal growth cones during axonal elongation or retraction, pathfinding, and target recognition.[2,16,18,45,76,77-81,84,89-93,95,96] The mature axonal growth cone is distinguished by a thin fan-like morphology, consisting of membranous sheets largely filled with a filamentous actin meshwork, termed the lamellipodium, interspersed with microspikes of bundled actin fibrils, or filopodia, arrayed with their growing tips extending distally.[79,89,95,97-99] Well-differentiated growth cones can be further subdivided into two distinct domains on the basis of key structural and functional attributes. Accordingly, the most peripheral region, termed the P-domain, represents the dynamic leading edge, comprising both actin monomers and a filamentous actin (F-actin) matrix, but few associated organelles.[79,89,95,97-99] In contrast, the growth cone region closest to the neuronal soma, or central (C)-domain, constitutes a complex and dense network of actin, neurofilaments, polarized plus-end distal microtubules, mitochondria, endosomes and clear vesicles, membrane-bound vesicle stacks, and polyribosomes. Thus, the C-domain alone provides cellular machinery critical for energy production and the local synthesis or transport of proteins required for ongoing axonal growth and maturation.[79,89,95,97-99]

As neuronal differentiation proceeds, axonal growth cone morphology changes dynamically as attractive and repulsive interactions between local microenvironmental cues and the underlying cytomatrix guide axons toward appropriate synaptic targets.[79-84,93,94,100-107] Complex growth cone shapes emerge at decision points during encounters with a target-rich extracellular environment,[94,104-107] and the most simple shapes appear as growth cones fasciculate and extend rapidly along other axons of long-distance projections.[94,104-107] Throughout the developing axonal fascicles of long-distance projections and short-distance tracts alike, nascent branches begin to extend and retract. Early studies of neuronal branching morphogenesis suggested that axonal branches could arise either through the simple bifurcation of a growth cone tip, or through filopodial outgrowth from an axon shaft to form collateral branches well behind an extending growth cone (Figs. 3, 5). In fact, most vertebrate neurons establish connections with multiple targets through this process of collateral branching, in which a neuron initially extends a pioneer axon to a primary target and only after a prolonged delay then generates secondary branches along the axon shaft to innervate secondary targets.[2,90,104-109]

Once the focus of controversy, the interrelationship between axonal growth cone bifurcation and collateral branching has been gradually clarified through a number of ongoing investigations.[2,94,104-115] Initial studies by O'Leary and collaborators, beginning in the 1980s, generated important data that began to advanced the understanding of collateral branch

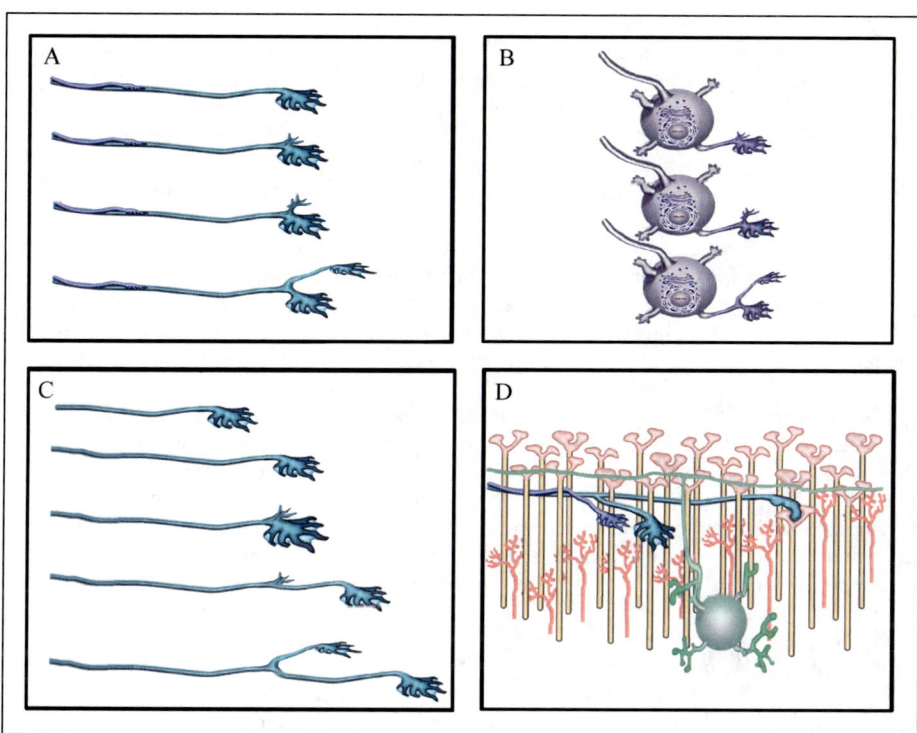

Figure 3. Schematic illustration of mechanisms proposed to underlie neuronal branching morphogenesis. A) Axonal branching produced through simple growth cone bifurcation, in which local microenvironmental cues stimulate cytoskeletal reorganization culminating in the formation of a second active growth cone. B) Dendritic branching produced through simple growth cone bifurcation. C) Axonal branching produced through growth cone pausing followed by collateral branch outgrowth, or "delayed interstitial branching," in which a side shoot extends from the axon shaft well after the primary growth cone has continued to elongate. D) Axonal branching produced as repellent molecule exposure (green neuron and vertical pink cells) triggers primary growth cone collapse (turquoise axon) and the subsequent sprouting of lamellipodia and filopodia from the axon shaft (turquoise axon branch). In addition, defasciculation of bundled axons (blue axon) often accompanies this form of branch induction. Figure reprinted with permission from K.M. Kollins, constructed from data presented in references 2, 15, 36, 38, 76, 79, 90, 98, 101, 106, 114, 115, 144, 279, 565, 636 and 637.

formation.[2,76,90,94,98,114,115] Examination of the axonal connectivity of cortical layer V neurons revealed that their stereotypical connections with the basilar pons in the hindbrain were not formed through simple growth cone bifurcation, but rather through "delayed interstitial branching," in which a side shoot is extended from the axon shaft well after the growth cone has continued to extend.[2,90,94,98,114,115] Similar delayed interstitial branching also underlies the establishment of a variety of developing neuronal circuits, including the collateral projection from the hippocampal formation to the mammillary bodies, the dorsal root ganglia projection to the spinal cord, and the retinal projection to the lateral geniculate nucleus and optic tectum.[2,106,110-115] More recent in vitro observations by Kalil and coworkers suggest that extension of branches from the middle of an axon shaft may result from bifurcation-related structural changes generated by an active growth cone that does not immediately bifurcate, but instead continues to elongate.[104-106] Another series of recent observations indicate that de novo branches can sprout from a developing axon in response to collapse of the primary growth cone,[101] or at positions along the axon shaft demarcating points where growth cones paused in response to microenvironmental cues.[106,108,109] For example, collapse of retinal ganglion cell (RGC) growth cones, produced either through mechanical manipulation or contact with re-pellent molecules, can trigger lateral extensions of lamellipodia and filopodia from an axon shaft in addition to promoting defasciculation of bundled axons.[101] In vivo, local induction of this interstitial back-branching may be critical for establishing appropriate connectivity be-tween RGC axons and their target optic tectal neurons during topographical mapping of the developing visual system.

Although several types of axonal branching have been identified, it remains unclear whether simple growth cone bifurcation, delayed interstitial branching, and collapse-induced branch-ing arise through separate mechanisms or are instead achieved through a common pathway. Thus, the cellular and molecular mechanisms responsible for generating and stabilizing axonal branches are a current focus of intense study. To this end, the recent work of Kalil and collabo-rators has proved particularly significant in clarifying the sequence of de novo branching dur-ing development.[104-106,108,109-112] For example, direct observation of living cortical brain slices revealed that, after reaching their synaptic target regions, axonal growth cones paused, flat-tened, and enlarged as microtubules formed robust loops characteristic of slow-growth states.[105,116] During these periods, paused growth cones demonstrated repeating cycles of col-lapse, retraction, and extension without overall forward growth, shown to be necessary for demarcating the sites of future axonal branch points.[105-109] Once the primary growth cone resumed steady forward growth several hours to days after pausing, filopodial and lamellipodial remnants remained along the axon shaft as varicosities that began to elongate into interstitial branches tipped with active growth cones.[105,106,108]

Initiation of branching strictly at those points where complex growth cone behaviors occur suggests that specific microenvironmental cues trigger the pause in growth cone migration, local modification of the underlying cytomatrix, and the establishment of a nascent branch point through localized targeting of subcellular components.[105,106,108] According to this novel model of branching morphogenesis, Kalil and colleagues argue that delayed interstitial branch-ing actually represents a form of target-induced growth cone bifurcation, since the original growth cone resumes forward advance while contributing to the formation of new growth cones along the trailing axon shaft (Fig. 3).[105,106] Microtubule reorganization is a critical com-ponent of this axonal branching process, involving the splaying apart of looped microtubules within the underlying axon shaft, accompanied by local microtubule fragmentation that allows short microtubules to begin invading the nascent branch.[2,85-87,93,104-106,117,118] Indeed, studies characterizing microtubule dynamics during cortical neuron axonogenesis demonstrated that when microtubules within an axon remain bundled, branches fail to form even when transient filopodia invested with microtubules arise along the axon shaft. Moreover, only those filopodia that capture sufficient numbers of invading short microtubules are able to develop into stable branches.[2,80,81,85-87,104-106,119,120]

Coincident with dynamic changes in microtubule arrangement, redistribution of F-actin is necessary for ongoing branching morphogenesis. For example, recent studies provided evidence that focal accumulation of F-actin is coordinated with microtubule splaying but precedes micro-tubule penetration into forming branches (Fig. 4).[104-106] In fact, time-lapse imaging of interac-tions between microtubules and F-actin within branching cortical axons revealed that co-localization of these cytoskeletal elements reflects their coordinated polymerization and depo-lymerization within an active growth cone, or at branch points.[105] Significantly, the inhibition of dynamic changes in either microtubule or F-actin polymerization abolishes branching, while allowing continued axonal elongation. Thus, bidirectional signaling between these cytoskeletal elements provides an essential pathway for regulating axonal growth and branching morphogen-esis.[105,121-123] Following the initiation of stable axonal branches, ongoing morphogenesis is charac-terized by the transport of microtubules and organelles into growing branches, which may elon-gate differentially through a process termed sibling bias.[124] For example, hippocampal neurons

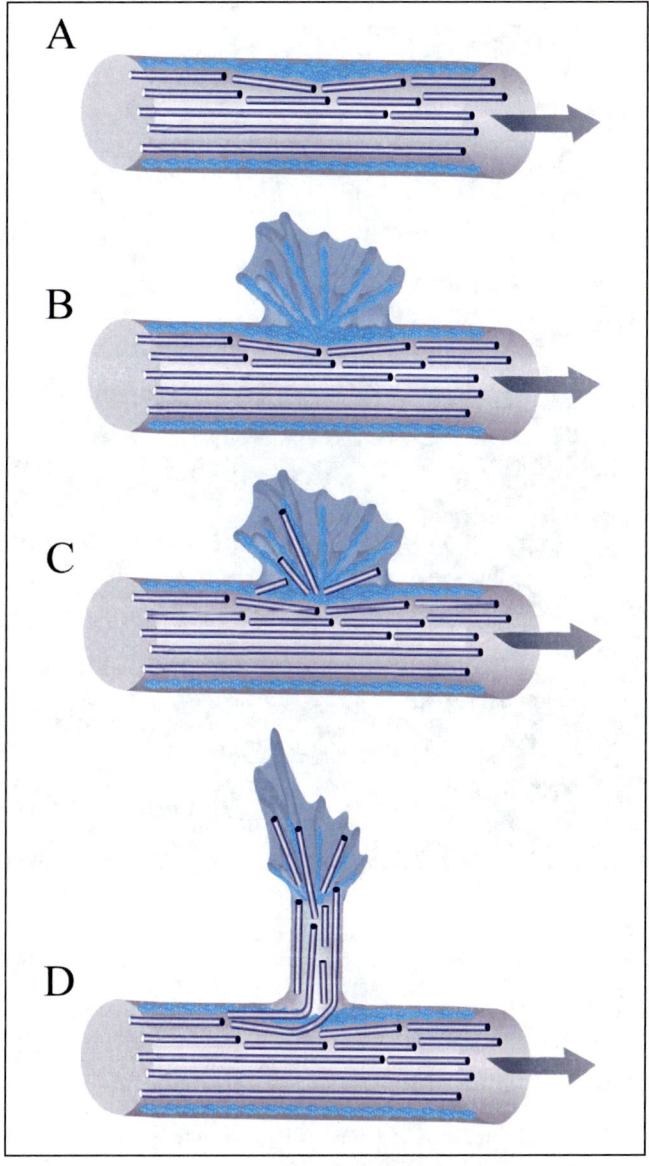

Figure 4. Schematic illustration of a model proposed for cytoskeletal reorganization occurring during axonal branching morphogenesis. A) The earliest cytoskeletal modifications during branching morphogenesis involve the splaying apart of microtubules (MTs) (blue tubules) within the underlying axon shaft, accompanied by local MT fragmentation and focal F-actin accumulation (turquoise filaments). B) Coincident with the emergence of an F-actin-enriched lamellipodium along the axon shaft, C) individual fragmented MTs begin to explore this nascent process through rapid cycles of extension and retraction. D) Short MTs continue to penetrate the nascent branch, while stabilized MTs already within the growing branch elongate through ongoing polymerization. As axonal branches mature, the colocalization of F-actin and MTs reflects their coordinated polymerization and depolymerization within active growth cones at branch points. Arrows indicate direction of axonal growth. Figure reprinted with permission from K.M. Kollins, constructed from data presented in references 2, 36, 52, 104-106, 108, 143, 156 and 157.

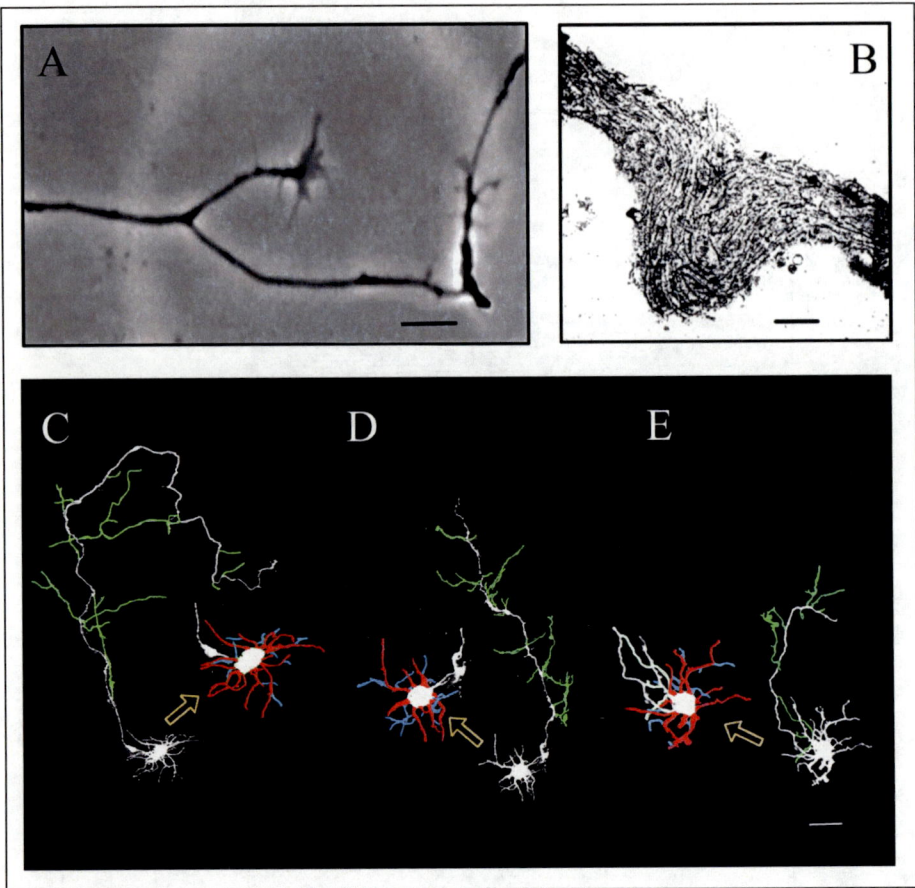

Figure 5. Examples of vertebrate neuronal branching. Branches can sprout from a developing axon in response to collapse of the primary growth cone, or at positions along the axon shaft demarcating points where growth cones paused, while dendritic branches arise through simple growth cone bifurcation. A) Temporal RGC axons encountering posterior optic tectal neurons in vitro exhibit growth cone collapse and sprouting of lateral extensions from the axon shaft. Scale bar = 10 μm. B) Temporal RGC axon sprouting lateral extensions in vitro was fixed and prepared for EM. The micrograph image reveals a kinked portion of the axon shaft filled with looped microtubules and associated organelles, characteristic of nascent branch points. Scale bar = 1 μm. C-E) Cerebellar granule neurons developing with chronic neurotrophin stimulation establish complex axonal and dendritic arbors in vitro. Granule neurons were maintained as high cell-density monolayer cultures seeded with lipophilic DiIC18$_{(3)}$ dye labeled neurons, fixed, and visualized with confocal microscopy. The digital montage comprises multiple low-magnification high-contrast confocal fields of mature granule neurons. Images are saturated to reveal thin neurites and branches, with axonal branches pseudo-colored green, dendrites pseudo-colored red, and dendritic branches pseudo-colored blue. Scale bar = 10 μm. Figure constructed with data obtained from K.M. Kollins and R.W. Davenport, unpublished observations, and references 36, 254, 446 and 565.

display characteristic patterns of alternating growth for separate branches extending from a single axon, presumably reflecting changes in the preferential transport of subcellular materials into one of several branches. As a consequence of the balance between various epigenetic influences and cell-autonomous constraints, axonal branch morphology retains some degree of plasticity throughout development and maturation, allowing ongoing sculpting of synaptic circuitry.

Mechanisms of Dendritic Branching Morphogenesis

For most vertebrate neuronal populations, axonal morphogenesis occurs under electrically silent conditions, or with low levels of spontaneous activity, whereas dendritic development instead proceeds in the context of various forms of neuronal activity. Consequently, a neuronal cell-type-dependent balance of specific forms and patterns of electrical activity is important for sculpting the architecture of dendritic branches.[2,15,38,125,126] Vaughn's synaptotropic dendritic branching hypothesis, formulated in response to static EM observations, proposes that branches form at regions of dendritic contact with afferent partners, and that these nascent branches are successively stabilized through synapse maturation.[19,126,127] More recently, a number of live-cell imaging experiments achieved through two-photon microscopy provided direct support for this hypothesized mechanism of dendritic branching.[642] For example, in zebrafish tectal neurons, new dendritic filopodia, the putative precursors to stable branches, are the favored sites for synapse formation, and dendritic filopodia that fail to make stabilizing synaptic contacts eventually retract.[642] Similarly, during hippocampal neuron development, branch formation is induced only in those dendrites receiving direct synaptic contact through afferent innervation,[128-133] and for ciliary ganglion cells the extent of dendritic arborization is directly proportional to the number of axonal contacts received.[12,128-133]

It follows then, that dendritic branches associated with immature, unstable, or weakly-active synaptic connections may retract passively through insufficient structural stabilization, or actively through neurotransmitter-based competition.[15,134-136] In fact, global activity blockade can influence the overall spread and branching pattern of developing dendritic arbors, while more local changes in neurotransmitter signaling can modulate the stabilization or retraction of individual dendritic branches.[134-139] For example, selective elimination of excitatory input to the ventral dendrites of chick nucleus laminaris results in retraction of these dendritic arbors, while innervated dorsal arbors are maintained.[137-139] Clearly, for numerous neuronal populations, the number of presynaptic contacts made with developing dendrites stimulates local dendritic branching and determines the resulting arbor complexity, while failure to establish mature and active synaptic contacts variously stunts neuronal differentiation or elicits dendritic arbor retraction.[15,38,125,133]

Dendritic arborization is a highly dynamic process for most differentiating neurons, characterized by filopodial-driven branch addition and retraction that is rapid during early developmental periods, but decreases with progressive neuronal maturation as arbors increasingly stabilize.[15,38,50,140-144] Neurotransmission plays a significant role in modulating these rapid changes in arbor remodeling. Consequently, the direction of dendritic filopodial extension is biased toward maximizing the number of contacts made with presynaptic axons from target neurons that display physiologically appropriate activity patterns or neurotransmitter release.[15,38,125,144,145,146] The sequence of subcellular reorganization within actively branching dendritic filopodia was recently characterized, facilitated by important technical advances in live-cell microscopy.[144-146] These time-lapse observations initially revealed that dendritic branching typically involves the bifurcation of an active growth cone, leading to the emergence of two similar, though smaller, growth cones that extend in divergent paths (Figs. 3, 5).[2,15,133,142,144] Additional developmental studies demonstrated that, during this process of dendritic branching, nascent branch points are first invaded by membrane-bound vesicles, forming membranous cisternae, and then successively by actin fibers and microtubules.[38,50] In fact, the molecular basis for cytoskeletal modifications underlying dendritic branching, similar to axonal branching, involves regulation of both the synthesis and stabilization of actin and microtubule networks.[15,38,50,125,133,144] Interestingly, ultrastructural observations have revealed that no microtubule nucleation sites are present within the bifurcation points of developing dendritic branches,[50] suggesting the contribution of other regulatory processes in directing branch elongation. Indeed, during the maturation of dendritic arbors, nascent actin-based branches are stabilized and lengthened through the proximo-distal invasion of microtubules, bent in relation to the orientation of microtubules within the primary dendrite shaft.[15,50,143] It remains unclear whether the growth cone bifurcation mechanism initiating dendritic branching is followed by localized microtubule debundling and splaying to allow elongation, similar to that observed for developing axonal branches. Given the relative disparity between branch lengths achieved by dendrites as compared to axons, in combination with characteristic differences in underlying microtubule polarity, it is likely that significant differences in cytoskeletal reorganization are also produced for arborizing axons and dendrites.

Closely related to dendritic branching in the CNS is the formation of small, highly-motile protrusions termed dendritic spines. Receiving the majority of excitatory synaptic input in the mature cerebral cortex, dendritic spines display both structural and functional heterogeneity that ultimately influences neuronal signaling properties.[533,643-646] For this reason, the initial formation and ongoing morphological plasticity of dendritic spines has been a focus of intense study. Structurally, dendritic spines comprise a bulbous actin-rich head attached to the dendrite shaft through a narrow neck, and each of these elements exhibits tremendous variation among spines.[553,643-646] Morphologically, dendritic spines can be subdivided into two groups: small spines, characterized as filopodial or thin, and large spines, characterized as stubby, fenestrated, or mushroom-like.[553,643-646] Although the regulation of dendritic spine formation is yet unresolved, the sequence of spine outgrowth is well characterized. During cortical neuron development, dendrites initially generate long, thin, filopodial protrusions, but as differentiation proceeds, these labile dendritic filopodia are replaced first with polymorphic 'protospines' and then with knobby actin-rich spines containing postsynaptic density (PSD-95) clusters.[141,553,642-648] Thus, one significant area of investigation has centered on clarifying the relationship between morphological changes in dendritic filopodia and the onset of spine formation during synaptogenesis.

Tremendous progress in clarifying the mechanism of spine formation has been achieved through an elegant series of EM imaging and three-dimensional structural analysis studies, and recent live-cell observations performed with two-photon time-lapse imaging.[642,643,645] Together, these investigations revealed that exploratory dendritic filopodia grow into the neuropil, locate suitable axonal partners, and typically consolidate these initial contacts into stable synapses while transforming into mature spines, although shaft synapses also arise. Thus, spine formation appears to follow the initial stages of synapse formation.[642-646] Indeed, within a short interval after contacting appropriate axonal partners, but before spine formation, dendritic filopodia begin to accumulate active vesicles and subsequently PSD-95 clustering occurs.[642] A critical role for afferent activity in ongoing dendritic spine formation is evident from observations that, for an individual neuron, spine density is significantly greater in regions of the dendritic arbor receiving high levels of innervation as compared to uninnervated dendrites.[647] Taken together, these findings suggest that neurotransmission may be necessary to trigger the morphological and physiological transformation of dendritic filopodia into dendritic spines. Interestingly, in these recent studies a second population of dendritic filopodia was observed to persist as stable branches throughout tectal neuron development, suggesting that additional controls may be involved in determining which filopodia become branches or spines.[642] Ultimately, the ultrastructural changes within dendritic filopodia that produce branches or spines are critical for triggering physiological changes thought to underlie fast synaptic transmission and the consolidation of learning and memory.

The Regulation of Neuronal Branching Morphogenesis through Intrinsic Mechanisms and Epigenetic Cues

Defining Intrinsic and Epigenetic Regulation of Branching Morphogenesis

The variation in neuronal morphology evident throughout the nervous system is generated by patterns of differentiation unique to each population, reflecting those structural specializations that allow particular physiological properties to emerge. While many of the developmental mechanisms underlying morphogenesis are likely to be conserved among neuronal populations, the tremendous diversity in neuronal cytoarchitecture suggests that cell-type-specific differences in the regulation of axonal and dendritic structure must also arise. Significant advances have been made toward characterizing the numerous genes, molecules, and cell-cell interactions that regulate early stages of axonal and dendritic outgrowth. However, much less is understood about the mechanisms involved in initiating and regulating the subsequent process of neurite arborization. In fact, the degree to which variation in branching architecture among neuronal populations reflects fundamental differences in the control of arborization, or instead represents cell context-dependent modulation of common regulatory pathways, remains largely unclear. Therefore, one essential aim of current research is to determine the relative contributions of cell-intrinsic and epigenetic regulation throughout branching morphogenesis.[2,8,26-29,38,49,50,54,85-87,156-163]

Generally, morphological characteristics that consistently appear during neuronal development, both in situ and in reduced culture conditions, are thought to represent invariant cell-type-specific attributes established through intrinsic regulation.[26-29] In one such pathway, a largely cell-autonomous program of gene expression can control intracellular signaling cascades and, in turn, the synthesis of molecular effectors and cytoskeletal components that generate neuronal form. In addition, intrinsic regulation of neuronal morphogenesis also reflects constraints imposed by cellular dimensions, the biophysical properties of microtubule polymers and the actin cytomatrix, and the corresponding exertion of various biomechanical forces upon changing neuronal structures.[8,18,26-29,38,44,50,52,76,77,106,125,144,164-170] In contrast, epigenetic influences on the pattern of neuronal morphogenesis arise from temporally or spatially constrained microenvironmental cues including homotypic and heterotypic cell-cell interactions, substrate-bound and diffusible chemoattractants and chemorepellents, hormones and growth factors, and patterns and levels of electrical activity.[8,29,85-87]

Although the distinction between intrinsic and epigenetic control of neuronal morphogenesis is a useful construct for examining certain stages of differentiation, in practice it is often difficult to separate cell-autonomous regulation from instructive or permissive effects induced by the environment. In fact, the weight of current evidence suggests that the structure attained by a mature neuron is largely determined by differential gene expression, produced by a combination of intrinsic developmental programs and epigenetic cues. Ultimately, set against a background of constitutively expressed genes and cell-autonomous transcription factor activity, cascades of gene induction are stimulated by an array of extrinsic cues. As such, the differentiation of each neuronal cell type reflects stereotyped transcriptional changes triggered by a sequence of epigenetic interactions, constituting developmental subroutines.[8,9,39,44,171,172] It follows that the complement of signals within the local microenvironment of a given neuron may influence the order in which these subroutines are induced during development. One important result of this type of regulation is that the onset of any given subroutine can limit the ability of a neuron to respond to other extrinsic cues, leading to progressive restriction of developmental potential.[8,171,172] As a consequence, the spatiotemporal control of exposure to individual epigenetic cues can orchestrate a unique program of branching morphogenesis for each developing neuronal population.

Control from the Inside Out: Intrinsic Regulation of Branching Morphogenesis

Intrinsic regulation of neuronal branching morphogenesis arises through a combination of structural constraints imposed by cellular dimensions, biophysical properties of intracellular cytoskeletal elements, and cell-autonomous control of arbor topology and branching probability (Fig. 6).

At the gross structural level, the size of the neuronal soma intrinsically delimits the number and area of primary axons and dendrites and associated branch projections, which can be described as the total amount of cross-sectional process area allowed.[29-33] Indeed, a direct correlation between the soma diameter and the combined cross-sectional dendrite area has been observed consistently within populations of motoneurons, hippocampal pyramidal cells, and both

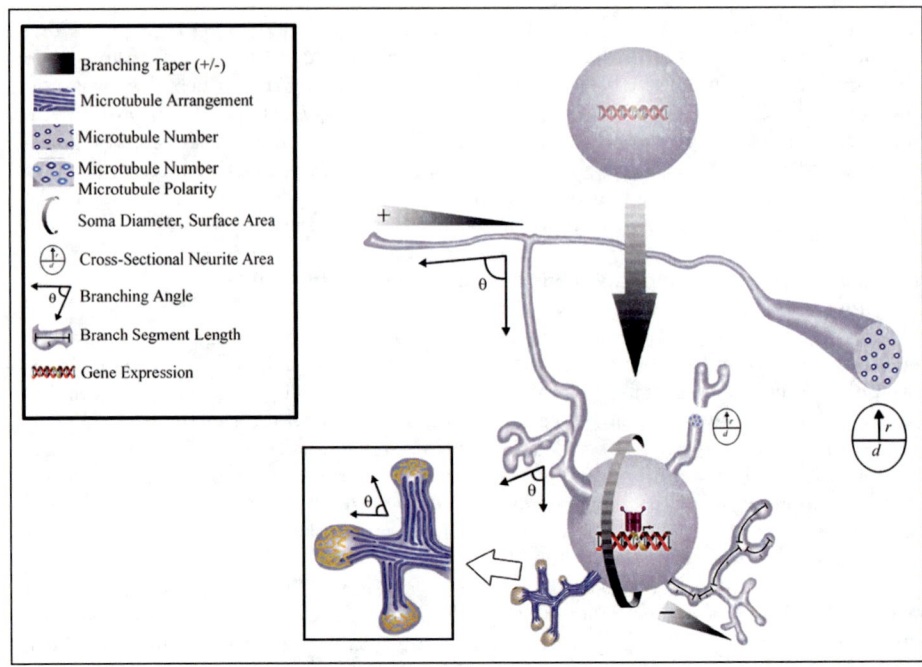

Figure 6. Intrinsic factors affecting the morphological differentiation of vertebrate neurons. The size of the neuronal soma dictates the total cross-sectional process area allowed to be partitioned into axons or dendrites (curving black arrow). The total cross-sectional area of the axonal domain may increase through increased branching proximally, or progressive proximo-distal enlargement of the axon (area symbol). In contrast, the cross-sectional area of dendrites may be conserved or instead decrease successively through partitioning into segments of decreasing diameter following the 3/2 power scaling law (area symbol). Neurite branching is constrained by the total cross-sectional process area, which may be partitioned into successive branch segments until a limiting size is met, representing the minimum number of microtubules required to maintain structural stability (blue circles). Developing neurites and their associated branches exhibit a characteristic projection orientation, termed the vector or branching angle, that results in the increased spread of the arbor through successive bifurcations along the neurite (angle symbol). Although the distance between branch points, or segment lengths, may vary for a given neuron, each neuronal cell-type exhibits a characteristic range of segment lengths for dendrites and the axon (segment length symbol). When the cross-sectional area decreases between branching segments, negative taper is produced, with positive taper instead resulting from increasing cross-sectional area between neurite branch points (gradient). The characteristic taper is established by the proportion of microtubules compartmentalized into successive branch segments, achieved in part through microtubule-associated protein binding and differential microtubule spacing or orientation (blue tubules). An additional form of intrinsic control over neuronal morphology during the development of polarity is produced through the background complement of genes expressed, independent of extracellular cues. Inset schematic represents an enlarged view of the typical cytoskeletal organization within a granule neuron dendrite. Figure reprinted with permission from K.M. Kollins, constructed from data presented in references 3, 8, 9, 26-29, 36, 49-54, 134, 164, 174-176, 189-191, 195, 196 and 343.

Purkinje cells and granule neurons of the cerebellum.[27-29] Further, the relative diameter, or caliber, of axons and dendrites differs significantly at their point of origin from the soma, and changes in cross-sectional area arising through subsequent branching follow distinctly different trends for these separate classes of neurites.[8] For example, the total cross-sectional area of the axonal domain may increase either through net increases in branching, progressive proximo-distal enlargement of the axon, or a combination of both parameters.[8,173-175] In contrast, the total cross-sectional area of the dendritic domain is either conserved between branch points, or is successively partitioned into segments of decreasing diameter following the 3/2 power scaling law, described as: $P^{3/2} = D_1^{3/2} + D_2^{3/2}$, where P and D represent the diameters of the parent and daughter branches, respectively.[8,26,29,176] In fact, measurements across various neuronal populations reveal that the exponent of this descriptive branch power equation actually varies between 3/2 and 2, suggesting some degree of plasticity in the intrinsic regulation of branching.[8]

The arrangement of neurites arising from the soma may take the form of a single large process or multiple smaller-caliber processes, producing the characteristic arborization pattern observed for each neuronal population.[8,26-29,159,160] Thus, while cerebellar basket/stellate interneurons produce up to five dendrites and an average of two branches per dendrite, Purkinje cells, the sole output neuron of the cerebellum, instead extend one or two dendrites but may develop over 100 secondary and tertiary branches and more than 80,000 dendritic spines.[3,177-179] It is likely that the strict intrinsic control of neurite area reflects the limited plasma membrane surface area, and metabolic constraints on the rate and number of microtubules produced, together establishing the architecture of emerging processes.[26-29] For example, analysis of the cytoskeletal structure of rat Purkinje cell dendritic arbors has revealed that a consistent maximum number of approximately 800 microtubules are extended from the soma into the primary dendrite, which are then distributed into branches as this stem process arborizes.[26-29] Indeed, an essential property of neuronal branching is that the total amount of cross-sectional process area is partitioned into successive branch segments until a limiting size is met, which represents the minimum number of microtubules required to maintain structural stability.[8,29,159,160] For dendritic branches, sufficient stability is maintained to a minimal diameter of 0.5-1.0 μm, indicating a viable limit of three microtubules at distal dendritic segments.[8,29,159,160,180] Regulation of microtubule partitioning at axonal branch points appears to be accomplished, in part, through competition between regions of an extending filopodium within stimulatory or permissive regions of the microenvironment, leading to differential microtubule distribution. However, the lower limit of microtubules required to maintain axonal branch stability has not been adequately characterized.[8,26,29,86,87,106,181]

While it is generally accepted that neurite branching is mediated through complex interactions involving epigenetic cues within the microenvironment, each neuronal population is predisposed to develop particular arbor characteristics through cell-type-specific constraints on branching.[2,38,85,86,106,125,182,183] Consequently, a surprising degree of similarity in mature arborization patterns has been observed within homogeneous populations of spinal cord neurons,[184] cortical neurons,[185] or hippocampal pyramidal neurons,[4] allowed to develop on patternless substrates in reduced culture conditions. Moreover, the characteristic arbor pattern developed by each neuronal cell type in vitro largely recapitulates the branching pattern observed for the corresponding population in situ.[4,182-185] This cell-autonomous control of neuronal arbor topology is evident in the angle of branch extension, the average distance between branches, and the trend toward increasing or decreasing branch diameter between branch segments.[26,29,161,162,182,183] Further, each of these structural parameters can be regulated differently for developing axons and dendrites as neurons gradually attain cell-type-specific arbor characteristics.

Developing neurites exhibit a characteristic projection orientation, often described as a vector or branching angle, resulting in increasing arbor spread through successive bifurcations along the neurite length.[29,161,162] Interestingly, axons and dendrites differ fundamentally with respect to their outgrowth patterns and branching angles, possibly reflecting intrinsic differences in cytoskeletal composition.[29,161,162] Thus, in an essentially patternless microenvironment, axons tend to grow in straight paths and form minor bends of less than 17°, as measured through fractal analysis, attributed to the inherent "stiff elastic" properties of microtubule polymer bundles.[182,183,186-188] Significantly, the presence of local epigenetic cues during axonogenesis in situ allows regulated axon pathfinding through regions of complex topology, and consequently axons have been observed to form angles of 90° or more.[182,183,189] Similarly, axonal branch extension is generally characterized by obtuse angles.[14,26-29,40] In contrast, dendritic outgrowth generally does not entail the long-distance pathfinding characteristic of axonogenesis, and while the range of dendritic branch angles varies across neuronal cell types, branching typically produces acute angles.[14,40]

A second morphological parameter useful for characterizing developing neurons is the distance between successive branches, or segment length. Typically, segment lengths are greater in axons than dendrites, and greater between secondary branches than tertiary or higher order branches, although tremendous variation is evident within the developing nervous system.[8,9,29,180] In fact, segment lengths between successive axonal or dendritic branches also can vary significantly within the arbor of a given neuron.[29,180] Remarkably, despite such plasticity in branch distribution, each neuronal population exhibits a characteristic range of segment lengths, with the greatest dendritic segment lengths observed in hippocampal pyramidal neurons.[29,180]

An additional morphometric parameter that describes changes in neurite structure through successive branching is the taper, which is characterized as positive when the cross-sectional area increases between segments, and negative when the cross-sectional area decreases between branch points.[26,29-32] In fact, both positive and negative taper may develop along the extent of a given axon or dendrite, such that cross-sectional area can be differentially distributed between each of the branch segments generated.[26,29-32] The specific direction of neurite taper is established by the relative proportion of the microtubule population compartmentalized into successive branch segments, in combination with microtubule-associated protein (MAP) molecules that space these microtubule bundles.[29,49,50,53,190-193] Furthermore, by virtue of the differing subcellular complement of plus- and minus-end distal microtubules, and the segregation of various MAP isoforms, axons and dendrites demonstrate intrinsic constraints on the degree of taper which may result.[46,49,50,59,65,156,157,190,191,193-198] Indeed, the reorganization of various cytoskeletal elements throughout branching morphogenesis is critical for generating those structural and functional characteristics that define well-polarized neurons.

During neuronal development, dendritic trees expand through a cell-type-specific pattern of branching and branch elongation or retraction.[124,133,199,200] Accordingly, the overall rate of dendritic arborization depends on both the total number of dendrites and the branching probability for each dendrite as a function of time.[133] In fact, the rate of individual dendritic growth cone branching events may not remain constant during development, as limited intracellular constituents are increasingly partitioned among a growing number of branch segments.[124,133] For example, recent examination of dendritic branching in rat cortical neurons revealed that the branching probability for an individual dendritic segment decreases with increasing numbers of dendritic branches, likely reflecting increased competition for cytoskeletal elements.[133] Indeed, a variety of earlier studies demonstrated that cortical neurons display a pattern of robust dendritic branching during the earliest stages of development, followed by a protracted period of branch elongation without increases in branching.[133] This rapidly decreasing baseline branching rate suggests that the largest drive for dendritic branching occurs during the first phase of differentiation for at least one neuronal population, possibly reflecting significant developmental events.[133,144,199,200] Two non-mutually exclusive scenarios may account for these observations: first, cell-autonomous regulation, or exposure to certain epigenetic factors, may initiate a program of biochemical changes that gradually restricts neuronal sensitivity to branching cues; second, the progressive stabilization of individual branches may bias ongoing microtubule transport into these more mature processes. Similar analysis of axonal branching probability will be necessary to determine whether changes in the rate of branching morphogenesis follows stereotyped developmental trends for both neurite classes. In addition, since neurons employ a variety of cell-autonomous constraints on morphogenesis, it will be interesting to determine whether the rate and probability of branching are regulated differently in separate neuronal populations.

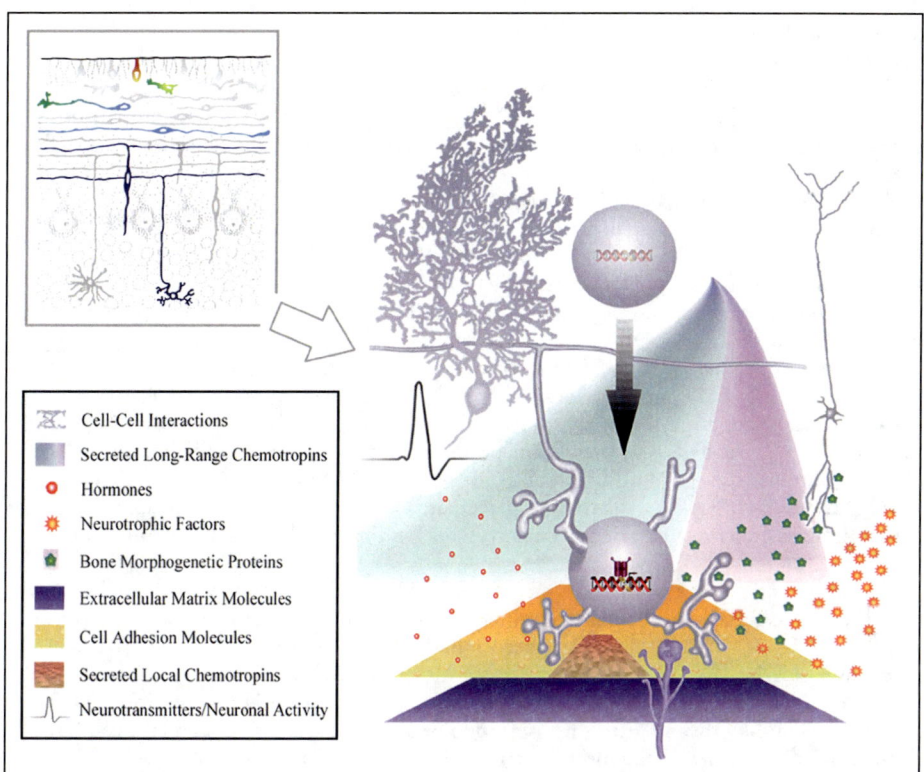

Figure 7. Epigenetic factors affecting the morphological differentiation of vertebrate neurons. While the more general morphological characteristics of developing neurons are regulated in large part though intrinsic mechanisms, attainment of mature neuronal morphology and appropriate physiology requires complex epigenetic interactions. Such epigenetic factors include: extracellular matrix components (blue plane) and cell adhesion molecules (yellow plane), secreted long-range chemoattractant and chemorepellent molecules (multi-color gradient), secreted local chemotropins (textured plane), cell-cell interactions (neurite matrix), hormones and hormone-like factors (red spheres), secreted bone morphogenetic proteins (green stars), various secreted neurotrophic factors (orange spheres), and neurotransmitters and distinct forms of neuronal activity (action potential symbol). Inset schematic illustration of the spatiotemporal sequence of cerebellar granule neuron differentiation depicts one representative neuron for each successive stage of morphogenesis. Figure reprinted with permission from K.M. Kollins, constructed from data presented in references 3, 8, 9, 18, 36, 38, 39, 44, 85, 86, 106, 125, 164, 201-203, 279, 324, 333, 360, 370, 484 and 638-641.

Control from the Outside In: Epigenetic Regulation of Branching Morphogenesis

Since the phenotype attained by a neuron is partly determined through signaling cascades triggered by extracellular cues, recent studies have sought to characterize the epigenetic factors involved in regulating branching morphogenesis. Experimental evidence suggests that many of the extracellular cues known to serve early developmental functions during neurite outgrowth and pathfinding may also modulate the more advanced stages of neuronal arborization. Such epigenetic factors include: extracellular matrix factors and cell-surface adhesion molecules, secreted long-range chemoattractant and chemorepellent molecules, hormones and hormone-like factors, bone morphogenetic proteins, neurotrophic factors, and neuronal activity (Fig. 7).[2,8,18,38,39,44,76,85-87,106,125,144,164,201]

Extracellular Matrix and Cell-Surface Adhesion Molecules

Morphological changes that accompany the development of branches involve dynamic cytoskeletal reorganization, in part regulated through interactions between neuronal cell-surface receptors and the local extracellular matrix.[44,201-210] Indeed, recent structural comparisons of neuronal populations maintained in the presence of soluble or substratum-localized extracellular matrix (ECM) molecules have implicated differential adhesion as an early mechanism underlying neurite outgrowth and arborization.[84,201,211-219]

Within an active growth cone, actin filaments respond to a variety of secreted ECM substrate components through direct interactions with cell-surface adhesion molecules.[84,202,219-223] As such, axonal outgrowth, guidance, and the establishment of appropriate neuronal connectivity, are critically dependent on the spatial distribution of ECM cues within the extracellular milieu. In fact, activation of cell-surface receptors through local or global exposure to ECM components can produce distinctly different regulatory effects on neurite extension and arborization.[8,44,224-229] For example, a local source of collagen IV promotes the outgrowth of rat superior cervical ganglion cell axonal arbors, while a substrate of the complex combination of ECM molecules in the basement membrane protein extract, matrigel, stimulates axonal elaboration and also increases the production of dendrites.[8,225,229] Most recently, the Kallmann syndrome gene product, KAL-1, was shown to mediate axonal branching for *C. elegans* sensory neurons through a novel regulatory pathway involving binding with heparin sulfate proteoglycans thought to localize to the ECM.[230]

Classic cell and substrate adhesion molecules (CAMs and SAMs) were originally characterized as cell-surface molecules that mediate both cell-cell and cell-ECM interactions.[203,206-209,215,216,223,231-239] In fact, accumulating evidence has implicated CAM/SAM activity in regulating diverse aspects of neuronal morphogenesis including stimulation of axonal and dendritic arborization.[38,200,215,216,220,221,231] Cell-surface adhesion receptors fall into four separate families: the integrins, the immunoglobulin gene superfamily, and the cadherins, all of which influence branching morphogenesis, and the selectins, which do not.[232,238,240-243] Although these CAM/SAM families are united by similar functions in regulating cell adhesion during the establishment of neuronal morphology, the specific developmental actions mediated by each adhesion molecule are quite distinct and cell context-dependent.

Integrins

Integrins constitute membrane-spanning noncovalently-bound heterodimers that facilitate local transmembrane linkages between the ECM and underlying cytoskeletal structures, such as are generated in focal adhesion complexes.[220,221,232,233,238,239,244-246] Expression of individual integrin receptor subunits can be selectively compartmentalized to a particular subcellular domain, such as the concentration of $\alpha 5$ integrin within the primary apical dendrites of hippocampal pyramidal neurons, and exclusion from basal dendritic arbors.[237,238,247-253] In this way, morphogenetic changes also can be selectively compartmentalized to particular axonal or dendritic domains. For example, the apical and basal dendritic arbors of hippocampal neurons generate strikingly different morphologies in response to local adhesive cues, due to the differential distribution of integrin receptors.[237-239]

Immunoglobulin Gene Superfamily

Members of the immunoglobulin gene superfamily (IgSF) of cell adhesion molecule/ECM receptors are widely expressed throughout the nervous system, and mediate their diverse functions, in part, through their specific membrane-linkages and cell-surface distributions.[44,206-209,254-259] Typically, IgSF members are plasma membrane-spanning molecules differentially localized to the axonal domain, such as L1, or the somatodendritic domain, as evidenced by the neural cell adhesion molecule 180 isoform ($NCAM_{180}$).[206-210,255,256] However, several IgSF members instead are tethered to the noncytoplasmic surface of the neuronal plasma membrane through a glycosyl-phosphatidylinositol (GPI) anchor, suggesting the potential for unique cell-surface localization and downstream signaling properties.[255,256,259,260]

In the case of transmembrane NCAM family members, recent studies have demonstrated that corticospinal axons elaborate interstitial branches in response to polysialic acid (PSA) modification of NCAMs that normally promote fasciculation.[261] Interestingly, PSA follows a characteristic spatiotemporal distribution pattern during development, initially restricted to distal axon segments in parallel with the onset of collateral branching, and expanding along the axon shaft in conjunction with increased branching.[261] Analysis of the physical properties of PSA during branching morphogenesis suggests that PSA modification attenuates cell-cell and cell-substrate interactions directly mediated by NCAMs, thereby reducing axonal fasciculation and enhancing axonal exposure to local branching cues.[261]

Cadherins

Cadherins comprise a superfamily of Ca^{2+}-sensitive adhesion proteins, widely distributed within the adult CNS, but characterized by distinct region-specific expression patterns for certain isoforms during nervous system development.[44,262] At least 80 members of the cadherin superfamily have been identified to date, including classic cadherins, desmogleins, desmocollins, protocadherins, cadherin-related neuronal receptors, Fats, seven-pass transmembrane cadherins, and Ret tyrosine kinases.[262,264,265] Indeed, the combinatorial expression and localization of cadherin superfamily members allows tremendous variation in the adhesive affinities of developing neurons during branching morphogenesis and the establishment of functional connectivity.[44,262-266] Within the developing cerebellum, for example, N-cadherin indiscriminately stimulates Ca^{2+}-dependent cell adhesion and neurite outgrowth among various neuronal populations. However, M-cadherin instead produces changes in neuronal adhesive affinity that selectively mediates the formation of synaptic glomeruli between granule neuron dendrites and mossy fiber axonal termini.[262,264,265]

Secreted Long-Range and Local Chemoattractants and Chemorepellents

Members of the ephrin, semaphorin, and Slit chemotropic families mediate neuronal morphogenesis through direct chemoattraction or chemorepulsion of axonal or dendritic processes in a cell-type-specific manner. Originally characterized for their function as guidance cues during axonal outgrowth, additional roles for these regulatory molecules have been described recently, including stimulation of axonal and dendritic branching morphogenesis.[2,87,106,217,267-279] Indeed, Kalil and colleagues suggest that axon guidance and interstitial branch formation might be similarly regulated by chemotropic molecules that modulate growth cone dynamics,[104-106] implicating these molecules in a variety of morphological pathways necessary for establishing functional neuronal circuits.

Ephrins

Ephrins comprise a family of multifunctional chemotropic signaling molecules, some of which serve critical roles in establishing appropriate topographic patterning of retinotectal connections during visual system development.[217,218,273-282] In addition to their well-characterized functions in the visual system, particular ephrins also have been implicated in axon repulsion within the developing hippocampus, thalamus, midbrain dopaminergic system, striatum, and olfactory cortex,[267,283,284] enhanced neurite outgrowth for sympathetic ganglion neurons,[217,267] and stimulation of axonal and dendritic branching for a variety of neuronal populations.[2,38,44,279,281] Taken together, such findings demonstrate that chemotropic ephrin molecules can function bi-directionally as either attractive or repulsive signals in a cell-type-specific manner during nervous system development.[273,276,278,279,285] Ephrin proteins are ligands for the Eph family of receptor tyrosine kinases,[273,279,286,287] currently known to comprise 14 receptor isoforms.[217,218,273,279] Ephrin ligands can be subdivided into two distinct classes based on their particular membrane linkages and Eph receptor binding affinities, which are thought to confer specific functional properties during neuronal development.[283,288-290] While ephrin-A ligands are GPI-anchored, binding preferentially to the EphA class of receptors, ephrin-B ligands instead are transmembrane proteins, and bind preferentially to cognate EphB receptors.[288] Interestingly, only GPI-anchored ephrin-A signaling stimulates morphological changes that do not directly contribute to initial axon guidance.[273,276-279,288] For example, ephrin-A5 signaling has been found to produce local changes in cortical neuron adhesion by activating β1 integrin expression, in turn stimulating changes in morphology consistent with the initial stages of neurite branching.[290-292] In contrast, EphA3 receptor activity stimulated by GPI-linked ephrin ligands decreases both apical and basal dendrite branching for developing cortical pyramidal neurons.[273,283,291] In combination, these recent studies demonstrate that ephrins serve numerous roles during neuronal morphogenesis, collectively regulating axonal or dendritic outgrowth, fasciculation, and arborization.

Semaphorins

Semaphorins constitute a large family of secreted and transmembrane proteins variously involved in growth cone collapse (Collapsin-1/Semaphorin IIIA[107]), axon guidance through chemorepulsion (Collapsin-1/Semaphorin IIIA, IIID[293-295]), stimulation or inhibition of axonal arborization (Semaphorin IIIA[85-87,296]), and neuronal apoptosis (Semaphorin IIIA[297]). Semaphorin-mediated signaling occurs in a cell-type-specific manner through the differential activation of neuropilin (NP) family receptors. Consequently, the multifunctionality of semaphorin activity may be determined chiefly through the complement of available NP receptors and downstream effector molecules as a function of neuronal development.[107,268,272,294,295,298-300] Within the mammalian cortex, Semaphorin (Sema) IIIA activity serves as a chemorepellent cue for cortical axons and inhibits branching morphogenesis, mediated by localized NP-1 receptor signaling.[295] In contrast, RGC axons in the developing *Xenopus* visual system respond to SemaIIIA/NP-1 activity with transient growth cone collapse followed by robust axonal branching upon recovery.[296,301,302] In fact, specific RGC growth cone collapse, turning, or branching responses are age-dependent and produced through varying levels of cGMP second messenger signaling, indicating changing neuronal sensitivity to SemaIIIA activity.[291,296] When immature *Xenopus* RGCs are aged in culture, they develop SemaIIIA responsiveness in parallel with age-matched RGCs developing in vivo, corresponding with the onset of NP-1 expression.[296] Expression of Plexin family receptor molecules, also critical for NP-1 signaling, follows a similar developmental time course both in vitro and in vivo.[296,303] Therefore, an intrinsic "molecular clock" mechanism has been postulated to regulate RGC growth cone sensitivity to SemaIIIA, which may be critical for confining the growth cone collapse-associated branching effects mediated by this molecule to the correct terminal position within the optic tectum.[296] Ongoing constitutive expression of SemaIIIA observed in mature populations of olfactory neurons, cerebellar Purkinje cells, and both cranial and spinal motoneurons, may indicate an additional function for this molecule in regulating branching plasticity within the adult CNS.[268,293,304]

Slit Proteins

Slit proteins are secreted factors with multifunctional roles during neuronal development, including growth cone chemorepulsion at the CNS midline during axon pathfinding, stimulation of axonal elongation in some cell contexts, and the recently characterized promotion of branching.[270,274,305-311] Slit2 is a mammalian homolog of *Drosophila* Slit, encoding a large (190-200 kDa) secreted protein that is proteolytically cleaved into two bioactive fragments, the 140 kDa amino terminus and the smaller carboxyl terminus.[270,292,305,307,308] Mammalian Slit protein signaling is mediated in turn by homologs of *Drosophila* Robo receptor proteins, Robo1-3, which demonstrate an expression pattern complementary to that of Slit isoforms during CNS development.[270] Similar to ephrin and semaphorin signaling, the morphoregulatory effects produced through Slit/Robo signaling are also cell-type-specific and depend on the developmental context of each neuronal population. For example, the purified amino-terminal fragment, Slit2-N, but not the full-length molecule, Slit2, stimulates axonal arborization for dorsal root ganglion (DRG) sensory neurons by increasing both the number of branchpoints and individual branch lengths.[270,292] Recent studies indicate that full-length Slit2 may actually function as an antagonist of Slit2-N in many contexts,[292] suggesting that the competing processes of axon guidance and axonal arborization may be developmentally regulated through post-translational enzymatic cleavage of Slit proteins.[314] However, in certain cell contexts full-length Slit proteins also stimulate branching morphogenesis. For example, early exposure to Slit2 induces premature arborization for central trigeminal axons entering the brainstem, and generates robust axonal arborization for maturing neocortical interneurons.[312-314] In addition to Slit activity during axonal morphogenesis, recent studies have demonstrated a role for Slit1 in stimulating robust dendritic growth and branching for cortical neurons developing in vitro.[311] Together, these findings indicate not only that Slit proteins serve multifunctional roles, but also that the nature of the morphoregulatory response appears to be dependent on the neuronal cell type and developmental stage, potentially allowing individual populations within the same cortical environment to pursue separate developmental programs.

Hormones and Hormone-Like Factors

Hormonal influences on neuronal morphogenesis have been examined extensively, revealing dendrite-specific regulatory effects for both gonadal hormones and glucocorticoids, and pleiotropic actions for thyroid hormones. Collectively, hormones and hormone-like factors produce numerous cell-type-specific modifications to neuronal structure and function, often exhibiting a developmental critical period or characteristic periodic time course for these effects to emerge.[38,39,203,315-319]

Gonadal Hormones

Gonadal steroid hormones serve various morphoregulatory roles during neuronal development, and mediate synaptic connectivity within the adult central nervous system.[318,321] Through these pathways, gonadal hormones have been found to influence the circuitry of vertebrate neuronal populations within the arcuate nuclei, the lateral septum, the medial amygdala, the hypothalamic ventromedial nuclei, the preoptic area, and the hippocampus.[318,320-322] Examination of dendritic structure in CA1 hippocampal pyramidal neurons has revealed that significant morphological changes are produced through cyclic fluctuations in estradiol, estrogen, and progesterone.[318,320,321] For example, in adult hippocampal slice cultures estradiol promotes dendritic elaboration and spine outgrowth,[320,322] estrogen increases dendritic spine density, but progesterone reduces both dendritic arborization and synaptic contacts.[318,321] The functional significance of such striking gonadal hormone-mediated control of dendritic morphology in CA1 hippocampal neurons remains largely unclear, however. Given the role of the CA1 neuronal population in memory storage, it is possible that these structural changes may critically link reproductive processes with learning and memory.[318] Similar effects on dendritic elaboration induced through estradiol stimulation also have been reported for slice cultures of mature hypothalamus or preoptic area cortex, indicating that a variety of CNS neuronal arbors retain a great degree of plasticity not only throughout development, but also into adulthood.[321,323,324]

Glucocorticoids

In contrast to the largely stimulatory effects of gonadal hormones on dendritic morphology, glucocorticoids instead appear to chiefly inhibit morphogenesis, producing a range of structural and physiological deficits including diminished dendritic plasticity.[325-330] For example, chronic or acute cortisol administration during a critical developmental period reduces dendrite numbers and limits dendritic arborization for stellate neurons in the embryonic chick telencephalon.[327] Similarly, late embryonic adrenalectomy results in dendritic arborization failure for slow developing hippocampal granule neurons within the rat dentate gyrus, although rapidly developing pyramidal neurons within the CA1 and CA3 regions remain unaffected.[329] As such, it is likely that glucocorticoids affect dendritic branching morphogenesis most significantly during developmental stages characterized by the greatest degree of neuronal plasticity, representing a transient period of enhanced sensitivity to epigenetic perturbations.

Thyroid Hormones

Appropriate levels of the thyroid hormones T3 and T4 are essential for regulating a wide range of developmental processes within the CNS.[3,38,315-317,331-337] Consequently, regions of the brain characterized by postnatal differentiation, including the hippocampus, olfactory bulbs, and cerebellum, demonstrate significant defects in both neuronal proliferation and morphogenesis with restricted thyroid hormone exposure.[316,317,332-337] For example, neuronal populations developing in hypothyroid animals generally exhibit reduced axonal and dendritic arbors, contrasting with greatly enhanced dendritic arborization produced for neurons developing within hyperthyroid animals.[38,335,336] Interestingly, treatment of normal neonatal rats with thyroid hormone rapidly stimulates dendritic arborization for hippocampal pyramidal neurons; however, adult hippocampal neurons exposed to thyroid hormone display only moderate dendritic spine extension.[38,335,336] Thus, it is likely that a critical period exists during which developing neurons are maximally responsive to the influence of thyroid hormone activity on branching morphogenesis.

Bone Morphogenetic Proteins

Members of the bone morphogenetic protein (BMP) subfamily of the transforming growth factor-β (TGF-β) superfamily are secreted morphogens originally characterized for their role in osteogenesis.[339,341,343] More recently, BMPs have been implicated in regulating various aspects of dendritic morphogenesis for PNS and CNS neuronal populations.[338-344] Similar regulation of axonal morphogenesis has not been reported, however, suggesting that BMP-mediated morphogenetic effects may be dendrite-specific. The particular effects these proteins exert upon dendritic differentiation appear to be independent of the extent to which the target neurons rely upon intrinsic or epigenetic cues during development. Thus, for PNS populations, in which dendritic arborization critically depends on extrinsic factors, and for CNS populations, which elaborate dendrites in reduced conditions, BMPs similarly mediate robust increases in dendritic length or branching.[339,341,342] For example, BMP-7, also called osteogenic protein-1 (OP-1), has been reported to stimulate both dendritic elongation and branching for populations of hippocampal pyramidal neurons developing in vitro.[340,345-347] Within the cerebellum, BMP-5 expression is sustained at high levels throughout development, suggesting a possible role for this factor during dendritic branching morphogenesis in early-differentiating Purkinje cells and basket/stellate neurons, as well as for later-differentiating granule neurons.[341,348]

The regulatory effects of BMP activity during branching morphogenesis are mediated through long-range signaling as gradient morphogens, and through short-range cell-cell communication. Both forms of BMP-mediated signaling are initiated through ligand-dependent association of hetero-oligomeric complexes of type I/II serine/threonine kinase receptors.[341,343] These heteromeric receptor complexes act in a combinatorial manner to induce distinctly different signaling responses, depending on the identity of the associated type I and II receptors.[341,343] To date, however, those molecular mechanisms downstream of BMP activity that mediate aspects of dendritic differentiation remain largely uncharacterized. Interestingly, OP-1 has been found to regulate expression of the cell adhesion molecules L-1 and N-CAM in a neural cell line, suggesting that one important pathway affected by BMPs during dendritic morphogenesis may involve modulation of neuronal adhesion.[201,349] Members of the TGF-β family regulate synthesis of homeodomain-containing transcription factors; therefore, it is possible that the morphoregulatory signaling mediated by diverse BMPs also converges at the level of differential gene expression.[339,341,350,351]

Neurotrophic Factors

Neurotrophic factors, a family of structurally related homodimeric proteins originally identified through their ability to sustain neuronal survival, also regulate diverse aspects of neuronal differentiation and morphogenesis. These epigenetic proteins, widely expressed throughout both the developing and mature PNS and CNS, include the prototypic neurotrophic factor, nerve growth factor (NGF), brain-derived neurotrophic factor (BDNF), neurotrophin-3 (NT-3), neurotrophin-4/5 (NT-4/5), and novel neurotrophins recently cloned from fish (NT-6 and NT-7).[86,125,168,169,352-369] In many regions of the mammalian CNS, neurotrophin expression is developmentally regulated, producing a characteristic pattern of maximum NT-3 synthesis and release during embryonic periods, with BDNF availability instead largely restricted to periods of postnatal differentiation.[370-372] In turn, both the synthesis and release of neurotrophins can be upregulated dynamically through positive feedback signaling triggered either by additional neurotrophin stimulation, or by various forms of neuronal activity.[367,368,370,373-388] Thus, precise spatiotemporal control of neurotrophin signaling through regulated expression and release may underlie distinct stages of neuronal morphogenesis.

Signal transducing receptors for the neurotrophins constitute two distinct classes: the solitary low-affinity p75 receptor (p75^{LNTR}), a member of the tumor necrosis factor receptor/fas/CD40 superfamily, and a structurally related group of high-affinity proto-oncogene receptor tyrosine kinases, collectively referred to as Trk receptors. These include TrkA, the cognate receptor for NGF, TrkB, which binds both BDNF and NT-4/5 with high affinity and NT-3 with low affinity, and TrkC, the high-affinity receptor for NT-3.[367,368,389-406] Trk receptor signaling influences neuronal differentiation by eliciting both short-term changes in protein phosphorylation, and long-term changes in gene expression through the activation of transcription factors such as cyclic AMP-response-element-binding protein (CREB).[163,168,353,366,384,399,400,405,407-412] The modulation of particular aspects of differentiation by neurotrophin activity appears to depend on the cell-surface distribution of Trk receptor isoforms, the activation of individual transduction pathways, the availability of downstream signaling substrates, and the developmental context of exposed neurons. In this way, neurotrophins can regulate neuronal precursor proliferation and initiate neurite outgrowth, as well as stimulating processes associated with terminal differentiation, including branching morphogenesis and synapse formation.[19,125,166-169,397,402-404,413,414]

The numerous structural changes underlying axonal and dendritic branching morphogenesis ultimately arise through the combined effects of differential gene induction and reorganization of specific cytoskeletal proteins, processes known to be downstream targets of neurotrophins.[86,125,163,168] For example, actin, microtubules, and numerous microtubule-associated proteins have been reported to be regulated by NT-3 and BDNF at the level of both transcription and post-translational modification.[415-421] Despite their ability to exert similar influences on the cytomatrix of axons and dendrites, neurotrophic factors also can produce separate and distinct structural changes for each of these neurite classes during branching morphogenesis.

Neurotrophin Modulation of Axonal Branching

Throughout the nervous system, target-derived neurotrophins regulate axonal outgrowth and branching morphogenesis during neuronal development, and mediate aspects of cytoskeletal reorganization that underlie presynaptic plasticity changes in mature neurons.[19,86,125,168,312,358,359,422-424] These effects are particularly well characterized in the vertebrate visual system.[144,382,407,425-442] For example, BDNF promotes rapid and prolonged changes in both the complexity and dynamic state of *Xenopus* RGC axonal arbors in vitro, with similar effects induced through neurotrophin microinjection or delivery through microspheres localized to the optic tectum in vivo.[144,431-434,443] Interestingly, this role for BDNF in preferentially stimulating RGC axonal arborization is achieved, in part, due to the localization of TrkB receptors to the axonal cell-surface domain of these neurons.[144,433,442,443] Within the primary visual cortex, infusion of BDNF or NT-4/5 into layer IV blocks the appropriate formation of ocular dominance columns during mammalian visual system development, presumably through exuberant branching of axonal terminals arising from the lateral geniculate nucleus, the functional counterpart of the non-mammalian optic tectum.[444,445] Similarly, chronic BDNF exposure during cerebellar granule neuron development in vitro stimulates atypical early increases in axonal branching and branch elongation.[446] Thus, the strict spatiotemporal regulation of Trk receptor expression, and neurotrophin release, is likely to impose direct local control on axonal branching morphogenesis necessary for the formation of appropriate circuitry during CNS development.

Neurotrophin Modulation of Dendritic Branching

Neurotrophic factors act on dendrites in a cell-type-specific manner to modulate the initial extent of arborization, regulate branch stabilization or retraction, and mediate ongoing dynamic changes in branch structure that underlie synaptogenesis.[38,125,435-453] While neurotrophic factors influence dendritic branching morphogenesis throughout the developing nervous system, these effects are often manifested differently for PNS and CNS populations, partially reflecting differences in the complement of neurotrophins and Trk receptors expressed.[38,125,435-453] For autonomic ganglion cells of the PNS, there is a clear correlation between the number of afferent inputs received through the process of preganglionic convergence, and the maximum dendritic arborization achieved, suggesting a critical role for limiting levels of target-derived neurotrophin.[19,454,455] Consistent with this model, increased NGF signaling in sympathetic ganglion cells acts instructively to induce rapid increases in both the number and length of primary dendrites, and enhance dendritic arborization.[19,352,454-457]

Within the CNS, each morphological attribute of dendritic arbors, including the number and length of primary and higher-order branches, can be differentially regulated by separate neurotrophic factors.[38,125,418,458,459] The range of dendritic branching responses elicited by individual neurotrophins has been particularly well described for pyramidal neurons developing within the visual cortex.[125,144,407,408,425-427,443] In this system, each neurotrophin produces distinct structural changes in either the apical or basal dendritic arbors of pyramidal neurons as a function of their laminar position within the cortex.[125] Thus, for pyramidal neurons developing within cortical layer IV, basal dendritic branching is stimulated with exogenous BDNF and NT-4, but not NT-3, while in layer VI, NT-4 increases dendritic arborization, NT-3 elicits no changes, and BDNF instead decreases dendritic branching.[125] Significantly, these lamina-specific responses demonstrate exquisite local control of instructive morphological changes within functional groups of neurons. It is likely that this degree of autonomy is necessary for pyramidal neurons to develop phenotypically appropriate dendritic arbors, allowing the establishment of particular synaptic input patterns.[125] In contrast to the segregated activity of particular neurotrophins within the developing visual cortex, other CNS populations instead require combinatorial neurotrophin signaling during branching morphogenesis. For example, establishment of appropriate dendritic arbors within the cerebellum necessitates cooperative BDNF and NT-3 signaling for Purkinje cells,[370,460-462] although BDNF alone mediates dendritic elongation and branching for their synaptic partners, the stellate/basket cells and granule neurons.[179,370,414,424,446,463-471]

Endogenous neurotrophins also produce antagonistic effects on dendritic branching in a cell context-dependent manner.[86,87,125,144] In fact, neurotrophins are thought to serve as potent triggers for dendritic branch retraction under certain conditions, possibly by inducing a requirement for stabilization of nascent branches, mediating destabilization of actin or tubulin polymers, or upregulating expression of repellent molecules.[86,125,459] For example, BDNF stimulation of developing *Xenopus* RGCs significantly increases axonal arborization, but paradoxically reduces dendritic branching through local inhibition of both branch formation and stabilization.[430-434,443] These antagonistic branching effects are especially interesting given the cell-surface concentration of TrkB receptors along the axon,[144,433,442,443] suggesting that changes in dendritic structure might arise indirectly through downstream signaling cascades triggered at the axonal domain. Alternatively, low-level BDNF signaling achieved through sparsely distributed TrkB receptors within immature dendrites might generate a distinctly different complement of downstream events. It is also possible that, in certain neuronal populations, neurotrophin signaling alternately stimulates or inhibits branching depending on the changing developmental context in which axons and then dendrites are elaborated. Such differential effects could arise through changes in the association of particular neurotrophins with ECM microdomains,[472] through localized neurotrophin delivery by target axons, through changes in Trk receptor expression or cell-surface distribution,[358,359,377,442,473-476] or through changes in downstream signaling substrates or background gene expression.[477-479]

Consistent with this developmental model, recent data suggests that distinct morphological effects elicited by neurotrophin activity are partly contingent upon the unique ontogenetic history of each neuron.[125,477-479] For example, BDNF-mediated TrkB signaling is required during early stages of cerebellar Purkinje cell development to allow complete dendritic differentiation during much later stages of maturation.[460,461,463,464] In fact, phenotypically mature dendritic arbors fail to develop in postnatal Purkinje cells cultured in the absence of BDNF-secreting granule neurons, and appropriate dendritic differentiation is not rescued with supplemental BDNF.[406,460,461] Instead, exogenous BDNF selectively increases dendritic spine density for these Purkinje cells, demonstrating that individual aspects of dendrite arbor structure may be independently regulated by neurotrophins during development.[460,461] When considered together, recent findings indicate that a variety of pathways contribute to the spatiotemporal regulation of neurotrophin signaling, enabling the local control of dendritic branching that is necessary for establishing appropriate circuitry during nervous system development.

Neurotransmitters and Neuronal Activity

Within the developing nervous system, a variety of neurotransmitter classes directly regulate morphological differentiation, and organize local circuit connections, by generating specific patterns and levels of electrical activity.[38,39,125,144,480-484] In tandem, neuronal activity can variously enhance or restrict the expression of numerous epigenetic molecules implicated in branching morphogenesis, modulating the physiological effects produced by these cues.[70,231,378,379,471,473,485] As such, a limited number of active signaling pathways can generate a complex array of effects during neuronal differentiation, contingent upon the complement of molecular effectors recruited.[70,231,378,379,471,473,485] Collectively, the processes regulated or modulated by neuronal activity that influence branching morphogenesis involve: selective induction of downstream second messenger signaling cascades, regulated changes in intracellular Ca^{2+} levels, posttranslational protein modifications, and differential gene expression.[2,18,38,39,44,76,106,125,144,164,195,196,480-482,484,486,487]

For most CNS populations, axonal differentiation occurs under electrically silent conditions, or with low-levels of spontaneous activity, whereas dendritic development proceeds in the context of specific forms and patterns of electrical activity.[2,15,38,106,125,144,488,489] Not surprisingly, electrical activity serves important roles in sculpting dendritic architecture at the earliest stages of neuronal development, and both establishing and refining synaptic connections at maturity.[15,38,106,125,144,346] In fact, individual neuronal populations can respond to various sources of electrical activity in a cell context-dependent manner, illustrating some degree of cell-autonomous control during dendritic branching morphogeneisis.[2,15,38,125,144,346] For example, tetanic electrical stimulation increases both the rate and number of dendritic filopodia extended from hippocampal pyramidal neurons developing in slice cultures, while excitatory N-methyl-$_D$-aspartate (NMDA) glutamate receptor signaling instead abolishes filopodial dynamics.[2,38,125] In contrast, NMDA-mediated activity is necessary for appropriate extension of RGC dendritic filopodia.[15,38,125,144,490] Significantly, inhibitory γ-aminobutyric acid (GABA) receptor activation reverses the effects of glutamatergic signaling during dendritic morphogenesis for a number of neuronal populations, suggesting that physiological sculpting of neuronal architecture may occur through variable patterns of excitatory and inhibitory activity.[2,15,38,125,144,164,203,484,491,492]

During later stages of neuronal development, the processes of dendritic arborization and synaptogenesis are closely linked, such that appropriate patterns and levels of activity serve to stimulate or restrict branching and synapse formation in parallel.[2,15,19,38,106,125-127,135,144,346,493-495] Accordingly, Vaughn's synaptotropic hypothesis proposes that dendritic outgrowth and branching may be modulated directly by synaptic contacts, with arborization occurring preferentially in target regions retaining the greatest availability of presynaptic elements.[126,127,135,493,494] Indeed, developmental studies have shown that increasing the number of axonal contacts made with dendritic filopodia increasingly stimulates dendritic branching and arbor complexity, while activity blockade variously stunts branching or elicits dendrite retraction.[15,38,125,134,136,144,496] Moreover, activity-mediated enhancement or maintenance of dendritic arbors is often localized to the discrete regions of contact between presynaptic axonal termini and postsynaptic dendritic branches or spines.[15,38,125,144] Another important observation is that the direction of dendritic arborization is governed by a tendency toward maximizing the number of contacts made with presynaptic axons from specific target regions that display physiologically relevant activity patterns.[15,38,125,133,144] For example, in the developing visual system, ocular dominance columns are established and maintained through the segregation of eye-specific afferent inputs that are characterized by distinct patterns of electrical activity. In this way, the dendritic growth properties for neurons positioned within individual columns can be differentially regulated as a function of location, generating eye-specific circuits.[38,125,145,146,437,438] Consequently, dendrites near an ocular dominance column border region elongate and branch only within the confines of that discrete functional unit, and do not extend inappropriate arbors into the adjacent column.[38,125,145,146,437,438] It has been postulated that nascent branches may "test" the local microenvironment for appropriate target axons displaying particular patterns or levels of activity, with initial contacts developing into synapses only when physiological activity is maintained, in some contexts triggering stop-growing pathways.[15,144] Because specific activity patterns also have been found to direct selective pruning of dendritic arbors, neuronal activity is thought to regulate the overall degree and rate of dendritic branch additions and retractions as neurons mature and form synaptic contacts.[15,38,125,144]

Throughout the developing nervous system, dendritic branches unable to compete successfully for limiting amounts of neurotransmitter, or axonal surface area, are often stunted or eliminated through pruning. For example, during cerebellar development, Purkinje cell dendrites that fail to establish functional synapses with granule neurons differentiate incompletely, suggesting a critical role for glutamatergic activity in the morphogenesis of these arbors.[463,464,482,486,487,497-501] Although eliminating endogenous electrical activity prevents dendritic branching for immature Purkinje cells, their dendrites continue to elongate, suggesting that functional synapses may be required to switch from a default pathway of linear growth to a developmental pattern of branching.[463,464,482] Interestingly, such alterations in dendritic growth patterns occur in parallel with changing intracellular Ca^{2+} regulation, indicating that the onset of electrical activity may produce shifts in Ca^{2+} homeostasis, thereby stimulating various forms of Ca^{2+}-dependent cytoskeletal remodeling important for dendritic arborization.[15,38,125,144,482] Since embryonic exposure to BDNF is also necessary for subsequent elaboration of Purkinje cell dendrites,[460-464] the chief role of neuronal activity in modulating branching morphogenesis may be to integrate or amplify downstream signaling cascades stimulated by a range of epigenetic cues. Alternatively, early neurotrophin signaling may be required to prime developing neurons to respond appropriately to subsequent electrical activity by inducing selective gene expression or stimulating post-translational protein modifications.[460-464] Context-specific morphological responses also are produced through activity-dependent targeting and translation of TrkB mRNA within stimulated dendrites for a variety of neuronal populations, providing a complementary mechanism for dynamic local control of arbor plasticity.[383,442,475,476] Taken together, these findings indicate that the onset of neuronal activity during later stages of dendrite maturation may enable a greater degree of control over arbor development, and the orchestration of synaptogenesis, through the activation of diverse signaling pathways.

As neuronal development continues, specific forms and patterns of activity are thought to mediate the stabilization of dendritic arbors by inhibiting dynamic cycles of branch elongation and retraction, and strengthening nascent synaptic contacts.[15,38,125,144,171,172,502] One mechanism thought to underlie this activity-dependent control of arbor maturation is the differential expression of particular combinations of glutamate receptors as a function of development.[503-507] For example, in the visual system, nascent contacts between RGCs and optic tectal neurons are characterized by a large proportion of silent synapses that are exclusively mediated by NMDA receptors, potentiating dendritic arbor expansion. However, as these initial contacts are strengthened and mature as synapses, α-amino-3-hydroxy-5-methyl-4-isoxazole propionic acid (AMPA)-type receptors are additionally recruited, resulting in more robust activity and the corresponding stabilization of dendritic arbors.[15,144,508,509] Similarly, during hippocampal pyramidal neuron development, synaptically-released glutamate stabilizes dendritic spines through AMPA receptor activation, and failure to recruit sufficient AMPA receptors results in dendritic branch retraction.[15,144] In conjunction with changing glutamate receptor populations, certain cytoskeletal elements, channel proteins, and vesicular proteins, appear to be initially targeted to dendritic arbors by activity-independent mechanisms, with subsequent retention in functionally significant microdomains regulated by the onset of synaptic activity.[39,54,66-75,510] Thus, a number of important subcellular modifications are triggered by emerging neuronal activity as development proceeds, collectively promoting dendritic branching morphogenesis. When considered together, the various roles served by neuronal activity demonstrate exquisite spatiotemporal control over the arborization of individual dendrites, allowing the establishment of physiologically appropriate circuitry.

Molecular Regulation of Branching Morphogenesis

Since neurons differentiate as part of a developing circuit, instructive cues provided within the local microenvironment, and through cell-cell signaling, are critical for synchronizing the progression of branching morphogenesis among neuronal partners.[8,9,15,16,18,39,45,76,77,144,164,195,196] Ultimately, these complex signaling pathways converge at the level of the cytoskeleton, with the structural changes characteristic of neuronal branching arising through dynamic regulation of the actin cytomatrix, microtubules, and a variety of microtubule-associated proteins. Within the last decade, tremendous progress has been made toward identifying molecular mechanisms that orchestrate the cytoskeletal reorganization underlying both axonal and dendritic branching.[2,15,38,49-51,104-106,125,144] These recent studies have revealed that persistent morphological remodeling may arise in one of two ways: as the consequence of a short-duration signaling event that stimulates local post-translational modifications to certain cytoskeletal proteins, or one that activates an epigenetic sequence of long-term changes in gene expression.[15,38,125,144,163,172,399,400,511-516] Both forms of morphological regulation involve rapid changes in local intracellular free Ca^{2+} levels that subsequently produce downstream changes in protein phosphorylation. In this way, multiple epigenetic cues may be integrated through Ca^{2+}-mediated signaling pathways acting on different time scales to influence branching morphogenesis.

Calcium Signaling

During neuronal development, rapid changes in intracellular Ca^{2+} levels can be achieved through several non-mutually exclusive pathways. First, signaling through molecular asymmetries within the microenvironment can stimulate local Ca^{2+} influx or release from intracellular stores. Second, with a relatively homogeneous distribution of ECM molecules, or secreted and substrate-bound factors, local Ca^{2+} signaling can result from a cell-surface gradient of receptors and ion channels, or subcellular asymmetries in signal transduction molecules. Third, in a related mechanism, preferential membrane insertion at the growth cone leading edge through exocytosis can serve an active role in morphogenesis through the local insertion of voltage-gated or ligand-gated Ca^{2+} ion channels. Finally, the onset of depolarizing electrical activity can stimulate Ca^{2+} influx through NMDA-type glutamate receptors and other voltage-gated Ca^{2+} channels.[2,8,16,45,76,77,164,196,220,221,517,518]

Endoplasmic reticulum and mitochondria have been reported to localize at nascent branch points in developing neurons, suggesting that these organelles are spatially oriented to mediate dynamic Ca^{2+} release or uptake necessary for branching morphogenesis, in addition to regulating metabolic processes.[18,164,196,519,520] For branching achieved through simple growth cone bifurcation, induction of a local intracellular Ca^{2+} gradient can trigger F-actin depolymerization and also increase microtubule protrusion into the growth cone, in tandem with the activation of Ca^{2+}-binding proteins that modulate microtubule stability.[76,77,85-87,165,196,244,245,521-527] Thereafter, the rate of branch elongation is inversely proportional to the intracellular accumulation of free Ca^{2+}, requiring the establishment of a cytosolic Ca^{2+} gradient with the lowest levels underlying developing neurites.[8,16,18,39,45,76,77,164,196] Similar local Ca^{2+} influx and sequestration mechanisms are likely to underlie the formation of interstitial collateral branches, which also arise through coordinated dynamic changes in actin and microtubule polymerization.[2,85,87,106]

Downstream of rapid intracellular Ca^{2+} elevation, the activation of specific Ca^{2+}-dependent proteins can produce a wide range of cytoskeletal modifications that drive branching morphogenesis.[15,18,38,39,125,144,164,195,196,227,228,528] For example, studies examining developing hippocampal neurons indicate that activation of cyclic AMP (cAMP)-dependent protein kinases or Ca^{2+}/calmodulin-dependent serine/threonine kinases (CaMKs) is required for both axonal and dendritic branching, while increased protein kinase C (PKC) and protein kinase A (PKA) activation enhance dendritic arborization alone.[529,530] This redundancy in molecular control of branching morphogenesis suggests that cell context may critically determine downstream cytoskeletal changes, owing to developmental shifts in the complement of specific protein kinases expressed, the availability of second messengers, and background gene expression.

In addition to regulating branching morphogenesis through direct control of cytoskeletal plasticity, the activation of Ca^{2+} signaling cascades also provides an important pathway for modulating neuronal structure through selective changes in gene transcription.[386,388,409,410,531-533] For example, elevation of intracellular Ca^{2+} is essential for activating cAMP-response-element-binding protein (CREB), a transcription factor involved in regulating long-term changes in gene expression through activation of the cAMP-response-element (CRE) DNA target.[2,409,410,412,531,532,534] As such, a variety of extracellular molecules can induce specific immediate-early genes, such as *c-fos* and *c-jun*, and subsequently upregulate late-response genes, together orchestrating the differential protein expression required for structural and functional changes underlying branching morphogenesis.

Rho-Family GTPases

The earliest epigenetic control of neuronal branching appears to trigger rapid local modulation of the actin cytomatrix through one pathway, while inducing differential gene expression through a more protracted downstream signaling cascade. Recent findings suggest that at least one class of molecular effectors, the Rho-family GTPases, commonly mediates dynamic actin remodeling during these early stages of branching, induced by a variety of extracellular cues including: ECM interactions, secreted and substrate-bound chemoattractants and chemorepellents, neurotrophins, and electrical activity.[88,535-546] Through Rho GTPase signaling, significant changes in the local microenvironment may be rapidly translated into specific patterns of cytoskeletal remodeling which drive the competing processes of branch extension and retraction during development.[140,538-546]

Rho-family GTPases are members of the Ras superfamily of small GTP-binding proteins, and include: RhoA, RhoG, Rac1, Rac2, Cdc42, and the recently identified Rnd proteins Rnd1, Rnd2, and Rnd3/RhoE.[538-547] Acting as molecular checkpoints during neuronal development, each of these Rho-family GTPases serves a distinct function in regulating branching morphogenesis by transducing a complex array of extracellular signals into specific changes in actin polymerization.[538-547] Typically, while Rho activity increases neurite retraction or abolishes branch outgrowth, Rac and Cdc42 instead stimulate robust neurite extension and arborization.[6,45,77,88,312,313,536,545-549] For example, during *Xenopus* RGC development, Rac1 and Cdc42 promote robust dendritic branching whereas RhoA functions as a strong negative regulator for branch induction.[550] Axonal arborization also is regulated in a cell non-autonomous manner by Rho-family GTPase signaling, with high levels of Rac activity required to stimulate appropriate branching geometry for a variety of neuronal populations.[551] Characteristically, the majority of Rho-family GTPases involved in early axonal morphogenesis demonstrate overlapping roles in regulating similar aspects of dendritic differentiation during successive stages of neuronal development.

Rho-family GTPase signaling efficacy is regulated in turn by GTPase-activating proteins (GAPs) and guanine nucleotide exchange factors (GEFs), which accelerate the process of GDP/GTP exchange.[535-537,547-549] For example, Trio is a newly characterized GEF that selectively activates the Rac pathway via RhoG through one binding domain, GEFD1, and instead activates the RhoA pathway through the separate GEFD2 binding domain.[549] Recent studies also have identified a novel effector protein specifically involved in modulating the activity of Rnd GTPases, called Rapostlin.[552] Binding to activated brain-specific Rnd2 in a GTP-dependent manner, Rapostlin next binds directly to neuronal microtubules from the amino-terminal region, which has been observed to induce rapid neurite branching for PC12 cells.[552] Characterization of the developmental role of Rapostlin will be necessary in order to understand the functional consequences of this unique pathway for affecting multiple cytoskeletal elements in tandem.

Numerous effector molecules that function downstream of Rho-family GTPases have been identified, including molecules of the p21-activated kinase (PAK) family of serine/threonine kinases, Rac1- and Cdc42- specific LIM domain-containing protein kinases, and IQGAP, Cdc42-specific neuronal Wiskott-Aldrich syndrome protein (N-WASP), and Rho kinases (ROCKs).[140,536,553-559] Downstream signaling through activated Rho-family GTPases begins with stimulation of Rho-dependent kinases (such as p65[PAK] or p160ROCK) that phosphorylate one of the two isoforms of LIM kinase (LIMk1 and LIMk2), which subsequently phosphorylate cofilin and actin-depolymerizing factor to regulate the state of actin polymerization.[140,536,553,560] In addition, recent reports suggest that RhoA-dependent inactivation of myosin phosphatase may enhance the early myosin-actin interactions that ultimately drive retrograde actin flow during filopodial extension, while RhoA-induced growth-associated protein 43 (GAP-43) activity instead promotes later stages of neurite stabilization.[556-558,561-564] Given the antagonistic effects produced by Rho GTPase signaling cascades, recent investigations have focused on context-dependent recruitment of separate Rho effectors and the interplay between individual Rho GTPases. Significantly, these studies have suggested that the actin-reorganizing effects of Rac and Cdc42 GTPases are either dominant over the stabilizing actions of RhoA in actively growing neurites, or instead downregulate RhoA activity, thus potentiating outgrowth.[16,45,77,88,565] Considered together, the ability of each Rho-family GTPase to effect distinct changes in the actin cytomatrix suggests that sequential expression, or activation, of these molecules may orchestrate specific stages of neuronal branching morphogenesis.

In the early stages of neuronal branch formation, nascent F-actin enriched filopodia extend from developing neurites, and these protrusions enlarge as they are subsequently invested with microtubules.[2,38,104-106,125,537] It is likely that epigenetic cues which trigger local changes in actin polymerization during branching may do so by activating or inactivating particular Rho GTPases, allowing subsequent redistribution of microtubules and providing multiple pathways for reinforcing cytoskeletal changes.[140,551,552,566-568] Several alternative scenarios may account for these regulatory interactions throughout neuronal development. First, branch-inducing cues localized to the ECM could directly activate Rac and Cdc42 GTPases accumulated within the underlying cytosol, thereby overriding Rho and allowing microfilament depolymerization and the commencement of neurite outgrowth. Second, cues present within an extracellular microdomain could instead indirectly modulate the efficacy of Rho-family GTPase signaling by directly regulating the expression, localization, or activation state of GAPs or GEFs. Third, simultaneous activation of Rho-family GTPases and other actin-regulating effectors could destabilize the cortical actin cytoskeleton through independent but parallel pathways.[76,140,282,547,569-572] Through these mechanisms, it is possible for the combined effects of multiple actin-regulatory molecules to reinforce key signaling cascades that underlie branch formation, while also allowing precise spatiotemporal control of the dynamic actin changes produced.

In order to clarify the regulation of Rho-family GTPase activity during branching morphogenesis, recent studies have examined interactions producing actin reorganization during axonal and dendritic arborization in a variety of neuronal populations.[537,539-542,546,550-560] During development of the visual system, physiological branching of RGC axons has been shown to be a critical event in retinotectal map formation,[90,98,114,115,279,282,285,569,573] and may be induced through repellent optic tectal cell signaling.[286,296,542,574-577] In vitro, chick temporal RGC axonal growth cones rapidly collapse and retract after encountering repellent cues presented by posterior optic tectal cells.[101,296,297,565,578] Subsequently, lateral extensions appear along the RGC axon tract, reflecting a combination of nascent interstitial branches and defasciculating trailing axons separating from the pioneer axon.[101,102,279,296,297,565] This stereotyped collapse-induced branching is associated with depolymerization and redistribution of actin filaments both within the growth cone and along the axon shaft, mediated through Rho-family GTPase activity.[579] For example, inhibiting RhoA or the RhoA effector, p160ROCK, inhibits the generation of lateral extensions from cultured chick RGC axons,[565] while overexpression of Rac1B stimulates robust axonal branching.[565,568] Rho-family GTPases also have been shown to play a role in SemaIIIA-induced growth cone collapse for chick DRG neurons, and in ephrin-A5-induced collapse of chick RGC growth cones, although it is unclear whether these triggered actin redistributions are necessarily linked to the formation of interstitial branches.[579-581] Further support for a model of axonal branching in which inhibitory stimuli initiate actin rearrangement has been provided by recent experiments revealing functional interactions between Rho-family GTPases and the Plexin class of chemorepellent semaphorin receptors. In these studies, active GTP-bound Rac was found to interact directly with the cytoplasmic domain of mammalian Plexin-B1, allowing localized signal transduction to the underlying cytoskeleton.[303,535,582] Despite advances in characterizing Rho-family activity during branching morphogenesis, it remains unclear whether axonal growth cone collapse, retraction, and the formation of interstitial branches are inherently regulated independently of each other, or may be linked via the same molecular mechanisms.

In contrast to the wealth of data describing Rho GTPase activity during axonal morphogenesis, the regulatory roles of Rho-family GTPases during dendritic branching have only recently begun to be characterized.[140,142,150,550,583-585] An elegant series of studies in the developing *Xenopus* visual system initially revealed that early dendritic morphogenesis for both tectal neurons and RGCs involves downstream activation of Rac1 or Cdc42 to stimulate dendritic branching, and activation of RhoA to restrict subsequent dendritic growth.[140,150,550] Similarly, direct time-lapse imaging of developing chick RGCs demonstrated that constitutively active Rac1 or Cdc42 GTPases promote exuberant dendritic branching, while RhoA signaling instead suppresses branch formation.[142] Constitutively active RhoA has additionally been shown to reduce both dendritic growth and arborization for developing hippocampal neurons in slice cultures.[566,585] Considered together, these recent findings suggest that selective activation of particular Rho-family GTPases may constitute a conserved molecular mechanism important for dendritic remodeling and synaptic pruning as dendrites mature. Moreover, the correlated activation of specific combinations of Rho GTPases in presynaptic axons and postsynaptic dendrites is likely to be necessary for orchestrating plasticity changes that underlie synapse formation.

Since branching morphogenesis involves the dynamic redistribution of a variety of cytoskeletal elements, as axonal and dendritic branches elongate, the protruding sites of concentrated actin become engorged with microtubules.[2,15,39,85,87,144] Recent studies describing a regulatory interaction between microtubule dynamics, Rho-family GTPase activity, and actin reorganization, revealed that the various cytoskeletal changes underlying neurite branching may, in fact, be interdependent.[140,547,566,585] For example, continuous growth of microtubule polymers provides a positive feedback mechanism for Rac GTPase activation, and dynamic changes within the actin matrix of developing fibroblasts.[547] For these cells, tubulin monomers preferentially associate with Rac1-GDP, and tubulin polymerization activates Rac1, in turn allowing Rac1-GTP to dissociate from a growing microtubule and interact with the local actin matrix.[547] If this regulatory mechanism is recapitulated in differentiating neurons, the presence of growing microtubule polymers within the region of destabilized actin might be expected to amplify or potentiate filopodial dynamics produced through Rac GTPase activity. In turn, cycling between periods of microtubule growth and depolymerization would be expected to regulate the activity of Rac and Rho GTPases, respectively, modulating local actin polymerization and providing a vector for orienting the protrusion or retraction of growth cones.[547]

Several recently discovered functions for RhoA, Rac1, and Cdc42 are novel activity-based roles, likely involved in regulating ongoing axonal and dendritic maturation during synaptogenesis. For example, Cdc42 expression is induced within the CNS following the onset of several forms of neuronal activity,[586] and in turn modulates voltage-dependent Ca^{2+}-currents in vitro, as does active Rac1.[587] In yet another study, excitatory NMDA receptor-mediated signaling was shown to reduce RhoA activity in developing optic tectal neurons, thereby increasing dendritic branching and branch elongation.[150] In fact, establishment of appropriate neuronal arborization is likely to involve the dynamic regulation of RhoA-GTPase activity, allowing early robust branch outgrowth followed by cytoskeletal stabilization as synaptic contacts are initiated. According to this model, local modulation of dendritic branching could be achieved through spontaneous glutamatergic activity at the onset of synaptogenesis, leading to exuberant branching through increased Rac1 and Cdc42 activity and downregulated RhoA signaling. However, as initial synaptic contacts mature, and AMPA-type glutamate receptors are recruited,[508,509] RhoA-GTPase expression or signaling may be upregulated relative to Rac1 and Cdc42, and contribute to the stabilization of both axonal and dendritic arbors. In parallel, individual branches that fail to receive appropriate stabilizing synaptic contacts may be stunted and subsequently retract through persistent RhoA stimulation.[584,585] Alternatively, the physiological activity of RhoA may predominate during the earliest stages of neurite morphogenesis, serving as a developmental 'block' to branch outgrowth, with the onset of Rac1 and Cdc42 activity then sculpting arborization that is ultimately stabilized through synapse formation. Clearly, strict spatiotemporal control of Rho-family GTPase expression and activation is necessary, both for neuronal morphogenesis and for the establishment of appropriate circuitry within the developing nervous system.

Microtubule-Associated Proteins and Microtubule Affinity-Regulating Kinases

The development of neuronal branches involves a number of processes acting at the level of the cytoskeleton, including regulation of actin assembly, microtubule polymerization, and microtubule transport.[2,15,38,45,50-52,77,85-87,104-106,125,140,144] Following initial actin-based stages of branching morphogenesis, modulated in large part through Rho-family GTPase activity, later stages involve the combined effects of structural and motor microtubule-associated proteins (MAPs), microtubule affinity-regulating kinases (MARKs), and Ca^{2+}/calmodulin-dependent kinases (CaMKs) and phosphatases (calcineurin/PP2B).[2,8,9,15,16,39,44,45,53,54,190,191,418,506,514,515,588,589]

As nascent branches begin to fill with microtubules, ongoing morphogenesis involves a combination of branch segment lengthening and further partitioning of these segments into smaller diameter branches until a limiting size is met.[8,29,159,160] Structural MAPs bind to microtubules in a reversible but static manner to perform a number of important roles during these later stages of arborization, which include: promoting microtubule assembly through enhanced tubulin polymerization, stabilizing new microtubules through the formation of polymer bundles, regulating the spacing between individual bundled microtubules, and regulating the overall plasticity or stability of developing branches.[53,590] In turn, structural MAP activity is regulated chiefly through the actions of specific MARKs, which trigger phosphorylation changes in response to localized epigenetic cues.[53,590] Enhanced phosphorylation at most sites along a structural MAP protein serves to weaken binding interactions with microtubules, which become more widely spaced, while dephosphorylation instead limits dynamic instability by promoting ongoing tubulin polymerization and bundling.[8,16,45,53,164,190,191,195,196,590,591] Recent studies indicate that the phosphorylation state of structural MAPs may play a critical role in regulating neuronal arborization through significant changes in microtubule spacing, with MAP phosphorylation increasing the probability of branching, while dephosphorylation instead favors neurite elongation.[322,529,530,588,592] Using this experimental data to generate a computational model of branching, Hely and colleagues showed that the rate of dendritic elongation or branching can be determined theoretically from the ratio of phosphorylated to dephosphorylated MAP isoforms alone.[588] These findings suggest that the wide variation in arborization geometries displayed by neuronal populations is likely to develop, in part, due to significant differences in the subcellular localization or activation of specific MAPs and MAP kinases or phosphatases. Ultimately, such asymmetries may arise through purely stochastic fluctuations in molecular components, or through targeted signaling produced by cues within the local microenvironment.

Individual structural MAP isoforms have been categorized primarily according to their molecular weight, with high molecular mass MAPs including MAP1A, MAP1B, MAP1C, and neuron-specific MAP2A and MAP2B, and lower molecular mass proteins comprising MAP2C, MAP2D, and tau.[8,9,40,44,54] In addition, structural MAPs can be classified into functional groups with respect to characteristic differences in their subcellular distribution between phenotypically mature axonal and dendritic domains. For example, mature axons are enriched with the dephosphorylated form of microtubule-associated tau protein, while mature dendrites instead localize MAP2 isoforms and exclude tau.[8,9,44,54,590,593-595] One important consequence arising from this asymmetrical MAP distribution is the emergence of directional organelle and vesicle transport due to differences in microtubule spacing, and thus steric hindrance, between the axonal and dendritic domains.[8,9,44,53,54,190,191,590,593,596] As a result, axonal and dendritic branches attain significantly different subcellular compositions, allowing functional specializations to develop in parallel with branching morphogenesis.[2,8,9,38,40,44,85-87,125] Despite retaining separate classes of MAP isoforms and organelles, presynaptic axons and postsynaptic dendrites regulate the relative degree of microtubule plasticity through similar means as development proceeds.

For developing axons, specific microenvironmental cues are thought to trigger a pause in growth cone migration, local modification of the cytoskeletal matrix, and the establishment of a nascent branch point through localized targeting of subcellular components.[104-109] Microtubule reorganization is a critical component of this branching process, requiring localized microtubule fragmentation, or debundling, concomitant with actin accumulation.[104-109] Consequently, when microtubules within an axon remain bundled, stable branches fail to form even when transient filopodia invested with microtubules arise along the axon shaft.[104,106] Recent observations of cerebellar granule neurons developing in vitro suggest that the modifications in microtubule bundling required for axonal branch formation may be selectively mediated through changes in tau MAP phosphorylation.[594,595] As such, the dephosphorylated form of tau is enriched for portions of the axon shaft with tightly bundled microtubules, but in discrete regions of splayed microtubules, presumably representing areas of nascent branch formation, the phosphorylated form of tau predominates.[594,595] It remains unclear whether the growth cone splitting mechanism underlying branching in dendrites similarly requires microtubule splaying, although the reported increase in dendritic branching with MAP2 phosphorylation suggests that this may be the case.[53,322,530,588,597-599]

Providing an additional level of complexity to the regulatory interactions underlying microtubule dynamics is the recent finding that MAP phosphorylation varies significantly as a function of development for some neuronal populations.[503-505,597-599] Moreover, separate MAP phosphorylation sites can be regulated independently as neuronal maturation proceeds.[503-505,597-599] Such characteristic developmental changes in MAP phosphorylation reflect, in part, the regulatory activity provided by the onset of neurotrophin signaling cascades and neuronal activity. For example, administration of exogenous BDNF and NT-3 rapidly increases both the expression level and phosphorylation state of several MAP2 isoforms for early embryonic cortical neurons developing in vivo.[421] Glutamate receptor activation also efficiently regulates the phosphorylation state of various MAP isoforms.[504,505,600,601] Interestingly, the ability of certain neuronal populations to couple glutamatergic activity to changes in MAP2 phosphorylation differs with maturation, largely established by the specific glutamate receptor subtypes expressed at particular developmental stages.[503-505] For example, neonatal hippocampal pyramidal neurons respond to glutamate receptor stimulation with MAP2 phosphorylation, whereas dephosphorylation of MAP2 is only inducible in mature neurons.[504,505] Thus, for at least some neuronal subtypes, aspects of branching morphogenesis may be regulated indirectly as immature neurons gradually attain a mature complement of biochemical machinery and the competence to dephosphorylate MAPs, prolonging arbor plasticity.[504,505] Since phosphorylation of tau and MAP2 is also regulated by exposure to certain epigenetic factors, the unique developmental functions performed by these MAPs reflects a dynamic combination of cell-autonomous regulation and direct control exerted through the changing microenvironment.

In vivo, the degree of neuronal MAP phosphorylation is tightly controlled by a number of protein phosphatases and kinases, including MARKs.[53] MARKs comprise a spatiotemporally regulated gene family including MARKs1-4, each of which may occur in several splice variant forms. Structurally, these proteins possess a highly conserved N-terminal catalytic domain containing two activating phosphorylation sites, a region near the C-terminus thought to represent a membrane-targeting motif, and an extended spacer domain which is also thought to contribute to membrane localization.[53] Functionally, all MARKs are activated through phosphorylation at their catalytic domain by upstream kinases, in turn promoting MAP activation and the modulation of microtubule organization and stability that underlies the establishment of neuronal branches.[53] The physiological importance of regulated MARK activity is evident from striking observations of increasing cytoskeletal disorganization and cellular dysfunction with MARK overexpression, presumably resulting from diminished microtubule stability following the dissociation of MAPs.[53,190,191] These results have prompted the hypothesis that differential MARK activity may serve as a 'molecular switch' for developmentally relevant changes in the microtubule-based transport of microtubule polymers, vesicles, or membranous organelles.[53] According to this theory, MARK-induced depletion of MAPs within a localized region could decrease steric hindrance and in turn promote microtubule-based transport, potentially driving branch elongation. Ultimately, dynamic regulation of branch stability or plasticity during neuronal development may be mediated through the complement of particular MAP isoforms expressed, the subcellular localization of these MAPs, or dynamic changes in the MAP phosphorylation state.[53,602,603]

Several additional regulatory molecules that link microenvironmental cues with rapid cytoskeletal reorganization through changes in MAP phosphorylation include cAMP - dependent kinases, CaMKs, and the Ca^{2+}/calmodulin-dependent protein phosphatase, calcineurin.[15,38,39,125,144,195,196,227,228,409,410,531,532,588] Indeed, many of the morphological changes that occur during neuronal development are mediated through changing Ca^{2+} levels, and thus CaMK or calcineurin activity.[15,144,514,515,588,604] For example, recent studies revealed the fundamental role for Ca^{2+}/calmodulin-dependent protein kinase II (CaMKII) in dendritic branching by demonstrating a 30% reduction in branching with CaMKII inhibitors and greater than a 200% increase in branching when CaMKII activity was enhanced.[322] CaMKII mRNA is targeted to dendrites, and both translation and activation of CaMKII are enhanced by various forms of activity,[39,66,70,605] providing a mechanism for the local dynamic regulation of dendritic arbor plasticity during synapse formation.[144,506,507,529,530,588,606-611] In the basal state CaMKII is inactive, but increased Ca^{2+} influx through voltage-gated or ligand-gated Ca^{2+} channels, or triggered Ca^{2+} release from intracellular stores, results in rapid kinase phosphorylation and activation.[514,515,604] Consequently, multiple neuronal inputs, ranging from diffusible neurotrophins and local membrane-bound molecules to patterns and levels of activity, can be integrated through these Ca^{2+}-dependent pathways.[220,221,409,410,514,515,531,532,612-616] Following branch outgrowth and the development of phenotypically appropriate arborization geometry, neuronal branches are typically stabilized during synaptogenesis to allow the maturation of functional neuronal circuits.[15,38,125,144,495,506,507,608] In the developing retinotectal system, the transition from dynamic tectal cell dendritic arbor growth to structural stabilization is mediated through local increases in αCaMKII,[608] which similarly restricts the outgrowth of presynaptic RGC axonal arbors.[15,144,506,608] In order to enhance the formation and stabilization of appropriate synaptic connections in this maturing circuit, patterns of activity both restrict and maintain dendritic arborization by modulating MAP-binding interactions through changing CaMKII activity. Ultimately, as functional connectivity develops within the nervous system, dynamic regulation of branching morphogenesis requires a cell-type-specific balance of microtubule stability and plasticity modulated through rapid changes in MAP phosphorylation.

Gene Expression Changes during Branching Morphogenesis

Many of the molecular signaling cascades that underlie neuronal branching morphogenesis ultimately produce long-term changes in gene expression. Converging at the level of the cytoskeleton, various forms of epigenetic regulation allow both short- and long-term modifications in the transcriptional or translational level of immediate-early genes (IEGs) and late-response genes (LRGs). While LRGs constitute a wide array of gene types, IEGs instead comprise two classes, the transcription factor genes and the effector genes, both of which are involved in neuronal morphogenesis and the modulation of synaptic plasticity.[15,18,38,125,144,163,164,196,409,410,491,492,515,531,532,534,617-620] In addition to well-characterized changes in the expression of various cytoskeletal element genes that occur in parallel with neuronal branching, significant changes in effector genes also have been recently identified. Of the novel effector genes known to be directly involved in neuronal branching morphogenesis, the best characterized include: the activity-regulated cytoskeletal-associated (*Arc*) gene, the closely related neuronal activity-regulated pentraxin (*Narp*) gene, and a variety of candidate plasticity genes.[73,151-155,502,508,509,621-624]

Developmental regulation of gene expression within the CNS often begins with excitatory glutamatergic signaling, providing one important pathway for modulating structural and functional plasticity changes through the activation of particular second messenger cascades.[490,515,533,534,612,618-620,625-627] For example, activity-mediated elevation of intracellular Ca^{2+} through voltage-gated ion channels is essential for inducing CREB, a transcription factor involved in generating long-term changes in gene expression through a Ca^{2+}/calmodulin-dependent kinase (CaMKIV) regulated pathway.[409,410,532-534] In fact, recent findings indicate that the activity-dependent regulation of protein synthesis and protein targeting necessary for neuronal morphogenesis can occur through several independent pathways. First, certain physiologically relevant patterns or levels of activity can induce transcription factor genes, such as *CREB*, in turn stimulating mRNA transcription and translation within the soma for transport to diverse neuronal sites.[39,54,66-68,70-75,510,622] Second, appropriate patterns or levels of activity can target newly transcribed mRNAs toward remote domains for local translation at their site of function, as occurs with dendritic *Arc* and *Narp* genes.[73-75,502,621,622] Third, the onset of physiologically relevant activity can selectively enhance the translation of mRNA previously localized to a particular subcellular region, as recently described for *Arc* and also certain neurotrophic factors.[39,54,66-75,621,622] In combination, these activity-mediated pathways allow rapid and selective induction of IEGs, and subsequent regulation of downstream LRGs, collectively producing the differential protein expression required for branching morphogenesis.[409,410,531,532,534,618-620]

The molecular control of activity-dependent dendritic remodeling critically involves *Arc* and *Narp* genes, both of which are present at low levels basally and induced by strong electrical activity.[502,621] Following robust excitatory stimulation, newly synthesized *Arc* mRNA is targeted specifically to those regions of the dendritic arbor receiving strong activity, aggregating at postsynaptic sites. In tandem, *Arc* mRNA is depleted from neighboring unstimulated dendrite segments, suggesting that electrical activity may trigger the redistribution of a preexisting pool of mRNA, in addition to stimulating gene transcription.[621,622,628,629] In fact, recent findings suggest that strong electrical stimulation effectively 'tags' a synaptic site for preferential accumulation of newly synthesized *Arc* mRNA.[628,629] Thereafter, translated Arc protein remains localized to the actin-rich matrix beneath the plasma membrane of activated dendrite segments, which may establish a spatially-regulated pathway linking robust NMDA-mediated activity with local cytoskeletal modifications.[73-75,502,621,622,628,629] Although the precise function of Arc protein during dendritic branching morphogenesis and synapse formation remains unclear, recent findings support a role for rapid *Arc* gene induction in memory consolidation. Accordingly, dendritic targeting and translation of *Arc* mRNA is associated with the stabilization of long-term potentiation (LTP) in the hippocampus, a specific inducible form of synaptic plasticity.[73-75,622,628,629]

Similar to *Arc*, the *Narp* immediate-early gene is induced by excitatory glutamatergic activity during synaptogenesis, and also modulates ongoing changes in dendritic arbor plasticity.[15,144,508,509,630-633,650] Widely expressed within the developing nervous system, Narp protein is one member of the pentraxin family of secreted Ca^{2+}-dependent lectins, and forms N-terminal covalently-linked complexes with NP1 pentraxin.[630-633,650] In fact, recent studies indicate that the ratio of these aggregated pentraxins depends on both the developmental and electrical history of a neuron, with *Narp* rapidly induced by robust activity and subsequently integrated into NP1 assemblies at excitatory synapses.[630-633,650] Once these mixed pentraxin assemblies form, their homologous C-terminal domains promote AMPA-type glutamate receptor clustering, effectively increasing synaptogenic activity within the stimulated dendrite segment.[650] Further, as a direct consequence of localized Narp-mediated changes in glutamate receptor distribution and activity, the potentiation of excitatory signaling drives dendritic branching morphogenesis and synapse strengthening in tandem.[15,144,154,508,617,630-633,650] For example, nascent contacts between RGCs and optic tectal neurons are initially characterized by a large proportion of silent synapses, mediated by NMDA receptors, which subsequently recruit AMPA-type receptors as activity increases and synapses are stabilized.[15,144] Since active Narp/NP1 assemblies co-cluster AMPA-type receptors with pre-existing NMDA receptors at these developing excitatory synapses, reinforcing intracellular signaling cascades can efficiently promote ongoing structural and functional maturation of neuronal circuits.

A third key group of immediate-early genes activated through glutamatergic signaling comprises the candidate plasticity genes (CPGs), which encode transcription factors important for establishing functional CNS circuits and modulating structural plasticity during synaptogenesis.[151-155,634] Of the CPGs identified to date, the most completely characterized is CPG15, or *neuritin*,[155] encoding a small activity-regulated protein anchored to the extracellular cell surface by a glycosyl-phosphatidylinositol (GPI) linkage. Because of this structural motif, CPG15 is highly mobile within the plasma membrane, allowing it to act as a local cell-surface growth-promoting molecule for closely apposed neurons.[151-154,634] CPG15 is widely expressed by neuronal populations within the visual, auditory, and olfactory systems during the sequential stages of dendritic branching, afferent innervation, and synaptogenesis.[151-155,634] For example, during visual system development, CPG15 is expressed by presynaptic RGCs and also by postsynaptic optical tectal neurons, suggesting a regulatory pathway which may coordinate the differentiation of multiple neuronal structures that comprise functional synapses.[151-154,634] Indeed, the current understanding of *Xenopus* visual system maturation proposes that visual activity induces local expression of CPG15 within the developing RGC axonal domain, in turn stimulating dendritic differentiation for target optic tectal neurons.[152,153] In parallel, CPG15 expressed by optic tectal neurons is transported to their axonal domains, and is thought to influence dendritic arborization for postsynaptic target neurons in a cell non-autonomous manner.[152,153] Moreover, postsynaptic tectal neuron expression of CPG15 has been observed to reciprocally enhance the elaboration of presynaptic RGC axons, promoting the formation and maturation of functional retinotectal synapses through the recruitment of non-NMDA glutamate receptors.[154]

Considered together, the various developmental roles served by activity-induced IEG expression demonstrate a similar high degree of spatiotemporal control. It is likely that such regulation is critical for orchestrating the local progression of branching morphogenesis and synaptogenesis through the integration of complex intracellular signaling cascades.

Future Directions

Within the developing vertebrate nervous system, strict control of morphogenesis is essential since the geometry of neuronal arbors critically influences the establishment of physiologically appropriate circuitry. As such, one aim of intense study within recent decades has been to identify and characterize the specific molecules and cellular mechanisms underlying spatiotemporal control of neuronal branching. Tremendous progress has been made toward identifying extracellular cues that stimulate or inhibit branching, and intracellular pathways leading to cytoskeletal reorganization as branching proceeds. However, many aspects of neuronal branching morphogenesis remain poorly understood. In the future it will be important to determine: (1) the relative importance of developmental critical periods versus spatiotemporal restriction of neuron exposure to epigenetic cues in producing appropriate arborization patterns, (2) the different subcellular effects stimulated by local versus global neuron exposure to epigenetic cues known to affect branching, (3) the role of cell-autonomous programs of background gene expression in conjunction with gene induction triggered by changing epigenetic cues, (4) the combinatorial branching effects generated by multiple cues within the extracellular milieu, and the mechanisms involved in integrating these complex signaling cascades, and (5) the pathways involved in synchronizing axonal and dendritic arbor plasticity changes throughout synaptogenesis, first as the CNS develops and later during learning and memory consolidation.

Facilitating these studies, recent advances in high-resolution real-time imaging of living neurons will allow direct observation of the formation and maturation of neuronal arbors. In this way, delayed interstitial branching and collapse-induced branching can be examined under a variety of conditions that enhance, perturb, or direct branch outgrowth. In addition, recent developments in microcontact printing and magnetic patterning techniques, designed to control the positioning of individual neurons forming cellular circuits in vitro,[649] may prove to be a powerful tool for studying cell-cell interactions involved in triggering or abolishing branching. The ability to employ green fluorescent protein (GFP)-tagged protein constructs, or inject fluorescent dyes and molecular probes, will greatly enhance visualization of the interplay between microtubules, MAPs, molecular motor proteins, and actin filaments throughout branching morphogenesis. Moreover, real-time imaging of these cytoskeletal elements will be imperative for unraveling the precise sequence of subcellular reorganization underlying discrete stages of branching morphogenesis. In combination with this technique, neurons undergoing behaviors of interest will need to be fixed and prepared for EM imaging and 3-D reconstruction in order to characterize the full complement of ultrastructural changes occurring during successive stages of branch formation. Elucidating the complex combination of molecular pathways regulating branching morphogenesis will be greatly aided by advances in proteomics and recombinant gene technology, together with the availability of a wide array of genetic mutants.

One of the greatest challenges in the field of developmental neuroscience is first to identify the molecules and mechanisms both necessary and sufficient for regulating branching in all neuronal cell-types, and then to determine the relative importance of cell context and ontogenetic history in sculpting cell-type-specific arbors. Future studies in these underrepresented areas of investigation will be critical for a more complete understanding of neuronal branching morphogenesis.

References

1. Ramón y Cajal S. Histologie du système nerveus de lí homme et des vertèbrès 1911. In: Azoulay L, ed. Translator. Madrid: Reprinted by Instituto Ramón y Cajal del CSIC, 1952-1955.
2. Acebes A, Ferrus A. Cellular and molecular features of axon collaterals and dendrites. Trends Neurosci 2000; 23(11):557-565.
3. Altman J, Bayer SA. Development of the cerebellar system: In relation to its evolution, structure and functions. In: New York: CRC Press, 1997.
4. Banker GA, Cowan WM. Further observations on hippocampal neurons in dispersed cell culture. J Comp Neurol 1979; 187(3):469-493.
5. Banker G, Goslin K. Developments in neuronal cell culture. Nature 1988; 336(6195):185-186.
6. Bartlett WP, Banker GA. An electron microscopic study of the development of axons and dendrites by hippocampal neurons in culture. II. Synaptic relationships. J Neurosci 1984; 4(8):1954-1965.
7. Bartlett WP, Banker GA. An electron microscopic study of the development of axons and dendrites by hippocampal neurons in culture. I. Cells which develop without intercellular contacts. J Neurosci 1984; 4(8):1944-1953.
8. Craig AM, Banker G. Neuronal polarity. Annu Rev Neurosci 1994; 17:267-310.
9. Craig AM, Jareb M, Banker G. Neuronal polarity. Curr Opin Neurobiol 1992; 2(5):602-606.
10. Mertz K, Schilling K. Differentiation and morphogenesis of cerebellar interneurons developing under controlled in vitro conditions. Ann Anat. 2001; 183(4):389-390.
11. Miller MW. Maturation of rat visual cortex: IV. The generation, migration, morphogenesis, and connectivity of atypically oriented pyramidal neurons. J Comp Neurol 1988; 274(3):387-405.
12. Purves D, Lichtman JW. Geometrical differences among homologous neurons in mammals. Science 1985; 228(4697):298-302.
13. Ventimiglia R, Jones BE, Moller A. A quantitative method for morphometric analysis in neuronal cell culture: Unbiased estimation of neuron area and number of branch points. J Neurosci Methods 1995; 57(1):63-66.
14. Hoff PR, Trapp BD, de Vellis J et al. The cellular components of nervous tissue. In: Zigmond MJ, Bloom FE, Landis SC, Roberts JL, Squire LR, eds. Fundamental neuroscience. San Diego: Academic Press, 1999:41-69.
15. Cline HT. Dendritic arbor development and synaptogenesis. Curr Opin Neurobiol 2001; 11(1):118-126.
16. Bradke F, Dotti CG. Establishment of neuronal polarity: Lessons from cultured hippocampal neurons. Curr Opin Neurobiol 2000; 10(5):574-581.
17. Dowling JE. The retina: An approachable part of the brain. Cambridge: Belknap Press of Harvard University Press, 1987.
18. Mattson MP. Establishment and plasticity of neuronal polarity. J Neurosci Res 1999; 57(5):577-589.
19. Snider WD, Lichtman JW. Are neurotrophins synaptotrophins? Mol Cell Neurosci 1996; 7(6):433-442.
20. Turrigiano GG. Homeostatic plasticity in neuronal networks: The more things change, the more they stay the same. Trends Neurosci 1999; 22(5):221-227.
21. Fohlmeister JF, Miller RF. Mechanisms by which cell geometry controls repetitive impulse firing in retinal ganglion cells. J Neurophysiol 1997; 78(4):1948-1964.
22. Schaefer AT, Larkum ME, Sakmann B et al. Coincidence detection in pyramidal neurons is tuned by their dendritic branching pattern. J Neurophysiol 2003; 89(6):3143-3154.
23. Sheasby BW, Fohlmeister JF. Impulse encoding across the dendritic morphologies of retinal ganglion cells. J Neurophysiol 1999; 81(4):1685-1698.
24. Spruston N. Branching out: A new idea for dendritic function. Focus on "Coincidence detection in pyramidal neurons is tuned by their dendritic branching pattern". J Neurophysiol 2003; 89(6):2887-2888.
25. Spruston N, Jaffe DB, Johnston D. Dendritic attenuation of synaptic potentials and currents: The role of passive membrane properties. Trends Neurosci 1994; 17(4):161-166.
26. Hillman DE. Neuronal shape parameters and substructures as a basis of neuronal form. In: Schmitt FO, Worden FG, eds. The neuroscience fourth study program. Cambridge: MIT Press, 1979:477-497.
27. Hillman DE, Chen S. Plasticity of synaptic size with constancy of total synaptic contact area on Purkinje cells in the cerebellum. Prog Clin Biol Res 1981:229-245.
28. Hillman DE, Chen S. Reciprocal relationship between size of postsynaptic densities and their number: Constancy in contact area. Brain Res 1984; 295(2):325-343.

29. Hillman DE. Parameters of dendritic shape and substructure: Intrinsic and extrinsic determination. In: Laske RJ, Black MM, eds. Intrinsic determinants of neuronal form and function. New York: Alan R. Liss Inc, 1988:83-113.
30. Rall W. Branching, dendritic trees and motoneuron membrane resistivity. Exp Neurol 1959; 1:491-527.
31. Rall W. Core conductor theory and cable properties of neurons. In: Kandel ER, ed. The nervous system: Cellular biology of neurons. Bethesda: Am Physiol Soc., 1977; 1:39-97.
32. Rall W. The theoretical foundation of dendritic function. Cambridge: MIT Press, 1995.
33. Rall W, Burke RE, Holmes WR et al. Matching dendritic neuron models to experimental data. Physiol Rev 1992; 72(4 Suppl):S159-186.
34. Spruston N, Jaffe DB, Johnston D. Dendritic attenuation of synaptic potentials and currents: The role of passive membrane properties. Trends Neurosci 1994; 17(4):161-166.
35. Spruston N, Stuart G, Häusser M. Dendritic integration. In: Stuart G, Spruston N, Häusser M, eds. Dendrites. New York: Oxford University Press, Inc., 1999:231-260.
36. Kollins KM. Development of cerebellar granule neuron polarity and the regulation of morphogenesis by brain derived neurotrophic factor. Ph.D diss University of Maryland College Park, 2003, ProQuest AAT3112643.
37. Mainen ZF, Sejnowski TJ. Influence of dendritic structure on firing pattern in model neocortical neurons. Nature 1996; 382(6589):363-366.
38. McAllister AK. Cellular and molecular mechanisms of dendrite growth. Cereb Cortex 2000; 10(10):963-973.
39. Barres BA, Barde Y. Neuronal and glial cell biology. Curr Opin Neurobiol 2000; 10(5):642-648.
40. Brady S, Colman DR, Brophy P. Subcellular organization of the nervous system. In: Zigmond MJ, Bloom FE, Landis SC, Roberts JL, Squire LR, eds. Fundamental neuroscience. San Diego: Academic Press, 1999:71-106.
41. Dotti CG, Simons K. Polarized sorting of viral glycoproteins to the axon and dendrites of hippocampal neurons in culture. Cell 1990; 62(1):63-72.
42. Winckler B, Poo MM. No diffusion barrier at axon hillock. Nature 1996; 379(6562):213.
43. Winckler B, Mellman I. Neuronal polarity: Controlling the sorting and diffusion of membrane components. Neuron 1999; 23(4):637-640.
44. Powell SK, Rivas RJ. The generation of polarity in neuronal cells. In: Bartles J, ed. Advances in molecular and cell biology. JAI Press, 1998; 26:157-180.
45. Bradke F, Dotti CG. The role of local actin instability in axon formation. Science 1999; 283(5409):1931-1934.
46. Baas PW, Deitch JS, Black MM et al. Polarity orientation of microtubules in hippocampal neurons: Uniformity in the axon and nonuniformity in the dendrite. Proc Natl Acad Sci USA 1988; 85(21):8335-8339.
47. Baas PW, Ahmad FJ. The transport properties of axonal microtubules establish their polarity orientation. J Cell Biol 1993; 120(6):1427-1437.
48. Baas PW, Ahmad FJ, Pienkowski TP et al. Sites of microtubule stabilization for the axon. J Neurosci 1993; 13(5):2177-2185.
49. Baas PW. Microtubules and axonal growth. Curr Opin Cell Biol 1997; 9(1):29-36.
50. Baas PW. Microtubules and neuronal polarity: Lessons from mitosis. Neuron 1999; 22(1):23-31.
51. Baas PW. Microtubule transport in the axon. Int Rev Cytol 2002; 212:41-62.
52. Baas PW, Ahmad FJ. Force generation by cytoskeletal motor proteins as a regulator of axonal elongation and retraction. Trends Cell Biol 2001; 11(6):244-249.
53. Drewes G, Ebneth A, Mandelkow EM. MAPs, MARKs and microtubule dynamics. Trends Biochem Sci 1998; 23(8):307-311.
54. Ginzburg I. Neuronal polarity: Targeting of microtubule components into axons and dendrites. Trends Biochem Sci 1991; 16(7):257-261.
55. Baas PW, Joshi HC. γ-tubulin distribution in the neuron: Implications for the origins of neuritic microtubules. J Cell Biol 1992; 119(1):171-178.
56. Ahmad FJ, Pienkowski TP, Baas PW. Regional differences in microtubule dynamics in the axon. J Neurosci 1993; 13(2):856-866.
57. Ahmad FJ, Hughey J, Wittmann T et al. Motor proteins regulate force interactions between microtubules and microfilaments in the axon. Nat Cell Biol 2000; 2(5):276-280.
58. Wang J, Yu W, Baas PW et al. Microtubule assembly in growing dendrites. J Neurosci 1996; 16(19):6065-6078.
59. Baas PW, Brown A. Slow axonal transport: The polymer transport model. Trends Cell Biol 1997; 7:380-384.

60. Foletti DL, Prekeris R, Scheller RH. Generation and maintenance of neuronal polarity: Mechanisms of transport and targeting. Neuron 1999; 23(4):641-644.
61. Schroer TA, Sheetz MP. Functions of microtubule-based motors. Annu Rev Physiol 1991; 53:629-652.
62. Schroer TA, Sheetz MP. Two activators of microtubule-based vesicle transport. J Cell Biol 1991; 115(5):1309-1318.
63. Hirokawa N. The mechanisms of fast and slow transport in neurons: Identification and characterization of the new kinesin superfamily motors. Curr Opin Neurobiol 1997; 7(5):605-614.
64. Hirokawa N, Noda Y, Okada Y. Kinesin and dynein superfamily proteins in organelle transport and cell division. Curr Opin Cell Biol 1998; 10(1):60-73.
65. Baas PW, Sinclair GI, Heidemann SR. Role of microtubules in the cytoplasmic compartmentation of neurons. Brain Res 1987; 420(1):73-81.
66. Mohr E. Subcellular RNA compartmentalization. Prog Neurobiol 1999; 57(5):507-525.
67. Steward O. Targeting of mRNAs to subsynaptic microdomains in dendrites. Curr Opin Neurobiol 1995; 5(1):55-61.
68. Steward O, Wallace CS. mRNA distribution within dendrites: Relationship to afferent innervation. J Neurobiol 1995; 26(3):447-449.
69. Fawcett JP, Aloyz R, McLean JH et al. Detection of brain-derived neurotrophic factor in a vesicular fraction of brain synaptosomes. J Biol Chem 1997; 272(14):8837-8840.
70. Wells DG, Richter JD, Fallon JR. Molecular mechanisms for activity-regulated protein synthesis in the synapto-dendritic compartment. Curr Opin Neurobiol 2000; 10(1):132-137.
71. Eberwine J, Job C, Kacharmina JE et al. Transcription factors in dendrites: Dendritic imprinting of the cellular nucleus. Results Probl Cell Differ 2001; 34:57-68.
72. Eberwine J, Miyashiro K, Kacharmina JE et al. Local translation of classes of mRNAs that are targeted to neuronal dendrites. Proc Natl Acad Sci USA 2001; 98(13):7080-7085.
73. Steward O, Worley P. Localization of mRNAs at synaptic sites on dendrites. Results Probl Cell Differ 2001; 34:1-26.
74. Steward O, Worley PF. Selective targeting of newly synthesized Arc mRNA to active synapses requires NMDA receptor activation. Neuron 2001; 30(1):227-240.
75. Steward O, Worley PF. A cellular mechanism for targeting newly synthesized mRNAs to synaptic sites on dendrites. Proc Natl Acad Sci USA 2001; 98(13):7062-7068.
76. Andersen SS, Bi GQ. Axon formation: A molecular model for the generation of neuronal polarity. Bioessays 2000; 22(2):172-179.
77. Bradke F, Dotti CG. Neuronal polarity: Vectorial cytoplasmic flow precedes axon formation. Neuron 1997; 19(6):1175-1186.
78. Turing AM. The chemical basis of morphogenesis. 1953. Bull Math Biol 1990; 52(1-2):153-197.
79. Bray D, Bunge MB. The growth cone in neurite extension. Ciba Found Symp 1973; 14:195-209.
80. Suter DM, Forscher P. An emerging link between cytoskeletal dynamics and cell adhesion molecules in growth cone guidance. Curr Opin Neurobiol 1998; 8(1):106-116.
81. Suter DM, Forscher P. Substrate-cytoskeletal coupling as a mechanism for the regulation of growth cone motility and guidance. J Neurobiol 2000; 44(2):97-113.
82. Goldberg DJ, Burmeister DW. Stages in axon formation: Observations of growth of Aplysia axons in culture using video-enhanced contrast-differential interference contrast microscopy. J Cell Biol 1986; 103(5):1921-1931.
83. Aletta JM, Greene LA. Growth cone configuration and advance: A time-lapse study using video-enhanced differential interference contrast microscopy. J Neurosci 1988; 8(4):1425-1435.
84. Burmeister DW, Rivas RJ, Goldberg DJ. Substrate-bound factors stimulate engorgement of growth cone lamellipodia during neurite elongation. Cell Motil Cytoskeleton 1991; 19(4):255-268.
85. Gallo G, Letourneau PC. Axon guidance: A balance of signals sets axons on the right track. Curr Biol 1999; 9(13):R490-492.
86. Gallo G, Letourneau PC. Neurotrophins and the dynamic regulation of the neuronal cytoskeleton. J Neurobiol 2000; 44(2):159-173.
87. Gallo G, Letourneau P. Axon guidance: Proteins turnover in turning growth cones. Curr Biol 2002; 12(16):R560-562.
88. Bito H, Furuyashiki T, Ishihara H et al. A critical role for a Rho-associated kinase, p160ROCK, in determining axon outgrowth in mammalian CNS neurons. Neuron 2000; 26(2):431-441.
89. Forscher P, Smith SJ. Actions of cytochalasins on the organization of actin filaments and microtubules in a neuronal growth cone. J Cell Biol 1988; 107(4):1505-1516.
90. O'Leary DD, Bicknese AR, De Carlos JA et al. Target selection by cortical axons: Alternative mechanisms to establish axonal connections in the developing brain. Cold Spring Harb Symp Quant Biol 1990; 55:453-468.

91. Kater SB, Mills LR. Regulation of growth cone behavior by calcium. J Neurosci 1991; 11(4):891-899.
92. Letourneau PC. The cytoskeleton in nerve growth cone motility and axonal pathfinding. Perspect Dev Neurobiol 1996; 4(2-3):111-123.
93. Lin CH, Thompson CA, Forscher P. Cytoskeletal reorganization underlying growth cone motility. Curr Opin Neurobiol 1994; 4(5):640-647.
94. Raper JA, Tessier-Lavigne M. Growth cones and axon pathfinding. In: Zigmond MJ, Bloom FE, Landis SC, Roberts JL, Squire LR, eds. Fundamental neuroscience. San Diego: Academic Press, 1999:519-546.
95. Lewis AK, Bridgman PC. Nerve growth cone lamellipodia contain two populations of actin filaments that differ in organization and polarity. J Cell Biol 1992; 119(5):1219-1243.
96. Bray D. Branching patterns of individual sympathetic neurons in culture. J Cell Biol 1973; 56:702-712.
97. Suter DM, Forscher P. Kalil K et al. Common mechanisms underlying growth cone guidance and axon branching. J Neurobiol 2000; 44(2):145-158.
98. O'Leary DD, Terashima T. Cortical axons branch to multiple subcortical targets by interstitial axon budding: Implications for target recognition and "waiting periods". Neuron 1988; 1(10):901-910.
99. Condeelis J. Life at the leading edge: The formation of cell protrusions. Annu Rev Cell Biol 1993; 9:411-444.
100. Goodman CS. Mechanisms and molecules that control growth cone guidance. Annu Rev Neurosci 1996; 19:341-377.
101. Davenport RW, Thies E, Cohen ML. Neuronal growth cone collapse triggers lateral extensions along trailing axons. Nat Neurosci 1999; 2(3):254-259.
102. Davenport RW. Functional domains and intracellular signalling: Clues to growth cone dynamics. In: McCaig CD, ed. Nerve Growth and Guidance. London: Portland Press, 1996:55-75.
103. Cooper MW, Smith SJ. A real-time analysis of growth cone-target cell interactions during the formation of stable contacts between hippocampal neurons in culture. J Neurobiol 1992; 23(7):814-828.
104. Dent EW, Callaway JL, Szebenyi G et al. Reorganization and movement of microtubules in axonal growth cones and developing interstitial branches. J Neurosci 1999; 19(20):8894-8908.
105. Dent EW, Kalil K. Axon branching requires interactions between dynamic microtubules and actin filaments. J Neurosci 2001; 21(24):9757-9769.
106. Kalil K, Szebenyi G, Dent EW. Common mechanisms underlying growth cone guidance and axon branching. J Neurobiol 2000; 44(2):145-158.
107. Raper JA, Kapfhammer JP. The enrichment of a neuronal growth cone collapsing activity from embryonic chick brain. Neuron 1990; 4(1):21-29.
108. Szebenyi G, Callaway JL, Dent EW et al. Interstitial branches develop from active regions of the axon demarcated by the primary growth cone during pausing behaviors. J Neurosci 1998; 18(19):7930-7940.
109. Szebenyi G, Dent EW, Callaway JL et al. Fibroblast growth factor-2 promotes axon branching of cortical neurons by influencing morphology and behavior of the primary growth cone. J Neurosci 2001; 21(11):3932-3941.
110. Kuang RZ, Kalil K. Specificity of corticospinal axon arbors sprouting into denervated contralateral spinal cord. J Comp Neurol 1990; 302(3):461-472.
111. Kuang RZ, Kalil K. Branching patterns of corticospinal axon arbors in the rodent. J Comp Neurol 1990; 292(4):585-598.
112. Kuang RZ, Kalil K. Development of specificity in corticospinal connections by axon collaterals branching selectively into appropriate spinal targets. J Comp Neurol 1994; 344(2):270-282.
113. Hogan D, Berman NE. Growth cone morphology, axon trajectory and branching patterns in the neonatal rat corpus callosum. Brain Res Dev Brain Res 1990; 53(2):283-287.
114. Bastmeyer M, O'Leary DD. Dynamics of target recognition by interstitial axon branching along developing cortical axons. J Neurosci 1996; 16(4):1450-1459.
115. Bastmeyer M, Daston MM, Possel H et al. Collateral branch formation related to cellular structures in the axon tract during corticopontine target recognition. J Comp Neurol 1998; 392(1):1-18.
116. Halloran MC, Kalil K. Dynamic behaviors of growth cones extending in the corpus callosum of living cortical brain slices observed with video microscopy. J Neurosci 1994; 14(4):2161-2177.
117. Tanaka E, Ho T, Kirschner MW. The role of microtubule dynamics in growth cone motility and axonal growth. J Cell Biol 1995; 128(1-2):139-155.
118. Tanaka EM, Kirschner MW. Microtubule behavior in the growth cones of living neurons during axon elongation. J Cell Biol 1991; 115(2):345-363.

119. Gordon-Weeks PR. Microtubules and growth cone function. J Neurobiol 2004; 58(1):70-83.
120. Gordon-Weeks PR, Fischer I. MAP1B expression and microtubule stability in growing and regenerating axons. Microsc Res Tech 2000; 48(2):63-74.
121. McNally FJ. Cytoskeleton: CLASPing the end to the edge. Curr Biol 2001; 11(12):R477-480.
122. Schuyler SC, Pellman D. Microtubule "plus-end-tracking proteins": The end is just the beginning. Cell 2001; 105(4):421-424.
123. Schuyler SC, Pellman D. Search, capture and signal: Games microtubules and centrosomes play. J Cell Sci 2001; 114(Pt 2):247-255.
124. Ruthel G, Hollenbeck PJ. Growth cones are not required for initial establishment of polarity or differential axon branch growth in cultured hippocampal neurons. J Neurosci 2000; 20(6):2266-2274.
125. McAllister AK, Katz LC, Lo DC. Neurotrophins and synaptic plasticity. Annu Rev Neurosci 1999; 22:295-318.
126. Vaughn JE. Fine structure of synaptogenesis in the vertebrate central nervous system. Synapse 1989; 3(3):255-285.
127. Vaughn JE, Barber RP, Sims TJ. Dendritic development and preferential growth into synaptogenic fields: A quantitative study of Golgi-impregnated spinal motor neurons. Synapse 1988; 2(1):69-78.
128. Purves D, Hume RI. The relation of postsynaptic geometry to the number of presynaptic axons that innervate autonomic ganglion cells. J Neurosci 1981; 1(5):441-452.
129. Purves D, Hadley RD. Changes in the dendritic branching of adult mammalian neurones revealed by repeated imaging in situ. Nature 1985; 315(6018):404-406.
130. Purves D, Hadley RD, Voyvodic JT. Dynamic changes in the dendritic geometry of individual neurons visualized over periods of up to three months in the superior cervical ganglion of living mice. J Neurosci 1986; 6(4):1051-1060.
131. Dailey ME, Smith SJ. The dynamics of dendritic structure in developing hippocampal slices. J Neurosci 1996; 16(9):2983-2994.
132. Jones EG. Microcolumns in the cerebral cortex. Proc Natl Acad Sci USA 2000; 97(10):5019-5021.
133. van Pelt J, Uylings HB. Branching rates and growth functions in the outgrowth of dendritic branching patterns. Network 2002; 13(3):261-281.
134. Katz LC, Shatz CJ. Synaptic activity and the construction of cortical circuits. Science 1996; 274(5290):1133-1138.
135. Vaughn JE, Barber RP, Sims TJ. Dendritic development and preferential growth into synaptogenic fields: A quantitative study of Golgi-impregnated spinal motor neurons. Synapse 1988; 2(1):69-78.
136. Goodman CS, Shatz CJ. Developmental mechanisms that generate precise patterns of neuronal connectivity. Cell 1993; 72(Suppl):77-98.
137. Benes FM, Parks TN, Rubel EW. Rapid dendritic atrophy following deafferentation: An EM morphometric analysis. Brain Res 1977; 122(1):1-13.
138. Deitch JS, Rubel EW. Afferent influences on brain stem auditory nuclei of the chicken: Time course and specificity of dendritic atrophy following deafferentation. J Comp Neurol 1984; 229(1):66-79.
139. Deitch JS, Rubel EW. Rapid changes in ultrastructure during deafferentation-induced dendritic atrophy. J Comp Neurol 1989; 281(2):234-258.
140. Redmond L, Ghosh A. The role of Notch and Rho GTPase signaling in the control of dendritic development. Curr Opin Neurobiol 2001; 11(1):111-117.
141. Ziv NE, Smith SJ. Evidence for a role of dendritic filopodia in synaptogenesis and spine formation. Neuron 1996; 17(1):91-102.
142. Wong WT, Faulkner-Jones BE, Sanes JR et al. Rapid dendritic remodeling in the developing retina: Dependence on neurotransmission and reciprocal regulation by Rac and Rho. J Neurosci 2000; 20(13):5024-5036.
143. Baas PW, Yu W. A composite model for establishing the microtubule arrays of the neuron. Mol Neurobiol 1996; 12(2):145-161.
144. Cline HT. Development of dendrites. In: Stuart G, Spruston N, Häusser M, eds. Dendrites. New York: Oxford University Press Inc., 1999:35-56.
145. Katz LC, Constantine-Paton M. Relationships between segregated afferents and postsynaptic neurones in the optic tectum of three-eyed frogs. J Neurosci 1988; 8(9):3160-3180.
146. Katz LC, Gilbert CD, Wiesel TN. Local circuits and ocular dominance columns in monkey striate cortex. J Neurosci 1989; 9(4):1389-1399.
147. Baird DH, Baptista CA, Wang LC et al. Specificity of a target cell-derived stop signal for afferent axonal growth. J Neurobiol 1992; 23(5):579-591.
148. Baird DH, Hatten ME, Mason CA. Cerebellar target neurons provide a stop signal for afferent neurite extension in vitro. J Neurosci 1992; 12(2):619-634.

149. Baird DH, Trenkner E, Mason CA. Arrest of afferent axon extension by target neurons in vitro is regulated by the NMDA receptor. J Neurosci 1996; 16(8):2642-2648.

150. Li Z, Van Aelst L, Cline HT. Rho GTPases regulate distinct aspects of dendritic arbor growth in Xenopus central neurons in vivo. Nat Neurosci 2000; 3(3):217-225.

151. Nedivi E, Hevroni D, Naot D et al. Numerous candidate plasticity-related genes revealed by differential cDNA cloning. Nature 1993; 363(6431):718-722.

152. Nedivi E, Wu GY, Cline HT. Promotion of dendritic growth by CPG15, an activity-induced signaling molecule. Science 1998; 281(5384):1863-1866.

153. Nedivi E, Javaherian A, Cantallops I et al. Developmental regulation of CPG15 expression in Xenopus. J Comp Neurol 2001; 435(4):464-473.

154. Cantallops I, Haas K, Cline HT. Postsynaptic CPG15 promotes synaptic maturation and presynaptic axon arbor elaboration in vivo. Nat Neurosci 2000; 3(10):1004-1011.

155. Naeve GS, Ramakrishnan M, Kramer R et al. Neuritin: A gene induced by neural activity and neurotrophins that promotes neuritogenesis. Proc Natl Acad Sci USA 1997; 94(6):2648-2653.

156. Baas PW, Slaughter T, Brown A et al. Microtubule dynamics in axons and dendrites. J Neurosci Res 1991; 30(1):134-153.

157. Black MM, Baas PW. The basis of polarity in neurons. Trends Neurosci 1989; 12(6):211-214.

158. Harvey AM. Johns Hopkins—the birthplace of tissue culture: The story of Ross G. Harrison, Warren Y. Lewis, and George O. Gey. Johns Hopkins Med J 1976; Suppl:114-123.

159. Hillman DE, Chen S. Neurotubule-initiating complex and formation of a dendritic cytoskeleton. Soc Neurosci Abst 1982; 8:787.

160. Hillman DE, Cuccio E. Evidence for continuity of microtubules in dendrites. Soc Neurosci Abst 1983; 9:337.

161. Lasek RJ. The dynamic ordering of neuronal cytoskeletons. Neurosci Res Program Bull 1981; 19(1):7-32.

162. Lasek RJ. Studying the intrinsic determinants of neuronal form and function. In: Lasek RJ, Black MM, eds. Intrinsic determinants of neuronal form and function. New York: Liss Inc., 1988:3-58.

163. Segal RA, Greenberg ME. Intracellular signaling pathways activated by neurotrophic factors. Annu Rev Neurosci 1996; 19:463-489.

164. Mattson MP. Neurotransmitters in the regulation of neuronal cytoarchitecture. Brain Res 1988; 472(2):179-212.

165. Borisy GG, Svitkina TM. Actin machinery: Pushing the envelope. Curr Opin Cell Biol 2000; 12(1):104-112.

166. Davies AM. Paracrine and autocrine actions of neurotrophic factors. Neurochem Res 1996; 21(7):749-753.

167. Davies AM. The neurotrophic hypothesis: Where does it stand? Philos Trans R Soc Lond B Biol Sci 1996; 351(1338):389-394.

168. Davies AM. Neurotrophins: Neurotrophic modulation of neurite growth. Curr Biol 2000; 10(5):R198-200.

169. Chao M, Casaccia-Bonnefil P, Carter B et al. Neurotrophin receptors: Mediators of life and death. Brain Res Brain Res Rev 1998; 26(2-3):295-301.

170. Chao MV. Trophic factors: An evolutionary cul-de-sac or door into higher neuronal function? J Neurosci Res 2000; 59(3):353-355.

171. Hatten ME, Heintz N. Mechanisms of neural patterning and specification in the developing cerebellum. Annu Rev Neurosci 1995; 18:385-408.

172. Hatten ME, Alder J, Zimmerman K et al. Genes involved in cerebellar cell specification and differentiation. Curr Opin Neurobiol 1997; 7(1):40-47.

173. Rall W, Shepherd GM, Reese TS et al. Dendrodendritic synaptic pathway for inhibition in the olfactory bulb. Exp Neurol 1966; 14(1):44-56.

174. Cullheim S, Kellerth JO. A morphological study of the axons and recurrent axon collaterals of cat sciatic alpha-motoneurons after intracellular staining with horseradish peroxidase. J Comp Neurol 1978; 178(3):537-557.

175. Pfeiffer G, Friede RL. The axon tree of rat motor fibres: Morphometry and fine structure. J Neurocytol 1985; 14(5):809-824.

176. Banker GA, Waxman AB. Hippocampal neurons generate natural shapes in cell culture. In: Lasek RJ, Black MM, eds. Intrinsic determination of neuronal form and function. New York: Alan R Liss Inc., 1988:61-82.

177. Chan-Palay V, Palay SL. High voltage electron microscopy of rapid golgi preparations. Neurons and their processes in the cerebellar cortex of monkey and rat. Z Anat Entwicklungsgesch 1972; 137(2):125-153.

178. Palay SL, Chan-Palay V. A guide to the synaptic analysis of the neuropil. Cold Spring Harbor Symp Quant Biol 1976; 40:1-16.
179. Mertz K, Koscheck T, Schilling K. Brain-derived neurotrophic factor modulates dendritic morphology of cerebellar basket and stellate cells: An in vitro study. Neuroscience 2000; 97(2):303-310.
180. Buell SJ, Coleman PD. Dendritic growth in the aged human brain and failure of growth in senile dementia. Science 1979; 206(4420):854-856.
181. Edds KT. Dynamic aspects of filopodial formation by reorganization of microfilaments. J Cell Biol 1977; 73(2):479-491.
182. Katz MJ. Axonal branch shapes. Brain Res 1985; 361(1-2):70-76.
183. Katz MJ. How straight do axons grow? J Neurosci 1985; 5(3):589-595.
184. Neale JH, Barker JL, Uhl GR et al. Enkephalin-containing neurons visualized in spinal cord cell cultures. Science 1978; 201(4354):467-469.
185. Kriegstein AR, Dichter MA. Morphological classification of rat cortical neurons in cell culture. J Neurosci 1983; 3(8):1634-1647.
186. Solomon F. Organizing microtubules in the cytoplasm. Cell 1980; 22:331-332.
187. Solomon F. Specification of cell morphology by endogenous determinants. J Cell Biol 1980; 90:547-553.
188. Katz MJ, George EB, Gilbert LJ. Axonal elongation as a stochastic walk. Cell Motil 1984; 4(5):351-370.
189. Tosney KW, Landmesser LT. Growth cone morphology and trajectory in the lumbosacral region of the chick embryo. J Neurosci 1985; 5(9):2345-2358.
190. Matus A. Microtubule-associated proteins: Their potential role in determining neuronal morphology. Annu Rev Neurosci 1988; 11:29-44.
191. Matus A. Stiff microtubules and neuronal morphology. Trends Neurosci 1994; 17(1):19-22.
192. Vouyiouklis DA, Brophy PJ. Microtubule-associated proteins in developing oligodendrocytes: Transient expression of a MAP2c isoform in oligodendrocyte precursors. J Neurosci Res 1995; 42(6):803-817.
193. Ferhat L, Cook C, Chauviere M et al. Expression of the mitotic motor protein Eg5 in postmitotic neurons: Implications for neuronal development. J Neurosci 1998; 18(19):7822-7835.
194. Ferhat L, Kuriyama R, Lyons GE et al. Expression of the mitotic motor protein CHO1/MKLP1 in postmitotic neurons. Eur J Neurosci 1998; 10(4):1383-1393.
195. Mattson MP. Effects of microtubule stabilization and destabilization on tau immunoreactivity in cultured hippocampal neurons. Brain Res 1992; 582(1):107-118.
196. Mattson MP. Calcium as sculptor and destroyer of neural circuitry. Exp Gerontol 1992; 27(1):29-49.
197. Ahmed S, Reynolds BA, Weiss S. BDNF enhances the differentiation but not the survival of CNS stem cell-derived neuronal precursors. J Neurosci 1995; 15(8):5765-5778.
198. Black MM, Slaughter T, Moshiach S et al. Tau is enriched on dynamic microtubules in the distal region of growing axons. J Neurosci 1996; 16(11):3601-3619.
199. Woldenberg MJ, O'Neill MP, Quackenbush LJ et al. Models for growth, decline and regrowth of the dendrites of rat Purkinje cells induced from magnitude and link-length analysis. J Theor Biol 1993; 162(4):403-429.
200. Parnavelas JG, Uylings HB. The growth of non-pyramidal neurons in the visual cortex of the rat: A morphometric study. Brain Res 1980; 193(2):373-382.
201. Prochiantz A. Neuronal polarity: Giving neurons heads and tails. Neuron 1995; 15(4):743-746.
202. Powell SK, Williams CC, Nomizu M et al. Laminin-like proteins are differentially regulated during cerebellar development and stimulate granule cell neurite outgrowth in vitro. J Neurosci Res 1998; 54(2):233-247.
203. Burgoyne RD, Cambray-Deakin MA. The cellular neurobiology of neuronal development: The cerebellar granule cell. Brain Res 1988; 472(1):77-101.
204. Rivas RJ, Burmeister DW, Goldberg DJ. Rapid effects of laminin on the growth cone. Neuron 1992; 8(1):107-115.
205. Kadmon G, Altevogt P. The cell adhesion molecule L1: Species- and cell-type-dependent multiple binding mechanisms. Differentiation 1997; 61(3):143-150.
206. Faivre-Sarrailh C, Falk J, Pollerberg E et al. NrCAM, cerebellar granule cell receptor for the neuronal adhesion molecule F3, displays an actin-dependent mobility in growth cones. J Cell Sci 1999; 112(Pt 18):3015-3027.
207. Faivre-Sarrailh C, Gennarini G, Goridis C et al. F3/F11 cell surface molecule expression in the developing mouse cerebellum is polarized at synaptic sites and within granule cells. J Neurosci 1992; 12(1):257-267.
208. Faivre-Sarrailh C, Rougon G. Are the glypiated adhesion molecules preferentially targeted to the axonal compartment? Mol Neurobiol 1993; 7(1):49-60.

209. Faivre-Sarrailh C, Rougon G. Axonal molecules of the immunoglobulin superfamily bearing a GPI anchor: Their role in controlling neurite outgrowth. Mol Cell Neurosci 1997; 9(2):109-115.
210. Sakurai T, Lustig M, Babiarz J et al. Overlapping functions of the cell adhesion molecules Nr-CAM and L1 in cerebellar granule cell development. J Cell Biol 2001; 154(6):1259-1273.
211. Chamak B, Prochiantz A. [Axons, dendrites and adhesion]. C R Acad Sci III 1989; 308(13):353-358.
212. Lafont F, Prochiantz A, Valenza C et al. Defined glycosaminoglycan motifs have opposite effects on neuronal polarity in vitro. Dev Biol 1994; 165(2):453-468.
213. Lafont F, Rouget M, Triller A et al. In vitro control of neuronal polarity by glycosaminoglycans. Development 1992; 114(1):17-29.
214. Lafont F, Rouget M, Rousselet A et al. Specific responses of axons and dendrites to cytoskeleton perturbations: An in vitro study. J Cell Sci 1993; 104(Pt 2):433-443.
215. Edelman GM, Chuong CM. Embryonic to adult conversion of neural cell adhesion molecules in normal and staggerer mice. Proc Natl Acad Sci USA 1982; 79(22):7036-7040.
216. Edelman GM, Jones FS. Gene regulation of cell adhesion: A key step in neural morphogenesis. Brain Res Brain Res Rev 1998; 26(2-3):337-352.
217. Flanagan JG, Vanderhaeghen P. The ephrins and Eph receptors in neural development. Annu Rev Neurosci 1998; 21:309-345.
218. Flanagan LA, Ju YE, Marg B et al. Neurite branching on deformable substrates. Neuroreport 2002; 13(18):2411-2415.
219. Luckenbill-Edds L. Laminin and the mechanism of neuronal outgrowth. Brain Res Brain Res Rev 1997; 23(1-2):1-27.
220. Bixby JL, Bookman RJ. Intracellular mechanisms of axon growth induction by CAMs and integrins: Some unresolved issues. Perspect Dev Neurobiol 1996; 4(2-3):147-156.
221. Bixby JL, Grunwald GB, Bookman RJ. Ca^{2+} influx and neurite growth in response to purified N-cadherin and laminin. J Cell Biol 1994; 127(5):1461-1475.
222. Powell SK, Kleinman HK. Neuronal laminins and their cellular receptors. Int J Biochem Cell Biol 1997; 29(3):401-414.
223. Sanes JR. Extracellular matrix molecules that influence neural development. Annu Rev Neurosci 1989; 12:491-516.
224. Ruoslahti E, Vaheri A. Cell-to-cell contact and extracellular matrix. Curr Opin Cell Biol 1997; 9(5):605-607.
225. Lein PJ, Higgins D. Laminin and a basement membrane extract have different effects on axonal and dendritic outgrowth from embryonic rat sympathetic neurons in vitro. Dev Biol 1989; 136(2):330-345.
226. Lein PJ, Banker GA, Higgins D. Laminin selectively enhances axonal growth and accelerates the development of polarity by hippocampal neurons in culture. Brain Res Dev Brain Res 1992; 69(2):191-197.
227. Esch T, Lemmon V, Banker G. Local presentation of substrate molecules directs axon specification by cultured hippocampal neurons. J Neurosci 1999; 19(15):6417-6426.
228. Esch T, Lemmon V, Banker G. Differential effects of NgCAM and N-cadherin on the development of axons and dendrites by cultured hippocampal neurons. J Neurocytol 2000; 29(3):215-223.
229. Lein PJ, Higgins D. Protein synthesis is required for the initiation of dendritic growth in embryonic rat sympathetic neurons in vitro. Brain Res Dev Brain Res 1991; 60(2):187-196.
230. Bulow HE, Berry KL, Topper LH et al. Heparan sulfate proteoglycan-dependent induction of axon branching and axon misrouting by the Kallmann syndrome gene kal-1. Proc Natl Acad Sci USA 2002; 99(9):6346-6351.
231. Fields RD, Itoh K. Neural cell adhesion molecules in activity-dependent development and synaptic plasticity. Trends Neurosci 1996; 19(11):473-480.
232. Hynes RO, Lander AD. Contact and adhesive specificities in the associations, migrations, and targeting of cells and axons. Cell 1992; 68(2):303-322.
233. Yamada KM, Miyamoto S. Integrin transmembrane signaling and cytoskeletal control. Curr Opin Cell Biol 1995; 7(5):681-689.
234. Hoffman S, Friedlander DR, Chuong CM et al. Differential contributions of Ng-CAM and N-CAM to cell adhesion in different neural regions. J Cell Biol 1986; 103(1):145-158.
235. Doherty P, Smith P, Walsh FS. Shared cell adhesion molecule (CAM) homology domains point to CAMs signalling via FGF receptors. Perspect Dev Neurobiol 1996; 4(2-3):157-168.
236. Doherty P, Walsh FS. CAM-FGF receptor interactions: A model for axonal growth. Mol Cell Neurosci 1996; 8(2-3):99-111.
237. Yanagida H, Tanaka J, Maruo S. Immunocytochemical localization of a cell adhesion molecule, integrin $\alpha5\beta1$, in nerve growth cones. J Orthop Sci 1999; 4(5):353-360.

238. Bi X, Lynch G, Zhou J et al. Polarized distribution of α5 integrin in dendrites of hippocampal and cortical neurons. J Comp Neurol 2001; 435(2):184-193.
239. Rohrbough J, Grotewiel MS, Davis RL et al. Integrin-mediated regulation of synaptic morphology, transmission, and plasticity. J Neurosci 2000; 20(18):6868-6878.
240. Ruoslahti E. Integrin signaling and matrix assembly. Tumour Biol 1996; 17(2):117-124.
241. Ruoslahti E, Obrink B. Common principles in cell adhesion. Exp Cell Res 1996; 227(1):1-11.
242. Ruoslahti E. Fibronectin and its integrin receptors in cancer. Adv Cancer Res 1999; 76:1-20.
243. Rosen SD, Bertozzi CR. The selectins and their ligands. Curr Opin Cell Biol 1994; 6(5):663-673.
244. Burridge K, Chrzanowska-Wodnicka M. Focal adhesions, contractility, and signaling. Annu Rev Cell Dev Biol 1996; 12:463-518.
245. Chrzanowska-Wodnicka M, Burridge K. Rho-stimulated contractility drives the formation of stress fibers and focal adhesions. J Cell Biol 1996; 133(6):1403-1415.
246. Sydor AM, Su AL, Wang FS et al. Talin and vinculin play distinct roles in filopodial motility in the neuronal growth cone. J Cell Biol 1996; 134(5):1197-1207.
247. Einheber S, Schnapp LM, Salzer JL et al. Regional and ultrastructural distribution of the α8 integrin subunit in developing and adult rat brain suggests a role in synaptic function. J Comp Neurol 1996; 370(1):105-134.
248. Pinkstaff JK, Detterich J, Lynch G et al. Integrin subunit gene expression is regionally differentiated in adult brain. J Neurosci 1999; 19(5):1541-1556.
249. Pinkstaff JK, Lynch G, Gall CM. Localization and seizure regulation of integrin β1 mRNA in adult rat brain. Brain Res Mol Brain Res 1998; 55(2):265-276.
250. Murase S, Hayashi Y. Expression pattern of integrin β1 subunit in Purkinje cells of rat and cerebellar mutant mice. J Comp Neurol 1996; 375(2):225-237.
251. Murase S, Hayashi Y. Concomitant expression of genes encoding integrin α5β5 heterodimer and vitronectin in growing parallel fibers of postnatal rat cerebellum: A possible role as mediators of parallel fiber elongation. J Comp Neurol 1998; 397(2):199-212.
252. Hayashi YK, Chou FL, Engvall E et al. Mutations in the integrin α7 gene cause congenital myopathy. Nat Genet 1998; 19(1):94-97.
253. Hayashi YK, Nagamatsu T, Ito M et al. Suppression of experimental crescent-type anti-glomerular basement membrane (GBM) nephritis by FK506 (tacrolimus hydrate) in rats. Jpn J Pharmacol 1996; 70(1):43-54.
254. Kollins KM, Powell SK, Rivas RJ. GPI-anchored human placental alkaline phosphatase has a nonpolarized distribution on the cell surface of mouse cerebellar granule neurons in vitro. J Neurobiol 1999; 39(1):119-141.
255. Brummendorf T, Rathjen FG. Axonal glycoproteins with immunoglobulin- and fibronectin type III-related domains in vertebrates: Structural features, binding activities, and signal transduction. J Neurochem 1993; 61(4):1207-1219.
256. Brummendorf T, Rathjen FG. Structure/function relationships of axon-associated adhesion receptors of the immunoglobulin superfamily. Curr Opin Neurobiol 1996; 6(5):584-593.
257. Stottmann RW, Rivas RJ. Distribution of TAG-1 and synaptophysin in the developing cerebellar cortex: Relationship to Purkinje cell dendritic development. J Comp Neurol 1998; 395(1):121-135.
258. Powell SK, Cunningham BA, Edelman GM et al. Targeting of transmembrane and GPI-anchored forms of N-CAM to opposite domains of a polarized epithelial cell. Nature 1991; 353(6339):76-77.
259. Furley AJ, Morton SB, Manalo D et al. The axonal glycoprotein TAG-1 is an immunoglobulin superfamily member with neurite outgrowth-promoting activity. Cell 1990; 61(1):157-170.
260. Harel R, Futerman AH. A newly-synthesized GPI-anchored protein, TAG-1/axonin-1, is inserted into axonal membranes along the entire length of the axon and not exclusively at the growth cone. Brain Res 1996; 712(2):345-348.
261. Daston MM, Bastmeyer M, Rutishauser U et al. Spatially restricted increase in polysialic acid enhances corticospinal axon branching related to target recognition and innervation. J Neurosci 1996; 16(17):5488-5497.
262. Yagi T, Takeichi M. Cadherin superfamily genes: Functions, genomic organization, and neurologic diversity. Genes Dev 2000; 14(10):1169-1180.
263. Arndt K, Nakagawa S, Takeichi M et al. Cadherin-defined segments and parasagittal cell ribbons in the developing chicken cerebellum. Mol Cell Neurosci 1998; 10(5-6):211-228.
264. Utton MA, Eickholt B, Howell FV et al. Soluble N-cadherin stimulates fibroblast growth factor receptor dependent neurite outgrowth and N-cadherin and the fibroblast growth factor receptor co-cluster in cells. J Neurochem 2001; 76(5):1421-1430.
265. Bahjaoui-Bouhaddi M, Padilla F, Nicolet M et al. Localized deposition of M-cadherin in the glomeruli of the granular layer during the postnatal development of mouse cerebellum. J Comp Neurol 1997; 378(2):180-195.

266. Inoue A, Sanes JR. Lamina-specific connectivity in the brain: Regulation by N-cadherin, neurotrophins, and glycoconjugates. Science 1997; 276(5317):1428-1431.

267. Gao PP, Sun CH, Zhou XF et al. Ephrins stimulate or inhibit neurite outgrowth and survival as a function of neuronal cell type. J Neurosci Res 2000; 60(4):427-436.

268. Giger RJ, Pasterkamp RJ, Holtmaat AJ et al. Semaphorin III: Role in neuronal development and structural plasticity. Prog Brain Res 1998; 117:133-149.

269. Van Vactor D. Axon guidance. Curr Biol 1999; 9(21):R797-799.

270. Van Vactor D, Flanagan JG. The middle and the end: Slit brings guidance and branching together in axon pathway selection. Neuron 1999; 22(4):649-652.

271. Van Vactor DV, Lorenz LJ. Neural development: The semantics of axon guidance. Curr Biol 1999; 9(6):R201-204.

272. Castellani V, Rougon G. Control of semaphorin signaling. Curr Opin Neurobiol 2002; 12(5):532-541.

273. Cutforth T, Harrison CJ. Ephs and ephrins close ranks. Trends Neurosci 2002; 25(7):332-334.

274. Guthrie S. Axon guidance: Starting and stopping with Slit. Curr Biol 1999; 9(12):R432-435.

275. Guthrie S. Ephrin cleavage: A missing link in axon guidance. Trends Neurosci 2000; 23(12):592.

276. Holmberg J, Frisen J. Ephrins are not only unattractive. Trends Neurosci. 2002; 25(5):239-243.

277. Knoll B, Drescher U. Ephrin-As as receptors in topographic projections. Trends Neurosci 2002; 25(3):145-149.

278. Mellitzer G, Xu Q, Wilkinson DG. Control of cell behaviour by signalling through Eph receptors and ephrins. Curr Opin Neurobiol 2000; 10(3):400-408.

279. O'Leary DD, Wilkinson DG. Eph receptors and ephrins in neural development. Curr Opin Neurobiol 1999; 9(1):65-73.

280. Davenport RW, Thies E, Nelson PG. Cellular localization of guidance cues in the establishment of retinotectal topography. J Neurosci 1996; 16(6):2074-2085.

281. Davenport RW, Thies E, Zhou R et al. Cellular localization of ephrin-A2, ephrin-A5, and other functional guidance cues underlies retinotopic development across species. J Neurosci 1998; 18(3):975-986.

282. Yates PA, Roskies AL, McLaughlin T et al. Topographic-specific axon branching controlled by ephrin-As is the critical event in retinotectal map development. J Neurosci 2001; 21(21):8548-8563.

283. Butler AK, Sullivan JM, McAllister AK et al. The role of ephrins in the development of intracortical circuitry. Soc Neurosci Abstr 1999; 25:2263.

284. Yue Y, Su J, Cerretti DP et al. Selective inhibition of spinal cord neurite outgrowth and cell survival by the Eph family ligand ephrin-A5. J Neurosci 1999; 19(22):10026-10035.

285. McLaughlin T, Hindges R, O'Leary DD. Regulation of axial patterning of the retina and its topographic mapping in the brain. Curr Opin Neurobiol 2003; 13(1):57-69.

286. Cheng HJ, Nakamoto M, Bergemann AD et al. Complementary gradients in expression and binding of ELF-1 and Mek4 in development of the topographic retinotectal projection map. Cell 1995; 82(3):371-381.

287. Tessier-Lavigne M. Eph receptor tyrosine kinases, axon repulsion, and the development of topographic maps. Cell 1995; 82(3):345-348.

288. Gale NW, Holland SJ, Valenzuela DM et al. Eph receptors and ligands comprise two major specificity subclasses and are reciprocally compartmentalized during embryogenesis. Neuron 1996; 17(1):9-19.

289. Karam SD, Burrows RC, Logan C et al. Eph receptors and ephrins in the developing chick cerebellum: Relationship to sagittal patterning and granule cell migration. J Neurosci 2000; 20(17):6488-6500.

290. Davy A, Robbins SM. Ephrin-A5 modulates cell adhesion and morphology in an integrin-dependent manner. Embo J 2000; 19(20):5396-5405.

291. Castellani V, Yue Y, Gao PP et al. Dual action of a ligand for Eph receptor tyrosine kinases on specific populations of axons during the development of cortical circuits. J Neurosci 1998; 18(12):4663-4672.

292. Wang KH, Brose K, Arnott D et al. Biochemical purification of a mammalian Slit protein as a positive regulator of sensory axon elongation and branching. Cell 1999; 96(6):771-784.

293. Rabacchi SA, Solowska JM, Kruk B et al. Collapsin-1/semaphorin-III/D is regulated developmentally in Purkinje cells and collapses pontocerebellar mossy fiber neuronal growth cones. J Neurosci 1999; 19(11):4437-4448.

294. Polleux F, Giger RJ, Ginty DD et al. Patterning of cortical efferent projections by semaphorin-neuropilin interactions. Science 1998; 282(5395):1904-1906.

295. Polleux F, Morrow T, Ghosh A. Semaphorin 3A is a chemoattractant for cortical apical dendrites. Nature 2000; 404(6778):567-573.

296. Campbell DS, Regan AG, Lopez JS et al. Semaphorin 3A elicits stage-dependent collapse, turning, and branching in Xenopus retinal growth cones. J Neurosci 2001; 21(21):8538-8547.
297. Jurney WM, Gallo G, Letourneau PC et al. Rac-1-mediated endocytosis during ephrin-A2 and semaphorin 3A-induced growth cone collapse. J Neurosci 2002; 22(14):6019-6028.
298. Chen H, Chedotal A, He Z et al. Neuropilin-2, a novel member of the neuropilin family, is a high affinity receptor for the semaphorins Sema E and Sema IV but not Sema III. Neuron 1997; 19(3):547-559.
299. Pasterkamp RJ, Verhaagen J. Emerging roles for semaphorins in neural regeneration. Brain Res Brain Res Rev 2001; 35(1):36-54.
300. Kolodkin AL. Growth cones and the cues that repel them. Trends Neurosci 1996; 19(11):507-513.
301. Chien CB, Harris WA. Axonal guidance from retina to tectum in embryonic Xenopus. Curr Top Dev Biol 1994; 29:135-169.
302. Dingwell KS, Holt CE, Harris WA. The multiple decisions made by growth cones of RGCs as they navigate from the retina to the tectum in Xenopus embryos. J Neurobiol 2000; 44(2):246-259.
303. Driessens MH, Hu H, Nobes CD et al. Plexin-B semaphorin receptors interact directly with active Rac and regulate the actin cytoskeleton by activating Rho. Curr Biol 2001; 11(5):339-344.
304. Catalano SM, Messersmith EK, Goodman CS et al. Many major CNS axon projections develop normally in the absence of semaphorin III. Mol Cell Neurosci 1998; 11(4):173-182.
305. Brose K, Tessier-Lavigne M. Slit proteins: Key regulators of axon guidance, axonal branching, and cell migration. Curr Opin Neurobiol 2000; 10(1):95-102.
306. Brose K, Bland KS, Wang KH et al. Slit proteins bind Robo receptors and have an evolutionarily conserved role in repulsive axon guidance. Cell 1999; 96(6):795-806.
307. Li HS, Chen JH, Wu W et al. Vertebrate Slit, a secreted ligand for the transmembrane protein Roundabout, is a repellent for olfactory bulb axons. Cell 1999; 96(6):807-818.
308. Ringstedt T, Braisted JE, Brose K et al. Slit inhibition of retinal axon growth and its role in retinal axon pathfinding and innervation patterns in the diencephalon. J Neurosci 2000; 20(13):4983-4991.
309. Takahashi T, Nakamura F, Jin Z et al. Semaphorins A and E act as antagonists of neuropilin-1 and agonists of neuropilin-2 receptors. Nat Neurosci 1998; 1(6):487-493.
310. Kramer SG, Kidd T, Simpson JH et al. Switching repulsion to attraction: Changing responses to Slit during transition in mesoderm migration. Science 2001; 292(5517):737-740.
311. Whitford KL, Marillat V, Stein E et al. Regulation of cortical dendrite development by Slit-Robo interactions. Neuron 2002; 33(1):47-61.
312. Ozdinler PH, Erzurumlu RS. Regulation of neurotrophin-induced axonal responses via Rho GTPases. J Comp Neurol 2001; 438(4):377-387.
313. Ozdinler PH, Erzurumlu RS. Slit2, a branching-arborization factor for sensory axons in the mammalian CNS. J Neurosci 2002; 22(11):4540-4549.
314. Sang Q, Wu J, Rao Y et al. Slit promotes branching and elongation of neurites of interneurons but not projection neurons from the developing telencephalon. Mol Cell Neurosci 2002; 21(2):250-265.
315. Nunez J, Couchie D, Aniello F et al. Regulation by thyroid hormone of microtubule assembly and neuronal differentiation. Neurochem Res 1991; 16(9):975-982.
316. Porterfield SP, Hendrich CE. The role of thyroid hormones in prenatal and neonatal neurological development—current perspectives. Endocr Rev 1993; 14(1):94-106.
317. Pasquini JM, Adamo AM. Thyroid hormones and the central nervous system. Dev Neurosci 1994; 16(1-2):1-8.
318. Woolley CS, McEwen BS. Estradiol regulates hippocampal dendritic spine density via an N-methyl-D-aspartate receptor-dependent mechanism. J Neurosci 1994; 14(12):7680-7687.
319. Lin X, Bulleit RF. Insulin-like growth factor I (IGF-I) is a critical trophic factor for developing cerebellar granule cells. Brain Res Dev Brain Res 1997; 99(2):234-242.
320. Toran-Allerand CD, Hashimoto K, Greenough WT et al. Sex steroids and the development of the newborn mouse hypothalamus and preoptic area in vitro: III. Effects of estrogen on dendritic differentiation. Brain Res 1983; 283(1):97-101.
321. McEwen BS, Woolley CS. Estradiol and progesterone regulate neuronal structure and synaptic connectivity in adult as well as developing brain. Exp Gerontol 1994; 29(3-4):431-436.
322. Audesirk T, Cabell L, Kern M et al. Enhancement of dendritic branching in cultured hippocampal neurons by 17β-estradiol is mediated by nitric oxide. Int J Dev Neurosci 2003; 21(4):225-233.

323. Jakab RL, Wong JK, Belcher SM. Estrogen receptor β immunoreactivity in differentiating cells of the developing rat cerebellum. J Comp Neurol 2001; 430(3):396-409.
324. Belcher SM. Regulated expression of estrogen receptor α and β mRNA in granule cells during development of the rat cerebellum. Brain Res Dev Brain Res 1999; 115(1):57-69.
325. Bohn MC, Lauder JM. Cerebellar granule cell genesis in the hydrocortisone-treated rats. Dev Neurosci 1980; 3(2):81-89.
326. Pavlik A, Buresova M. The neonatal cerebellum: The highest level of glucocorticoid receptors in the brain. Brain Res 1984; 314(1):13-20.
327. Stastny F, Pokorny J, Lisy V et al. A morphometric study of cortisol-induced changes in the development of neuronal process outgrowth in the corticoid zone of the embryonic chick telencephalon. Exp Clin Endocrinol 1986; 88(1):39-44.
328. Cameron HA, Gould E. Adult neurogenesis is regulated by adrenal steroids in the dentate gyrus. Neuroscience 1994; 61(2):203-209.
329. Sousa N, Madeira MD, Paula-Barbosa MM. Corticosterone replacement restores normal morphological features to the hippocampal dendrites, axons and synapses of adrenalectomized rats. J Neurocytol 1999; 28(7):541-558.
330. Persengiev SP. The neuroprotective and antiapoptotic effects of melatonin in cerebellar neurons involve glucocorticoid receptor and p130 signal pathways. J Steroid Biochem Mol Biol 2001; 77(2-3):151-158.
331. Nicholson JL, Altman J. The effects of early hypo- and hyperthyroidism on the development of rat cerebellar cortex. I. Cell proliferation and differentiation. Brain Res 1972; 44(1):13-23.
332. Nicholson JL, Altman J. The effects of early hypo- and hyperthyroidism on the development of the rat cerebellar cortex. II. Synaptogenesis in the molecular layer. Brain Res 1972; 44(1):25-36.
333. Lauder JM. Effects of early hypo- and hyperthyroidism on development of rat cerebellar cortex. IV. The parallel fibers. Brain Res 1978; 142(1):25-39.
334. Muller Y, Rocchi E, Lazaro JB et al. Thyroid hormone promotes BCL-2 expression and prevents apoptosis of early differentiating cerebellar granule neurons. Int J Dev Neurosci 1995; 13(8):871-885.
335. Gould E, Allan MD, McEwen BS. Dendritic spine density of adult hippocampal pyramidal cells is sensitive to thyroid hormone. Brain Res 1990; 525(2):327-329.
336. Gould E, Westlind-Danielsson A, Frankfurt M et al. Sex differences and thyroid hormone sensitivity of hippocampal pyramidal cells. J Neurosci 1990; 10(3):996-1003.
337. Sarafian T, Verity MA. Influence of thyroid hormones on rat cerebellar cell aggregation and survival in culture. Brain Res 1986; 391(2):261-270.
338. Fann MJ, Patterson PH. Depolarization differentially regulates the effects of bone morphogenetic protein (BMP)-2, BMP-6, and activin A on sympathetic neuronal phenotype. J Neurochem 1994; 63(6):2074-2079.
339. Kingsley DM. What do BMPs do in mammals? Clues from the mouse short-ear mutation. Trends Genet 1994; 10(1):16-21.
340. Lein P, Johnson M, Guo X et al. Osteogenic protein-1 induces dendritic growth in rat sympathetic neurons. Neuron 1995; 15(3):597-605.
341. Ebendal T, Bengtsson H, Soderstrom S. Bone morphogenetic proteins and their receptors: Potential functions in the brain. J Neurosci Res 1998; 51(2):139-146.
342. Guo X, Rueger D, Higgins D. Osteogenic protein-1 and related bone morphogenetic proteins regulate dendritic growth and the expression of microtubule-associated protein-2 in rat sympathetic neurons. Neurosci Lett 1998; 245(3):131-134.
343. Mehler MF, Mabie PC, Zhang D et al. Bone morphogenetic proteins in the nervous system. Trends Neurosci 1997; 20(7):309-317.
344. Augsburger A, Schuchardt A, Hoskins S et al. BMPs as mediators of roof plate repulsion of commissural neurons. Neuron 1999; 24(1):127-141.
345. Withers GS, Higgins D, Charette M et al. Bone morphogenetic protein-7 enhances dendritic growth and receptivity to innervation in cultured hippocampal neurons. Eur J Neurosci 2000; 12(1):106-116.
346. Higgins D, Burack M, Lein P et al. Mechanisms of neuronal polarity. Curr Opin Neurobiol 1997; 7(5):599-604.
347. Le Roux P, Behar S, Higgins D et al. OP-1 enhances dendritic growth from cerebral cortical neurons in vitro. Exp Neurol 1999; 160(1):151-163.
348. Martinez G, Loveland KL, Clark AT et al. Expression of bone morphogenetic protein receptors in the developing mouse metanephros. Exp Nephrol 2001; 9(6):372-379.
349. Perides G, Hu G, Rueger DC et al. Osteogenic protein-1 regulates L1 and neural cell adhesion molecule gene expression in neural cells. J Biol Chem 1993; 268(33):25197-25205.
350. Steinbeisser H, De Robertis EM, Ku M et al. Xenopus axis formation: Induction of goosecoid by injected Xwnt-8 and activin mRNAs. Development 1993; 118(2):499-507.

351. Alder J, Lee KJ, Jessell TM et al. Generation of cerebellar granule neurons in vivo by transplantation of BMP-treated neural progenitor cells. Nat Neurosci 1999; 2(6):535-540.

352. Levi-Montalcini R, Skaper SD, Dal Toso R et al. Nerve growth factor: From neurotrophin to neurokine. Trends Neurosci 1996; 19(11):514-520.

353. Barbacid M. Neurotrophic factors and their receptors. Curr Opin Cell Biol 1995; 7(2):148-155.

354. Barde YA. Neurotrophins: A family of proteins supporting the survival of neurons. Prog Clin Biol Res 1994; 390:45-56.

355. Gotz R, Koster R, Winkler C et al. Neurotrophin-6 is a new member of the nerve growth factor family. Nature 1994; 372(6503):266-269.

356. Nilsson AS, Fainzilber M, Falck P et al. Neurotrophin-7: A novel member of the neurotrophin family from the zebrafish. FEBS Lett 1998; 424(3):285-290.

357. Barde YA. Neurotrophic factors: An evolutionary perspective. J Neurobiol 1994; 25(11):1329-1333.

358. Thoenen H. Neurotrophins and neuronal plasticity. Science 1995; 270(5236):593-598.

359. Thoenen H. Neurotrophins and activity-dependent plasticity. Prog Brain Res 2000; 128:183-191.

360. Henderson CE. Role of neurotrophic factors in neuronal development. Curr Opin Neurobiol 1996; 6(1):64-70.

361. Ip NY, Yancopoulos GD. Neurotrophic factors and their receptors. Ann Neurol 1994; 35(Suppl):S13-16.

362. Vrbova G, Greensmith L, Nogradi A. Neurotrophic factors. Trends Neurosci 1999; 22(3):108-109.

363. Ip NY, Yancopoulos GD. The neurotrophins and CNTF: Two families of collaborative neurotrophic factors. Annu Rev Neurosci 1996; 19:491-515.

364. Ip NY. The neurotrophins and neuropoietic cytokines: Two families of growth factors acting on neural and hematopoietic cells. Ann NY Acad Sci 1998; 840:97-106.

365. Lewin GR, Barde YA. Physiology of the neurotrophins. Annu Rev Neurosci 1996; 19:289-317.

366. Meakin SO, Shooter EM. The nerve growth factor family of receptors. Trends Neurosci 1992; 15(9):323-331.

367. Barbacid M. The Trk family of neurotrophin receptors. J Neurobiol 1994; 25(11):1386-1403.

368. Barbacid M. Structural and functional properties of the TRK family of neurotrophin receptors. Ann NY Acad Sci 1995; 766:442-458.

369. Timmusk T, Belluardo N, Metsis M et al. Widespread and developmentally regulated expression of neurotrophin-4 mRNA in rat brain and peripheral tissues. Eur J Neurosci 1993; 5(6):605-613.

370. Lindholm D, Hamner S, Zirrgiebel U. Neurotrophins and cerebellar development. Perspect Dev Neurobiol 1997; 5(1):83-94.

371. Maisonpierre PC, Belluscio L, Friedman B et al. NT-3, BDNF, and NGF in the developing rat nervous system: Parallel as well as reciprocal patterns of expression. Neuron 1990; 5(4):501-509.

372. Rocamora N, Garcia-Ladona FJ, Palacios JM et al. Differential expression of brain-derived neurotrophic factor, neurotrophin-3, and low-affinity nerve growth factor receptor during the postnatal development of the rat cerebellar system. Brain Res Mol Brain Res 1993; 17(1-2):1-8.

373. Zafra F, Castren E, Thoenen H et al. Interplay between glutamate and γ-aminobutyric acid transmitter systems in the physiological regulation of brain-derived neurotrophic factor and nerve growth factor synthesis in hippocampal neurons. Proc Natl Acad Sci USA 1991; 88(22):10037-10041.

374. Zafra F, Lindholm D, Castren E et al. Regulation of brain-derived neurotrophic factor and nerve growth factor mRNA in primary cultures of hippocampal neurons and astrocytes. J Neurosci 1992; 12(12):4793-4799.

375. Bessho Y, Nakanishi S, Nawa H. Glutamate receptor agonists enhance the expression of BDNF mRNA in cultured cerebellar granule cells. Brain Res Mol Brain Res 1993; 18(3):201-208.

376. Leingartner A, Heisenberg CP, Kolbeck R et al. Brain-derived neurotrophic factor increases neurotrophin-3 expression in cerebellar granule neurons. J Biol Chem 1994; 269(2):828-830.

377. Wetmore C, Olson L, Bean AJ. Regulation of brain-derived neurotrophic factor (BDNF) expression and release from hippocampal neurons is mediated by non-NMDA type glutamate receptors. J Neurosci 1994; 14(3 Pt 2):1688-1700.

378. Canossa M, Griesbeck O, Berninger B et al. Neurotrophin release by neurotrophins: Implications for activity-dependent neuronal plasticity. Proc Natl Acad Sci USA 1997; 94(24):13279-13286.

379. Castren E, Berninger B, Leingartner A et al. Regulation of brain-derived neurotrophic factor mRNA levels in hippocampus by neuronal activity. Prog Brain Res 1998; 117:57-64.

380. Condorelli DF, Dell'Albani P, Timmusk T et al. Differential regulation of BDNF and NT-3 mRNA levels in primary cultures of rat cerebellar neurons. Neurochem Int 1998; 32(1):87-91.

381. Kruttgen A, Moller JC, Heymach JV Jr et al. Neurotrophins induce release of neurotrophins by the regulated secretory pathway. Proc Natl Acad Sci USA 1998; 95(16):9614-9619.

382. Lein ES, Shatz CJ. Rapid regulation of brain-derived neurotrophic factor mRNA within eye-specific circuits during ocular dominance column formation. J Neurosci 2000; 20(4):1470-1483.

383. Righi M, Tongiorgi E, Cattaneo A. Brain-derived neurotrophic factor (BDNF) induces dendritic targeting of BDNF and tyrosine kinase B mRNAs in hippocampal neurons through a phosphatidylinositol-3 kinase-dependent pathway. J Neurosci 2000; 20(9):3165-3174.

384. Yang F, He X, Feng L et al. PI-3 kinase and IP3 are both necessary and sufficient to mediate NT3-induced synaptic potentiation. Nat Neurosci 2001; 4(1):19-28.

385. Shieh PB, Hu SC, Bobb K et al. Identification of a signaling pathway involved in calcium regulation of BDNF expression. Neuron 1998; 20(4):727-740.

386. Fujita Y, Katagi J, Tabuchi A et al. Coactivation of secretogranin-II and BDNF genes mediated by calcium signals in mouse cerebellar granule cells. Brain Res Mol Brain Res 1999; 63(2):316-324.

387. Murray KD, Isackson PJ, Eskin TA et al. Altered mRNA expression for brain-derived neurotrophic factor and type II calcium/calmodulin-dependent protein kinase in the hippocampus of patients with intractable temporal lobe epilepsy. J Comp Neurol 2000; 418(4):411-422.

388. Tabuchi A, Nakaoka R, Amano K et al. Differential activation of brain-derived neurotrophic factor gene promoters I and III by Ca^{2+} signals evoked via L-type voltage-dependent and N-methyl-D-aspartate receptor Ca^{2+} channels. J Biol Chem 2000; 275(23):17269-17275.

389. Klein R, Nanduri V, Jing SA et al. The trkB tyrosine protein kinase is a receptor for brain-derived neurotrophic factor and neurotrophin-3. Cell 1991; 66(2):395-403.

390. Lamballe F, Klein R, Barbacid M. The trk family of oncogenes and neurotrophin receptors. Princess Takamatsu Symp 1991; 22:153-170.

391. Lamballe F, Klein R, Barbacid M. trkC, a new member of the trk family of tyrosine protein kinases, is a receptor for neurotrophin-3. Cell 1991; 66(5):967-979.

392. Lamballe F, Tapley P, Barbacid M. trkC encodes multiple neurotrophin-3 receptors with distinct biological properties and substrate specificities. Embo J 1993; 12(8):3083-3094.

393. Middlemas DS, Lindberg RA, Hunter T. trkB, a neural receptor protein-tyrosine kinase: Evidence for a full-length and two truncated receptors. Mol Cell Biol 1991; 11(1):143-153.

394. Squinto SP, Stitt TN, Aldrich TH et al. trkB encodes a functional receptor for brain-derived neurotrophic factor and neurotrophin-3 but not nerve growth factor. Cell 1991; 65(5):885-893.

395. Valenzuela DM, Maisonpierre PC, Glass DJ et al. Alternative forms of rat TrkC with different functional capabilities. Neuron 1993; 10(5):963-974.

396. Armanini MP, McMahon SB, Sutherland J et al. Truncated and catalytic isoforms of trkB are co-expressed in neurons of rat and mouse CNS. Eur J Neurosci 1995; 7(6):1403-1409.

397. Chao MV, Hempstead BL. p75 and Trk: A two-receptor system. Trends Neurosci 1995; 18(7):321-326.

398. Strohmaier C, Carter BD, Urfer R et al. A splice variant of the neurotrophin receptor trkB with increased specificity for brain-derived neurotrophic factor. Embo J 1996; 15(13):3332-3337.

399. Kaplan DR, Miller FD. Signal transduction by the neurotrophin receptors. Curr Opin Cell Biol 1997; 9(2):213-221.

400. Kaplan DR, Miller FD. Neurotrophin signal transduction in the nervous system. Curr Opin Neurobiol 2000; 10(3):381-391.

401. Bibel M, Hoppe E, Barde YA. Biochemical and functional interactions between the neurotrophin receptors trk and p75NTR. Embo J 1999; 18(3):616-622.

402. Casaccia-Bonnefil P, Gu C, Chao MV. Neurotrophins in cell survival/death decisions. Adv Exp Med Biol 1999; 468:275-282.

403. Casaccia-Bonnefil P, Gu C, Khursigara G et al. p75 neurotrophin receptor as a modulator of survival and death decisions. Microsc Res Tech 1999; 45(4-5):217-224.

404. Casaccia-Bonnefil P, Kong H, Chao MV. Neurotrophins: The biological paradox of survival factors eliciting apoptosis. Cell Death Differ 1998; 5(5):357-364.

405. Friedman WJ, Greene LA. Neurotrophin signaling via Trks and p75. Exp Cell Res 1999; 253(1):131-142.

406. Neveu I, Arenas E. Neurotrophins promote the survival and development of neurons in the cerebellum of hypothyroid rats in vivo. J Cell Biol 1996; 133(3):631-646.

407. Pizzorusso T, Fagiolini M, Gianfranceschi L et al. Role of neurotrophins in the development and plasticity of the visual system: Experiments on dark rearing. Int J Psychophysiol 2000; 35(2-3):189-196.

408. Pizzorusso T, Ratto GM, Putignano E et al. Brain-derived neurotrophic factor causes cAMP response element-binding protein phosphorylation in absence of calcium increases in slices and cultured neurons from rat visual cortex. J Neurosci 2000; 20(8):2809-2816.

409. Finkbeiner S, Tavazoie SF, Maloratsky A et al. CREB: A major mediator of neuronal neurotrophin responses. Neuron 1997; 19(5):1031-1047.

410. Finkbeiner S. CREB couples neurotrophin signals to survival messages. Neuron 2000; 25(1):11-14.

411. Bulleit RF, Hsieh T. MEK inhibitors block BDNF-dependent and -independent expression of GABA(A) receptor subunit mRNAs in cultured mouse cerebellar granule neurons. Brain Res Dev Brain Res 2000; 119(1):1-10.

412. Cavanaugh JE, Ham J, Hetman M et al. Differential regulation of mitogen-activated protein kinases ERK1/2 and ERK5 by neurotrophins, neuronal activity, and cAMP in neurons. J Neurosci 2001; 21(2):434-443.

413. Markus A, Patel TD, Snider WD. Neurotrophic factors and axonal growth. Curr Opin Neurobiol 2002; 12(5):523-531.

414. Segal RA. Selectivity in neurotrophin signaling: Theme and Variations. Annu Rev Neurosci 2003.

415. Paves H, Saarma M. Neurotrophins as in vitro growth cone guidance molecules for embryonic sensory neurons. Cell Tissue Res 1997; 290(2):285-297.

416. Zhang HL, Singer RH, Bassell GJ. Neurotrophin regulation of β-actin mRNA and protein localization within growth cones. J Cell Biol 1999; 147(1):59-70.

417. Zhang Y, Moheban DB, Conway BR et al. Cell surface Trk receptors mediate NGF-induced survival while internalized receptors regulate NGF-induced differentiation. J Neurosci 2000; 20(15):5671-5678.

418. Morfini G, DiTella MC, Feiguin F et al. Neurotrophin-3 enhances neurite outgrowth in cultured hippocampal pyramidal neurons. J Neurosci Res 1994; 39(2):219-232.

419. Coffey ET, Akerman KE, Courtney MJ. Brain derived neurotrophic factor induces a rapid upregulation of synaptophysin and tau proteins via the neurotrophin receptor TrkB in rat cerebellar granule cells. Neurosci Lett 1997; 227(3):177-180.

420. Tucker KL, Meyer M, Barde YA. Neurotrophins are required for nerve growth during development. Nat Neurosci 2001; 4(1):29-37.

421. Fukumitsu H, Ohashi A, Nitta A et al. BDNF and NT-3 modulate expression and threonine phosphorylation of microtubule-associated protein 2 analogues, and alter their distribution in the developing rat cerebral cortex. Neurosci Lett 1997; 238(3):107-110.

422. McCaig CD, Sangster L, Stewart R. Neurotrophins enhance electric field-directed growth cone guidance and directed nerve branching. Dev Dyn 2000; 217(3):299-308.

423. Schinder AF, Poo M. The neurotrophin hypothesis for synaptic plasticity. Trends Neurosci 2000; 23(12):639-645.

424. Conover JC, Yancopoulos GD. Neurotrophin regulation of the developing nervous system: Analyses of knockout mice. Rev Neurosci 1997; 8(1):13-27.

425. Bovolenta P, Frade JM, Marti E et al. Neurotrophin-3 antibodies disrupt the normal development of the chick retina. J Neurosci 1996; 16(14):4402-4410.

426. von Bartheld CS. Neurotrophins in the developing and regenerating visual system. Histol Histopathol 1998; 13(2):437-459.

427. von Bartheld CS, Butowt R. Expression of neurotrophin-3 (NT-3) and anterograde axonal transport of endogenous NT-3 by retinal ganglion cells in chick embryos. J Neurosci 2000; 20(2):736-748.

428. Akaneya Y, Tsumoto T, Hatanaka H. Brain-derived neurotrophic factor blocks long-term depression in rat visual cortex. J Neurophysiol 1996; 76(6):4198-4201.

429. Caleo M, Menna E, Chierzi S et al. Brain-derived neurotrophic factor is an anterograde survival factor in the rat visual system. Curr Biol 2000; 10(19):1155-1161.

430. Cohen-Cory S, Dreyfus CF, Black IB. Expression of high- and low-affinity nerve growth factor receptors by Purkinje cells in the developing rat cerebellum. Exp Neurol 1989; 105(1):104-109.

431. Cohen-Cory S, Fraser SE. BDNF in the development of the visual system of Xenopus. Neuron 1994; 12(4):747-761.

432. Cohen-Cory S, Fraser SE. Effects of brain-derived neurotrophic factor on optic axon branching and remodeling in vivo. Nature 1995; 378(6553):192-196.

433. Cohen-Cory S, Escandon E, Fraser SE. The cellular patterns of BDNF and trkB expression suggest multiple roles for BDNF during Xenopus visual system development. Dev Biol 1996; 179(1):102-115.

434. Cohen-Cory S. BDNF modulates, but does not mediate, activity-dependent branching and remodeling of optic axon arbors in vivo. J Neurosci 1999; 19(22):9996-10003.

435. Huber KM, Sawtell NB, Bear MF. Brain-derived neurotrophic factor alters the synaptic modification threshold in visual cortex. Neuropharmacology 1998; 37(4-5):571-579.

436. Huber K, Kuehnel F, Wyatt S et al. TrkB expression and early sensory neuron survival are independent of endogenous BDNF. J Neurosci Res 2000; 59(3):372-378.

437. Kossel A, Lowel S, Bolz J. Relationships between dendritic fields and functional architecture in striate cortex of normal and visually deprived cats. J Neurosci 1995; 15(5 Pt 2):3913-3926.

438. Kossel AH, Williams CV, Schweizer M et al. Afferent innervation influences the development of dendritic branches and spines via both activity-dependent and non-activity-dependent mechanisms. J Neurosci 1997; 17(16):6314-6324.

439. McAllister AK, Lo DC, Katz LC. Neurotrophins regulate dendritic growth in developing visual cortex. Neuron 1995; 15(4):791-803.

440. McAllister AK, Katz LC, Lo DC. Neurotrophin regulation of cortical dendritic growth requires activity. Neuron 1996; 17(6):1057-1064.

441. McAllister AK, Katz LC, Lo DC. Opposing roles for endogenous BDNF and NT-3 in regulating cortical dendritic growth. Neuron 1997; 18(5):767-778.

442. Kryl D, Yacoubian T, Haapasalo A et al. Subcellular localization of full-length and truncated Trk receptor isoforms in polarized neurons and epithelial cells. J Neurosci 1999; 19(14):5823-5833.

443. Lom B, Cohen-Cory S. Brain-derived neurotrophic factor differentially regulates retinal ganglion cell dendritic and axonal arborization in vivo. J Neurosci 1999; 19(22):9928-9938.

444. Cabelli RJ, Hohn A, Shatz CJ. Inhibition of ocular dominance column formation by infusion of NT-4/5 or BDNF. Science 1995; 267(5204):1662-1666.

445. Cabelli RJ, Shelton DL, Segal RA et al. Blockade of endogenous ligands of trkB inhibits formation of ocular dominance columns. Neuron 1997; 19(1):63-76.

446. Kollins KM, Rivas RJ. BDNF promotes dendritic development in cerebellar granule neurons in vitro. American Society for Cell Biology Abstr 1999; 39:1075.

447. Huang EJ, Reichardt LF. Neurotrophins: Roles in neuronal development and function. Annu Rev Neurosci 2001; 24:677-736.

448. Korte M, Carroll P, Wolf E et al. Hippocampal long-term potentiation is impaired in mice lacking brain-derived neurotrophic factor. Proc Natl Acad Sci USA 1995; 92(19):8856-8860.

449. Korte M, Griesbeck O, Gravel C et al. Virus-mediated gene transfer into hippocampal CA1 region restores long-term potentiation in brain-derived neurotrophic factor mutant mice. Proc Natl Acad Sci USA 1996; 93(22):12547-12552.

450. Korte M, Staiger V, Griesbeck O et al. The involvement of brain-derived neurotrophic factor in hippocampal long-term potentiation revealed by gene targeting experiments. J Physiol Paris 1996; 90(3-4):157-164.

451. Patterson SL, Abel T, Deuel TA et al. Recombinant BDNF rescues deficits in basal synaptic transmission and hippocampal LTP in BDNF knockout mice. Neuron 1996; 16(6):1137-1145.

452. D'Angelo E, Rossi P, Armano S et al. Evidence for NMDA and mGlu receptor-dependent long-term potentiation of mossy fiber-granule cell transmission in rat cerebellum. J Neurophysiol 1999; 81(1):277-287.

453. Yuzaki M, Furuichi T, Mikoshiba et al. A stimulus paradigm inducing long-term desensitization of AMPA receptors evokes a specific increase in BDNF mRNA in cerebellar slices. Learn Mem 1994; 1(4):230-242.

454. Snider WD. Nerve growth factor enhances dendritic arborization of sympathetic ganglion cells in developing mammals. J Neurosci 1988; 8(7):2628-2634.

455. Snider WD. Functions of the neurotrophins during nervous system development: What the knock-outs are teaching us. Cell 1994; 77(5):627-638.

456. Ruit KG, Osborne PA, Schmidt RE et al. Nerve growth factor regulates sympathetic ganglion cell morphology and survival in the adult mouse. J Neurosci 1990; 10(7):2412-2419.

457. Ruit KG, Snider WD. Administration or deprivation of nerve growth factor during development permanently alters neuronal geometry. J Comp Neurol 1991; 314(1):106-113.

458. Marty S, Carroll P, Cellerino A et al. Brain-derived neurotrophic factor promotes the differentiation of various hippocampal nonpyramidal neurons, including Cajal-Retzius cells, in organotypic slice cultures. J Neurosci 1996; 16(2):675-687.

459. Horch HW, Kruttgen A, Portbury SD et al. Destabilization of cortical dendrites and spines by BDNF. Neuron 1999; 23(2):353-364.

460. Shimada A, Mason CA, Morrison ME. TrkB signaling modulates spine density and morphology independent of dendrite structure in cultured neonatal Purkinje cells. J Neurosci 1998; 18(21):8559-8570.

461. Morrison ME, Mason CA. Granule neuron regulation of Purkinje cell development: Striking a balance between neurotrophin and glutamate signaling. J Neurosci 1998; 18(10):3563-3573.

462. Rabacchi SA, Kruk B, Hamilton J et al. BDNF and NT4/5 promote survival and neurite outgrowth of pontocerebellar mossy fiber neurons. J Neurobiol 1999; 40(2):254-269.

463. Schwartz PM, Borghesani PR, Levy RL et al. Abnormal cerebellar development and foliation in BDNF-/- mice reveals a role for neurotrophins in CNS patterning. Neuron 1997; 19(2):269-281.

464. Schwartz PM, Levy RL, Borghesani PR et al. Cerebellar pathology in BDNF -/- mice: The classic view of neurotrophins is changing. Mol Psychiatry 1998; 3(2):116-120.

465. Bao S, Chen L, Qiao X et al. Impaired eye-blink conditioning in Waggler, a mutant mouse with cerebellar BDNF deficiency. Learn Mem 1998; 5(4-5):355-364.
466. Bao S, Chen L, Qiao X et al. Transgenic brain-derived neurotrophic factor modulates a developing cerebellar inhibitory synapse. Learn Mem 1999; 6(3):276-283.
467. Qiao X, Chen L, Gao H et al. Cerebellar brain-derived neurotrophic factor-TrkB defect associated with impairment of eyeblink conditioning in Stargazer mutant mice. J Neurosci 1998; 18(17):6990-6999.
468. Minichiello L, Klein R. TrkB and TrkC neurotrophin receptors cooperate in promoting survival of hippocampal and cerebellar granule neurons. Genes Dev 1996; 10(22):2849-2858.
469. Lindholm D, Castren E, Tsoulfas P et al. Neurotrophin-3 induced by tri-iodothyronine in cerebellar granule cells promotes Purkinje cell differentiation. J Cell Biol 1993; 122(2):443-450.
470. Lindholm D, Dechant G, Heisenberg CP et al. Brain-derived neurotrophic factor is a survival factor for cultured rat cerebellar granule neurons and protects them against glutamate-induced neurotoxicity. Eur J Neurosci 1993; 5(11):1455-1464.
471. Lindholm D, Castren E, Berzaghi M et al. Activity-dependent and hormonal regulation of neurotrophin mRNA levels in the brain—implications for neuronal plasticity. J Neurobiol 1994; 25(11):1362-1372.
472. Fu WM, Liou HH, Wang CL. Collaboration of fibronectin matrix and neurotrophin in regulating spontaneous transmitter release at developing neuromuscular synapses in Xenopus cell cultures. Neurosci Lett 2001; 300(2):115-119.
473. Kohara K, Kitamura A, Morishima M et al. Activity-dependent transfer of brain-derived neurotrophic factor to postsynaptic neurons. Science 2001; 291(5512):2419-2423.
474. Rickman DW, Brecha NC. Expression of the proto-oncogene, trk, receptors in the developing rat retina. Vis Neurosci 1995; 12(2):215-222.
475. Tongiorgi E, Righi M, Cattaneo A. Subcellular localisation of neurotrophins and neurotrophin receptors: Implications for synaptic plasticity. Rev Bras Biol 1996; 56(Su 1 Pt 1):175-182.
476. Tongiorgi E, Righi M, Cattaneo A. Activity-dependent dendritic targeting of BDNF and TrkB mRNAs in hippocampal neurons. J Neurosci 1997; 17(24):9492-9505.
477. Song HJ, Ming GL, Poo MM. cAMP-induced switching in turning direction of nerve growth cones. Nature 1997; 388(6639):275-279.
478. Enokido Y, Wyatt S, Davies AM. Developmental changes in the response of trigeminal neurons to neurotrophins: Influence of birthdate and the ganglion environment. Development 1999; 126(19):4365-4373.
479. Bender RA, Lauterborn JC, Gall CM et al. Enhanced CREB phosphorylation in immature dentate gyrus granule cells precedes neurotrophin expression and indicates a specific role of CREB in granule cell differentiation. Eur J Neurosci 2001; 13(4):679-686.
480. Clendening B, Hume RI. Cell interactions regulate dendritic morphology and responses to neurotransmitters in embryonic chick sympathetic preganglionic neurons in vitro. J Neurosci 1990; 10(12):3992-4005.
481. Clendening B, Hume RI. Expression of multiple neurotransmitter receptors by sympathetic preganglionic neurons in vitro. J Neurosci 1990; 10(12):3977-3991.
482. Schilling K, Dickinson MH, Connor JA et al. Electrical activity in cerebellar cultures determines Purkinje cell dendritic growth patterns. Neuron 1991; 7(6):891-902.
483. Zigmond MJ, Castro SL, Keefe KA et al. Role of excitatory amino acids in the regulation of dopamine synthesis and release in the neostriatum. Amino Acids 1998; 14(1-3):57-62.
484. Burgoyne RD, Graham ME, Cambray-Deakin M. Neurotrophic effects of NMDA receptor activation on developing cerebellar granule cells. J Neurocytol 1993; 22(9):689-695.
485. Kojima M, Takei N, Numakawa T et al. Biological characterization and optical imaging of brain-derived neurotrophic factor-green fluorescent protein suggest an activity-dependent local release of brain-derived neurotrophic factor in neurites of cultured hippocampal neurons. J Neurosci Res 2001; 64(1):1-10.
486. Hirai H, Launey T. The regulatory connection between the activity of granule cell NMDA receptors and dendritic differentiation of cerebellar Purkinje cells. J Neurosci 2000; 20(14):5217-5224.
487. Vogel MW, Prittie J. Purkinje cell dendritic arbors in chick embryos following chronic treatment with an N-methyl-D-aspartate receptor antagonist. J Neurobiol 1995; 26(4):537-552.
488. Feller MB. Spontaneous correlated activity in developing neural circuits. Neuron 1999; 22(4):653-656.
489. Yuste R, Nelson DA, Rubin WW et al. Neuronal domains in developing neocortex: Mechanisms of coactivation. Neuron 1995; 14(1):7-17.
490. Waxham MN. Neurotransmitter receptors. In: Zigmond MJ, Bloom FE, Landis SC, Roberts JL, Squire LR, eds. Fundamental neuroscience. San Diego: Academic Press, 1999:235-267.

491. Nakanishi S. Molecular diversity of glutamate receptors and implications for brain function. Science 1992; 258(5082):597-603.
492. Santi MR, Ikonomovic S, Wroblewski JT et al. Temporal and depolarization-induced changes in the absolute amounts of mRNAs encoding metabotropic glutamate receptors in cerebellar granule neurons in vitro. J Neurochem 1994; 63(4):1207-1217.
493. Vaughn JE. Fine structure of synaptogenesis in the vertebrate central nervous system. Synapse 1989; 3(3):255-285.
494. Vaughn JE, Barber RP, Sims TJ. Dendritic development and preferential growth into synaptogenic fields: A quantitative study of Golgi-impregnated spinal motor neurons. Synapse 1988; 2(1):69-78.
495. Fiala JC, Feinberg M, Popov V et al. Synaptogenesis via dendritic filopodia in developing hippocampal area CA1. J Neurosci 1998; 18(21):8900-8911.
496. Koester SE, O'Leary DD. Functional classes of cortical projection neurons develop dendritic distinctions by class-specific sculpting of an early common pattern. J Neurosci 1992; 12(4):1382-1393.
497. Rakic P, Sidman RL. Organization of cerebellar cortex secondary to deficit of granule cells in weaver mutant mice. J Comp Neurol 1973; 152(2):133-161.
498. Rakic P, Sidman RL. Weaver mutant mouse cerebellum: Defective neuronal migration secondary to abnormality of Bergmann glia. Proc Natl Acad Sci USA 1973; 70(1):240-244.
499. Rakic P, Sidman RL. Sequence of developmental abnormalities leading to granule cell deficit in cerebellar cortex of weaver mutant mice. J Comp Neurol 1973; 152(2):103-132.
500. Rakic P. Role of cell interaction in development of dendritic patterns. Adv Neurol 1975; 12:117-134.
501. Sotelo C. Dendritic abnormalities of Purkinje cells in the cerebellum of neurologic mutant mice (Weaver and Staggerer). Adv Neurol 1975; 12:335-351.
502. Kunizuka H, Kinouchi H, Arai S et al. Activation of Arc gene, a dendritic immediate early gene, by middle cerebral artery occlusion in rat brain. Neuroreport 1999; 10(8):1717-1722.
503. Angenstein F, Buchner K, Staak S. Age-dependent differences in glutamate-induced phosphorylation systems in rat hippocampal slices. Hippocampus 1999; 9(2):173-185.
504. Quinlan EM, Halpain S. Emergence of activity-dependent, bidirectional control of microtubule-associated protein MAP2 phosphorylation during postnatal development. J Neurosci 1996; 16(23):7627-7637.
505. Quinlan EM, Halpain S. Postsynaptic mechanisms for bidirectional control of MAP2 phosphorylation by glutamate receptors. Neuron 1996; 16(2):357-368.
506. Wu GY, Cline HT. Stabilization of dendritic arbor structure in vivo by CaMKII. Science 1998; 279(5348):222-226.
507. Wu GY, Zou DJ, Rajan I et al. Dendritic dynamics in vivo change during neuronal maturation. J Neurosci 1999; 19(11):4472-4483.
508. Tsui CC, Copeland NG, Gilbert DJ et al. Narp, a novel member of the pentraxin family, promotes neurite outgrowth and is dynamically regulated by neuronal activity. J Neurosci 1996; 16(8):2463-2478.
509. O'Brien RJ, Xu D, Petralia RS et al. Synaptic clustering of AMPA receptors by the extracellular immediate-early gene product Narp. Neuron 1999; 23(2):309-323.
510. Kuhl D, Skehel P. Dendritic localization of mRNAs. Curr Opin Neurobiol 1998; 8(5):600-606.
511. Sherwood NT, Lo DC. Long-term enhancement of central synaptic transmission by chronic brain-derived neurotrophic factor treatment. J Neurosci 1999; 19(16):7025-7036.
512. Soderling TR, Derkach VA. Postsynaptic protein phosphorylation and LTP. Trends Neurosci 2000; 23(2):75-80.
513. Johnson BD, Byerly L. Ca^{2+} channel Ca^{2+}-dependent inactivation in a mammalian central neuron involves the cytoskeleton. Pflugers Arch 1994; 429(1):14-21.
514. Schulman H, Braun A. Ca^{2+}/calmodulin-dependent protein kinases. In: Carafoli E, Klee C, eds. Calcium as a cellular regulator. New York: Oxford University Press, 1998.
515. Schulman H, Hyman SE. Intracellular signaling. In: Zigmond MJ, Bloom FE, Landis SC, Roberts JL, Squire LR, eds. Fundamental neuroscience. San Diego: Academic Press, 1999:269-316.
516. Gao WQ, Hatten ME. Neuronal differentiation rescued by implantation of Weaver granule cell precursors into wild-type cerebellar cortex. Science 1993; 260(5106):367-369.
517. Goslin K, Banker G. Experimental observations on the development of polarity by hippocampal neurons in culture. J Cell Biol 1989; 108(4):1507-1516.
518. Goslin K, Banker G. Rapid changes in the distribution of GAP-43 correlate with the expression of neuronal polarity during normal development and under experimental conditions. J Cell Biol 1990; 110(4):1319-1331.
519. Berbel P, Innocenti GM. The development of the corpus callosum in cats: A light- and electron-microscopic study. J Comp Neurol 1988; 276(1):132-156.
520. Montero M, Alonso MT, Carnicero E et al. Chromaffin-cell stimulation triggers fast millimolar mitochondrial Ca^{2+} transients that modulate secretion. Nat Cell Biol 2000; 2(2):57-61.

521. Zheng J, Lamoureux P, Santiago V et al. Tensile regulation of axonal elongation and initiation. J Neurosci 1991; 11(4):1117-1125.

522. Zheng JQ, Felder M, Connor JA et al. Turning of nerve growth cones induced by neurotransmitters. Nature 1994; 368(6467):140-144.

523. Zheng JQ, Wan JJ, Poo MM. Essential role of filopodia in chemotropic turning of nerve growth cone induced by a glutamate gradient. J Neurosci 1996; 16(3):1140-1149.

524. Zheng JQ, Poo MM, Connor JA. Calcium and chemotropic turning of nerve growth cones. Perspect Dev Neurobiol 1996; 4(2-3):205-213.

525. Carlier MF. Actin: Protein structure and filament dynamics. J Biol Chem 1991; 266(1):1-4.

526. Carlier MF. Control of actin dynamics. Curr Opin Cell Biol 1998; 10(1):45-51.

527. Zhou K, Wang Y, Gorski JL et al. Guanine nucleotide exchange factors regulate specificity of downstream signaling from Rac and Cdc42. J Biol Chem 1998; 273(27):16782-16786.

528. Meberg PJ, Kossel AH, Williams CV et al. Calcium-dependent alterations in dendritic architecture of hippocampal pyramidal neurons. Neuroreport 1999; 10(3):639-644.

529. Cabell L, Audesirk G. Effects of selective inhibition of protein kinase C, cyclic AMP-dependent protein kinase, and Ca^{2+}-calmodulin-dependent protein kinase on neurite development in cultured rat hippocampal neurons. Int J Dev Neurosci 1993; 11(3):357-368.

530. Audesirk G, Cabell L, Kern M. Modulation of neurite branching by protein phosphorylation in cultured rat hippocampal neurons. Brain Res Dev Brain Res 1997; 102(2):247-260.

531. Finkbeiner S, Greenberg ME. Ca^{2+} channel-regulated neuronal gene expression. J Neurobiol 1998; 37(1):171-189.

532. Finkbeiner S. Calcium regulation of the brain-derived neurotrophic factor gene. Cell Mol Life Sci 2000; 57(3):394-401.

533. Crino P, Khodakhah K, Becker K et al. Presence and phosphorylation of transcription factors in developing dendrites. Proc Natl Acad Sci USA 1998; 95(5):2313-2318.

534. Watson FL, Heerssen HM, Moheban DB et al. Rapid nuclear responses to target-derived neurotrophins require retrograde transport of ligand-receptor complex. J Neurosci 1999; 19(18):7889-7900.

535. Hall A. Rho GTPases and the actin cytoskeleton. Science 1998; 279(5350):509-514.

536. Bishop AL, Hall A. Rho GTPases and their effector proteins. Biochem J 2000; 348(Pt 2):241-255.

537. Dickson BJ. Rho GTPases in growth cone guidance. Curr Opin Neurobiol 2001; 11(1):103-110.

538. Luo L, Liao YJ, Jan LY et al. Distinct morphogenetic functions of similar small GTPases: Drosophila Drac1 is involved in axonal outgrowth and myoblast fusion. Genes Dev 1994; 8(15):1787-1802.

539. Luo L, Hensch TK, Ackerman L et al. Differential effects of the Rac GTPase on Purkinje cell axons and dendritic trunks and spines. Nature 1996; 379(6568):837-840.

540. Luo L, Jan L, Jan YN. Small GTPases in axon outgrowth. Perspect Dev Neurobiol 1996; 4(2-3):199-204.

541. Luo L, Jan LY, Jan YN. Rho family GTP-binding proteins in growth cone signalling. Curr Opin Neurobiol 1997; 7(1):81-86.

542. Luo L. Rho GTPases in neuronal morphogenesis. Nat Rev Neurosci 2000; 1(3):173-180.

543. Mackay DJ, Hall A. Rho GTPases. J Biol Chem 1998; 273(33):20685-20688.

544. Mackay DJ, Nobes CD, Hall A. The Rho's progress: A potential role during neuritogenesis for the Rho family of GTPases. Trends Neurosci 1995; 18(11):496-501.

545. Ridley AJ. Rho: Theme and variations. Curr Biol 1996; 6(10):1256-1264.

546. Ridley A. Rho GTPases. Integrating integrin signaling. J Cell Biol 2000; 150(4):F107-109.

547. Waterman-Storer CM, Salmon E. Positive feedback interactions between microtubule and actin dynamics during cell motility. Curr Opin Cell Biol 1999; 11(1):61-67.

548. Estrach S, Schmidt S, Diriong S et al. The human Rho-GEF Trio and its target GTPase RhoG are involved in the NGF pathway, leading to neurite outgrowth. Curr Biol 2002; 12(4):307-312.

549. Schmidt S, Diriong S, Mery J et al. Identification of the first Rho-GEF inhibitor, TRIPα, which targets the RhoA-specific GEF domain of Trio. FEBS Let 2002; 523(1-3):35-42.

550. Ruchhoeft ML, Ohnuma S, McNeill L et al. The neuronal architecture of Xenopus retinal ganglion cells is sculpted by Rho-family GTPases in vivo. J Neurosci 1999; 19(19):8454-8463.

551. Ng J, Nardine T, Harms M et al. Rac GTPases control axon growth, guidance and branching. Nature 2002; 416(6879):442-447.

552. Fujita H, Katoh H, Ishikawa Y et al. Rapostlin is a novel effector of Rnd2 GTPase inducing neurite branching. J Biol Chem 2002; 277(47):45428-45543.

553. Ehlers MD. Molecular morphogens for dendritic spines. Trends Neurosci 2002; 25(2):64-67.

554. Tapon N, Hall A. Rho, Rac, and Cdc42 GTPases regulate the organization of the actin cytoskeleton. Curr Opin Cell Biol 1997; 9(1):86-92.

555. Tapon N, Nagata K, Lamarche N et al. A new Rac target POSH is an SH3-containing scaffold protein involved in the JNK and NF-κB signaling pathways. Embo J 1998; 17(5):1395-1404.
556. Kuhn TB, Schmidt MF, Kater SB. Laminin and fibronectin guideposts signal sustained but opposite effects to passing growth cones. Neuron 1995; 14(2):275-285.
557. Kuhn TB, Brown MD, Bamburg JR. Rac1-dependent actin filament organization in growth cones is necessary for β1-integrin-mediated advance but not for growth on poly-D-lysine. J Neurobiol 1998; 37(4):524-540.
558. Kuhn TB, Meberg PJ, Brown MD et al. Regulating actin dynamics in neuronal growth cones by ADF/cofilin and Rho family GTPases. J Neurobiol 2000; 44(2):126-144.
559. Boguski MS, McCormick F. Proteins regulating Ras and its relatives. Nature 1993; 366(6456):643-654.
560. Neumann H, Schweigreiter R, Yamashita T et al. Tumor necrosis factor inhibits neurite outgrowth and branching of hippocampal neurons by a Rho-dependent mechanism. J Neurosci 2002; 22(3):854-862.
561. Strittmatter SM. GAP-43 as a modulator of G protein transduction in the growth cone. Perspect Dev Neurobiol 1992; 1(1):13-19.
562. Strittmatter SM, Vartanian T, Fishman MC. GAP-43 as a plasticity protein in neuronal form and repair. J Neurobiol 1992; 23(5):507-520.
563. Fishman MC. GAP-43: Putting constraints on neuronal plasticity. Perspect Dev Neurobiol 1996; 4(2-3):193-198.
564. Benowitz LI, Routtenberg A. GAP-43: An intrinsic determinant of neuronal development and plasticity. Trends Neurosci 1997; 20(2):84-91.
565. Thies E, Davenport RW. Independent roles of Rho-GTPases in growth cone and axonal behavior. J Neurobiol 2003; 54(2):358-36.
566. Nakayama AY, Harms MB, Luo L. Small GTPases Rac and Rho in the maintenance of dendritic spines and branches in hippocampal pyramidal neurons. J Neurosci 2000; 20(14):5329-5338.
567. Morgan DL, Proske U. On the branching of motoneurons. Muscle Nerve 2001; 24(3):372-379.
568. Albertinazzi C, Gilardelli D, Paris S et al. Overexpression of a neural-specific Rho family GTPase, cRac1B, selectively induces enhanced neuritogenesis and neurite branching in primary neurons. J Cell Biol 1998; 142(3):815-825.
569. Roskies AL, O'Leary DD. Control of topographic retinal axon branching by inhibitory membrane-bound molecules. Science 1994; 265(5173):799-803.
570. Sakai T, Larsen M, Yamada KM. Fibronectin requirement in branching morphogenesis. Nature 2003; 423(6942):876-881.
571. Sang Q, Tan SS. Contact-associated neurite outgrowth and branching of immature cortical interneurons. Cereb Cortex 2003; 13(6):677-683.
572. Sata M, Moss J, Vaughn M. Structural basis for the inhibitory effect of brefeldin A on guanine nucleotide-exchange proteins for ADP-ribosylation factors. Proc Natl Sci USA 1999; 96(6):2752-2757.
573. Simon DK, O'Leary DD. Limited topographic specificity in the targeting and branching of mammalian retinal axons. Dev Biol 1990; 137(1):125-134.
574. Drescher U. The Eph family in the patterning of neuronal development. Curr Biol 1997; 7(12):R799-807.
575. Drescher U, Bonhoeffer F, Muller BK. The Eph family in retinal axon guidance. Curr Opin Neurobiol 1997; 7(1):75-80.
576. Drescher U. Eph receptor tyrosine kinases and their ligands in development. Ernst Schering Res Found Workshop 2000; (29):151-164.
577. Drescher U. Eph family functions from an evolutionary perspective. Curr Opin Genet Dev 2002; 12(4):397-402.
578. Monschau B, Kremoser C, Ohta K et al. Shared and distinct functions of RAGS and ELF-1 in guiding retinal axons. Embo J 1997; 16(6):1258-1267.
579. Vastrik I, Eickholt BJ, Walsh FS et al. Sema3A-induced growth-cone collapse is mediated by Rac1 amino acids 17-32. Curr Biol 1999; 9(18):991-998.
580. Jin Z, Strittmatter SM. Rac1 mediates collapsin-1-induced growth cone collapse. J Neurosci 1997; 17(16):6256-6263.
581. Wahl S, Barth H, Ciossek T et al. Ephrin-A5 induces collapse of growth cones by activating Rho and Rho kinase. J Cell Biol 2000; 149(2):263-270.
582. Driessens MH, Hu H, Nobes CD et al. Plexin-B semaphorin receptors interact directly with active Rac and regulate the actin cytoskeleton by activating Rho. Curr Biol 2001; 11(5):339-344.
583. Lee T, Winter C, Marticke SS et al. Essential roles of Drosophila RhoA in the regulation of neuroblast proliferation and dendritic but not axonal morphogenesis. Neuron 2000; 25(2):307-316.
584. Threadgill R, Bobb K, Ghosh A. Regulation of dendritic growth and remodeling by Rho, Rac, and Cdc42. Neuron 1997; 19(3):625-634.

585. Nakayama AY, Luo L. Intracellular signaling pathways that regulate dendritic spine morphogenesis. Hippocampus 2000; 10(5):582-586.
586. Gong TW, Hegeman AD, Shin JJ et al. Identification of genes expressed after noise exposure in the chick basilar papilla. Hear Res 1996; 96(1-2):20-32.
587. Wilk-Blaszczak MA, Singer WD, Quill T et al. The monomeric G-proteins Rac1 and/or Cdc42 are required for the inhibition of voltage-dependent calcium current by bradykinin. J Neurosci 1997; 17(11):4094-4100.
588. Hely TA, Graham B, Ooyen AV. A computational model of dendrite elongation and branching based on MAP2 phosphorylation. J Theor Biol 2001; 210(3):375-384.
589. Morfini G, Quiroga S, Rosa A et al. Suppression of KIF2 in PC12 cells alters the distribution of a growth cone nonsynaptic membrane receptor and inhibits neurite extension. J Cell Biol 1997; 138(3):657-669.
590. Goedert M, Crowther RA, Garner CC. Molecular characterization of microtubule-associated proteins tau and MAP2. Trends Neurosci 1991; 14(5):193-199.
591. Ainsztein AM, Purich DL. Stimulation of tubulin polymerization by MAP-2. Control by protein kinase C-mediated phosphorylation at specific sites in the microtubule-binding region. J Biol Chem 1994; 269(45):28465-28471.
592. Friedrich P, Aszodi A. MAP2: A sensitive cross-linker and adjustable spacer in dendritic architecture. FEBS Lett 1991; 295(1-3):5-9.
593. Binder LI, Frankfurter A, Kim H et al. Heterogeneity of microtubule-associated protein 2 during rat brain development. Proc Natl Acad Sci USA 1984; 81(17):5613-5617.
594. Zmuda JF, Rivas RJ. Actin disruption alters the localization of tau in the growth cones of cerebellar granule neurons. J Cell Sci 2000; 113(Pt 15):2797-2809.
595. Zmuda JF, Rivas RJ. Actin filament disruption blocks cerebellar granule neurons at the unipolar stage of differentiation in vitro. J Neurobiol 2000; 43(4):313-328.
596. Lopez LA, Sheetz MP. Steric inhibition of cytoplasmic dynein and kinesin motility by MAP-2. Cell Motil & Cytoskeleton 1993; 24:1-16.
597. Diez-Guerra FJ, Avila J. MAP2 phosphorylation parallels dendrite arborization in hippocampal neurones in culture. Neuroreport 1993; 4(4):419-422.
598. Diez-Guerra FJ, Avila J. Rapid dephosphorylation of microtubule-associated protein 2 in the rat brain hippocampus after pentylenetetrazole-induced seizures. Eur J Biochem 1993; 215(1):181-187.
599. Diez-Guerra FJ, Avila J. An increase in phosphorylation of microtubule-associated protein 2 accompanies dendrite extension during the differentiation of cultured hippocampal neurones. Eur J Biochem 1995; 227(1-2):68-77.
600. Guo Y, Sanchez C, Udin SB. MAP2 phosphorylation and visual plasticity in Xenopus. Brain Res 2001; 905(1-2):134-141.
601. Philpot BD, Lim JH, Halpain S et al. Experience-dependent modifications in MAP2 phosphorylation in rat olfactory bulb. J Neurosci 1997; 17(24):9596-9604.
602. Labelle C, Leclerc N. Exogenous BDNF, NT-3 and NT-4 differentially regulate neurite outgrowth in cultured hippocampal neurons. Brain Res Dev Brain Res 2000; 123(1):1-11.
603. Fukuda M, Gotoh Y, Tachibana T et al. Induction of neurite outgrowth by MAP kinase in PC12 cells. Oncogene 1995; 11(2):239-244.
604. Braun AP, Schulman H. The multifunctional calcium/calmodulin-dependent protein kinase: From form to function. Annu Rev Physiol 1995; 57:417-445.
605. Ouyang Y, Kantor D, Harris KM et al. Visualization of the distribution of autophosphorylated calcium/calmodulin-dependent protein kinase II after tetanic stimulation in the CA1 area of the hippocampus. J Neurosci 1997; 17(14):5416-5427.
606. Lisman J. The CaM kinase II hypothesis for the storage of synaptic memory. Trends Neurosci 1994; 17(10):406-412.
607. Mitchison T, Kirschner M. Cytoskeletal dynamics and nerve growth. Neuron 1988; 1(9):761-772.
608. Zou DJ, Cline HT. Postsynaptic calcium/calmodulin-dependent protein kinase II is required to limit elaboration of presynaptic and postsynaptic neuronal arbors. J Neurosci 1999; 19(20):8909-8918.
609. Walaas SI, Nairn AC. Multisite phosphorylation of microtubule-associated protein 2 (MAP-2) in rat brain: Peptide mapping distinguishes between cyclic AMP-, calcium/calmodulin-, and calcium/phospholipid-regulated phosphorylation mechanisms. J Mol Neurosci 1989; 1(2):117-127.
610. Leonard AS, Lim IA, Hemsworth DE et al. Calcium/calmodulin-dependent protein kinase II is associated with the N-methyl-D-aspartate receptor. Proc Natl Acad Sci USA 1999; 96(6):3239-3244.
611. Bayer KU, De Koninck P, Leonard AS et al. Interaction with the NMDA receptor locks CaMKII in an active conformation. Nature 2001; 411(6839):801-805.
612. Deutch AY, Roth RH. Neurotransmitters. In: Zigmond MJ, Bloom FE, Landis SC, Roberts JL, Squire LR, eds. Fundamental neuroscience. San Diego: Academic Press, 1999:193-234.

613. Berninger B, Garcia DE, Inagaki N et al. BDNF and NT-3 induce intracellular Ca^{2+} elevation in hippocampal neurones. Neuroreport 1993; 4(12):1303-1306.
614. Berninger B, Poo M. Exciting neurotrophins. Nature 1999; 401(6756):862-863.
615. Fukunaga K, Soderling TR, Miyamoto E. Activation of Ca^{2+}/calmodulin-dependent protein kinase II and protein kinase C by glutamate in cultured rat hippocampal neurons. J Biol Chem 1992; 267(31):22527-22533.
616. Fukunaga K, Stoppini L, Miyamoto E et al. Long-term potentiation is associated with an increased activity of Ca^{2+}/calmodulin-dependent protein kinase II. J Biol Chem 1993; 268(11):7863-7867.
617. Ripellino JA, Neve RL, Howe JR. Expression and heteromeric interactions of non-N-methyl-D-aspartate glutamate receptor subunits in the developing and adult cerebellum. Neuroscience 1998; 82(2):485-497.
618. Halegoua S, Armstrong RC, Kremer NE. Dissecting the mode of action of a neuronal growth factor. Curr Top Microbiol Immunol 1991; 165:119-170.
619. Gaiddon C, Loeffler JP, Larmet Y. Brain-derived neurotrophic factor stimulates AP-1 and cyclic AMP-responsive element dependent transcriptional activity in central nervous system neurons. J Neurochem 1996; 66(6):2279-2286.
620. Thakker-Varia S, Alder J, Crozier RA et al. Rab3A is required for brain-derived neurotrophic factor-induced synaptic plasticity: Transcriptional analysis at the population and single-cell levels. J Neurosci 2001; 21(17):6782-6790.
621. Lyford GL, Yamagata K, Kaufmann WE et al. Arc, a growth factor and activity-regulated gene, encodes a novel cytoskeleton-associated protein that is enriched in neuronal dendrites. Neuron 1995; 14(2):433-445.
622. Steward O, Wallace CS, Lyford GL et al. Synaptic activation causes the mRNA for the IEG Arc to localize selectively near activated postsynaptic sites on dendrites. Neuron 1998; 21(4):741-751.
623. Corriveau RA, Shatz CJ, Nedivi E. Dynamic regulation of CPG15 during activity-dependent synaptic development in the mammalian visual system. J Neurosci 1999; 19(18):7999-8008.
624. Udin S. CPG15 and the dynamics of retinotectal synapses. Nat Neurosci 2000; 3(10):971-972.
625. Bar-Peled O, Ben-Hur H, Biegon A et al. Distribution of glutamate transporter subtypes during human brain development. J Neurochem 1997; 69:2571-2580.
626. Nakanishi N, Shneider NA, Axel R. A family of glutamate receptor genes: Evidence for the formation of heteromultimeric receptors with distinct channel properties. Neuron 1990; 5:569-581.
627. Nakanishi S. Metabotropic glutamate receptors: Synaptic transmission, modulation, and plasticity. Neuron 1994; 13:1031-1037.
628. Guzowski JF, McNaughton BL, Barnes CA et al. Environment-specific expression of the immediate-early gene Arc in hippocampal neuronal ensembles. Nat Neurosci 1999; 2(12):1120-1124.
629. Guzowski JF, Lyford GL, Stevenson GD et al. Inhibition of activity-dependent Arc protein expression in the rat hippocampus impairs the maintenance of long-term potentiation and the consolidation of long-term memory. J Neurosci 2000; 20(11):3993-4001.
630. Rajan I, Cline HT. Glutamate receptor activity is required for normal development of tectal cell dendrites in vivo. J Neurosci 1998; 18(19):7836-7846.
631. Rajan I, Witte S, Cline HT. NMDA receptor activity stabilizes presynaptic retinotectal axons and postsynaptic optic tectal cell dendrites in vivo. J Neurobiol 1999; 38(3):357-368.
632. Sanchez C, Ulloa L, Montoro RJ et al. NMDA-glutamate receptors regulate phosphorylation of dendritic cytoskeletal proteins in the hippocampus. Brain Res 1997; 765(1):141-148.
633. McKinney RA, Capogna M, Durr R et al. Miniature synaptic events maintain dendritic spines via AMPA receptor activation. Nat Neurosci 1999; 2(1):44-49.
634. Nedivi E, Fieldust S, Theill LE et al. A set of genes expressed in response to light in the adult cerebral cortex and regulated during development. Proc Natl Acad Sci USA 1996; 93(5):2048-2053.
635. Baptista CA, Hatten ME, Blazeski R et al. Cell-cell interactions influence survival and differentiation of purified Purkinje cells in vitro. Neuron 1994; 12(2):243-260.
636. Sato M, Lopez-Mascaraque L, Heffner CD et al. Action of a diffusible target-derived chemoattractant on cortical axon branch induction and directed growth. Neuron 1994; 13(4):791-803.
637. Sato F, Parent M, Levesque M et al. Axonal branching pattern of neurons of the subthalamic nucleus in primates. J Comp Neurol 2000; 424(1):142-152.
638. Moran J, Patel AJ. Stimulation of the N-methyl-D-aspartate receptor promotes the biochemical differentiation of cerebellar granule neurons and not astrocytes. Brain Res 1989; 486(1):15-25.
639. Moran J, Patel AJ. Effect of potassium depolarization on phosphate-activated glutaminase activity in primary cultures of cerebellar granule neurons and astroglial cells during development. Brain Res Dev Brain Res 1989; 46(1):97-105.
640. Kuhar SG, Feng L, Vidan S et al. Changing patterns of gene expression define four stages of cerebellar granule neuron differentiation. Development 1993; 117(1):97-104.

641. Dandenault ME, Rivas RJ. Cessation of cell surface TAG-1 expression occurs by a non-cell autonomous mechanism in developing cerebellar granule neurons. American Society for Cell Biology Abstr 1999; 39:1199.
642. Niell CM, Meyer MP, Smith SJ. In vivo imaging of synapse formation on a growing dendritic arbor. Nature Neurosci 2004; 7(3):254-260.
643. Harris KM. Structure, development, and plasticity of dendritic spines. Curr Opin Neurobiol 1999; 9:343-348.
644. Matus A. Moving molecules make synapses. Nature Neurosci 2001; 4(10):967-968.
645. Fiala JC, Spacek J, Harris KM. Dendritic spine pathology: Cause or consequence of neurological disorders? Brain Res Rev 2002; 39:29-54.
646. Kasai H, Matsuzaki M, Noguchi J et al. Structure-stability-function relationships of dendritic spines. Trends Neurosci 2003; 26(7):360-368.
647. Maletic-Savatic M, Svoboda K. Rapid dendritic morphogenesis in CA1 hippocampal dendrites induced by synaptic activity. Science 1999; 283:1923-1926.
648. Roelandse M, Welman A, Wagner U et al. Focal motility determines the geometry of dendritic spines. Neurosci 2003; 121:39-49.
649. Lelkes PI, Unsworth BR, Saporta S et al. Culture of neuroendocrine and neuronal cells for tissue engineering. In: Vunjak-Novakovic G, Freshney RI, eds. Culture of cells for tissue engineering. Wiley Inc., 2004.
650. Xu D, Hopf C, Reddy R et al. Narp and NP1 form heterocomplexes that function in developmental and activity-dependent synaptic plasticity. Neuron 2003; 39:513-528.

CHAPTER 3

Branching of Single Cells in *Arabidopsis*

Daniel Bouyer and Martin Hülskamp

Abstract

Branching of single cells is controlled by intracellular or extracellular cues that lead to the establishment of a polarity axis and subsequently to the local activation of growth activity. Three model cell types in *Arabidopsis*, that elucidate different mechanisms of branch formation in single cells, are considered in this review. Trichomes serve as a model to study how multiple branches are formed in a predictable manner. Epidermal pavement cells enable study of pathways that integrate extracellular signals and facilitate coordinated growth within a tissue. The analysis of root hairs reveals information about how branch formation can be inhibited.

The mechanisms involved in branch formation at the level of single plant cells seem to be entirely different from those leading to branching at the level of tissues or the whole organism. Whereas the latter ultimately control the temporal and spatial coordination of cell divisions, at least in plants, branching at the single cell level regulates localized cell growth or expansion. This is achieved in several steps. Initially, the integration of external and internal cues creates the positional information for the proper initiation of branches. This information is best understood in three epidermal cell types in *Arabidopsis* (Fig. 1): leaf trichomes, leaf pavement cells and root hairs.[1] Leaf trichomes exemplify a cell type in which the spatial orientation of the branches is strictly correlated with respect to the leaf axis indicating the existence of a three-dimensional prepattern.[2] Leaf pavement cells are initially round or square and then initiate branches to adopt the shape of a piece of a jigsaw puzzle. Where these branches form is not predictable but their formation is always coordinated with morphogenesis in neighbouring cells. *Arabidopsis* root hairs, which are also composed of single cells, are normally not branched but certain mutants and drug treatments lead to their branching. Root hairs therefore serve as a model to study how branching is suppressed. In this chapter we will summarize the current view of how branch formation is controlled in these three cell types.

Branching of Trichomes

Leaf trichomes in *Arabidopsis* are single polyploid cells with a stereotyped branching pattern (Fig. 1).[3,4] The orientation of the three branches is highly predictable not only with respect to each other but also with respect to the leaf axis.[5,6] The two branches of the lower branch point are coaligned with the leaf axis and arranged in an angle of approximately 120 degrees. Only the branch pointing to the leaf tip continues to branch. The plane of the second branching event is parallel to the leaf plane and the two branches are arranged in an angle of about 90 degrees.

The branching pattern is laid down during the ontogenesis of the trichome.[7,8] Trichomes originate from protodermal cells. Concomitant with the decision to become a trichome, the cell stops dividing but continues to replicate its DNA (by endoreduplication). After four endoreduplication cycles the mature trichome cell has a DNA content of 32 C.[4] During this process, the incipient trichome cell increases in size and begins to grow out of the leaf surface.

Branching Morphogenesis, edited by Jamie A. Davies.

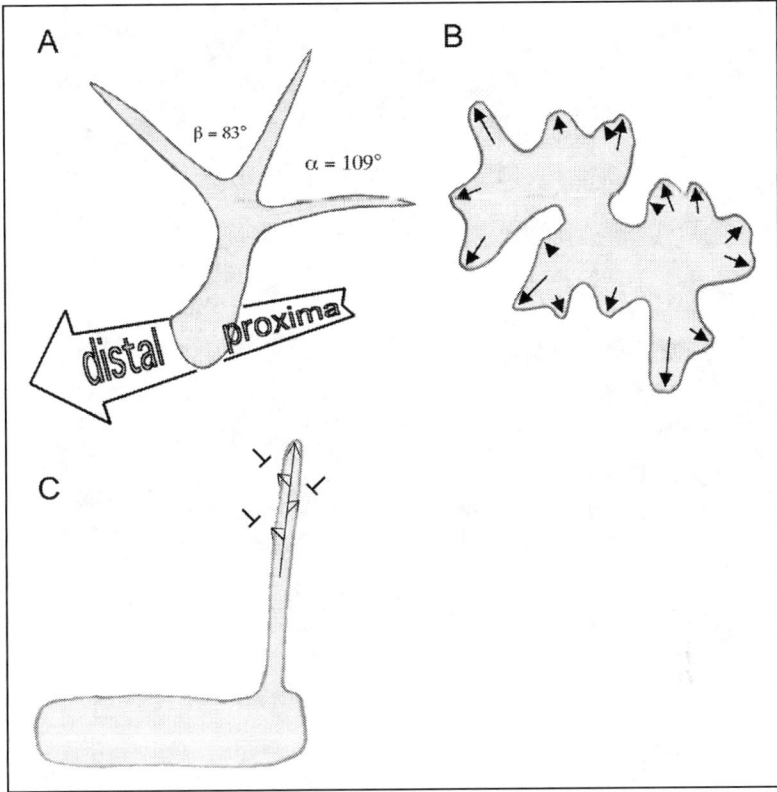

Figure 1. Model cell types to study single cell branching. Schematic drawings to illustrate the branched phenotype of trichomes (A), epidermal pavement cells (B) and root hairs (C).

When the cell reaches a DNA content of about 8C, the first branching event occurs. Shortly thereafter, the second branching event becomes visible as a small bulge on that branch that points towards the leaf tip. After branch initiation is completed, the trichome cell undergoes rapid cell expansion.

Trichome Branching Mutants

The facts that trichomes can be monitored very easily and that trichomes are not essential for the good health of *Arabidopsis* plants under laboratory conditions greatly facilitate a genetic approach, by which genes involved in trichome branching can be identified based on their mutant phenotype. More than 15 branching genes have been identified that function either as suppressors or activators of branch initiation (Fig. 2).[2,9-11] Among these genes, one class of genes appears to control branch number indirectly as a consequence of altered ploidy levels.[12] Mutants with a reduced DNA content such as *glabra3* or constitutive pathogene *response5* (*cpr5*) have smaller and less branched trichomes. Trichomes with a higher ploidy level as found in *triptychon* (*try*), *kaktus* (*kak*), *polychome* (*pym*), *spindly* (*spy*) and *rastafari* (*rfi*) mutants are larger and exhibit more branches.[4,11,12] As such a correlation between the ploidy level and branch number is also observed when treatments with DNA replication inhibitors reduce the DNA content and branch number (Schwab and Hülskamp, unpublished observations), and in situations in which higher ploidy levels in tetraploid plants lead to overbranched trichomes,[11] it is conceivable that the branching phenotype in endoreduplication mutants is caused indirectly.

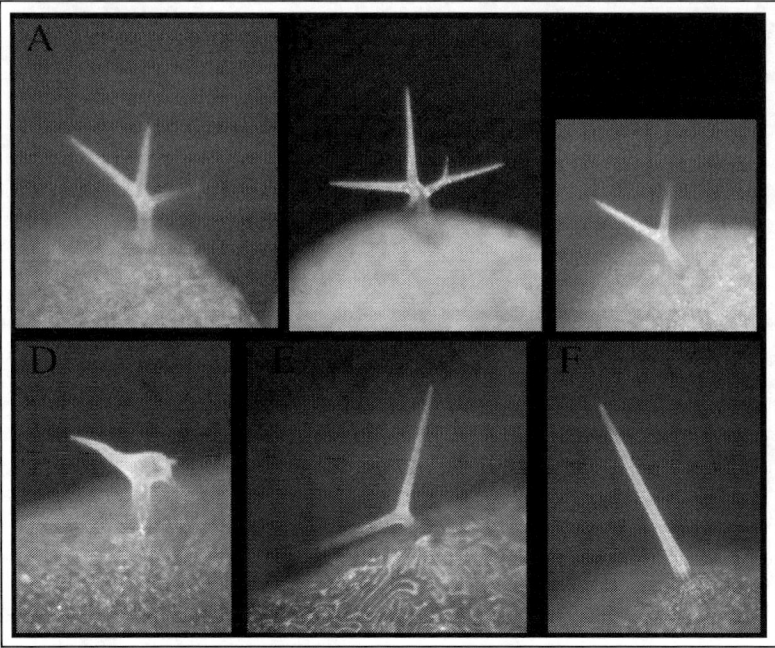

Figure 2. Trichome branching. A) Wild-type trichome with three branches. B) *Triptychon* mutant trichome with increased ploidy levels and increased branch number. C) *Glabra3* mutant trichome with lower ploidy levels and fewer branches. D) *Distorted* mutant trichome. Extension growth is irregular, compare with Figure 3C. E) *Angustifolia* mutant trichome with two branches. F) Unbranched *stichel* mutant trichome.

The second class of branching mutants comprises mutants exhibiting trichomes with a reduced or increased branch number but with a normal ploidy level. Only one such mutant, *noeck* (*nok*), is known, that has an increased number of branches—of up to seven.[5] All other mutants—*stichel* (*sti*), *angustifolia* (*an*), *stachel* (*sta*), *zwichel* (*zwi*), *furca1* (*frc1*), *frc2*, *frc3*, *frc4*, *tubulin folding cofactorA* (*tfcA*), *tfcC*, *spike* (*spk1*), *fragile fiber2* (*fra2*) (allelic to ectopic root hairs 3, botero1, fat root) and *fass/tonneau2* (*ton2*)—have two or no branches.[5,13,38,39,40,42,43,44,45] In addition, three mutants, suppressor of zwichel-1 (*suz1*), *suz2* and *suz3*, were found that suppress the branching phenotype of *zwi-3* in allele specific manner.[14] Except for *suz2*, all of these mutants display no trichome phenotype in a wild-type background. In a thorough genetic analysis, most of these genes have been placed into one interaction network.[13] From this analysis two major conclusions could be drawn: (1) Most branching genes act in parallel pathways. (2) Three genes, *AN, ZWI* and *FRC4*, act in the same pathway as the endoreduplication mutants. The molecular analysis of most of the branching genes has revealed a number of different processes participating in branch initiation.

Trichome Branching: The Cell Division/Endoreduplication Pathway

The genetic finding that three genes are required to mediate the function of the endoreduplication genes suggests that branch initiation might be mechanistically linked to the cell division machinery. This hypothesis is supported by the molecular analysis of two of these genes, *ZWI* and *AN*, and some other indirect evidence.

The *ZWI* gene was shown to encode the biochemically well-characterized calmodulin-binding KCBP gene[15-18] which is a member of the kinesin superfamily of motor proteins.[19] Kinesins are microtubule-based motors that are important in transport processes and microtubule organization. The biochemical function of *ZWI*, however, is still unclear. Although *zwi* mutants appear to have no obvious cell cycle defects, two lines of evidence suggest that the function of *ZWI* is linked to the cell cycle. First, it was shown that ZWI protein is localized to the preprophase band and the phragmoplast, two structures that mark in plants the division plane.[20] Second, the injection of *ZWI* antibodies in multi-cellular stamen hair cells of *Tradescantia* cause metaphase arrest and abnormal cell plate formation.[21]

The *AN* gene encodes a protein that shows sequence similarities to brefeldin A-induced ADP-ribosylated substrate (BARS) proteins and C-terminal binding proteins (CtBP) that act as transcriptional cofactors.[22,23] BARS were biochemically identified in rats as proteins that become ADP-ribosylated upon treatments with fungal toxin brefeldin A which is known to block Golgi dynamics.[24,25] CtBPs are corepressors that directly bind to the zinc-finger transcription factors Krüppel, Knirps and Snail.[26] Despite the quite different biochemical functions, these proteins appear to be homolog proteins as they can take over each other's function such that the rat BARS can bind to transcription factors such as the C-terminus of adenovirus protein E1A, and CtBPs become ribosylated after brefeldin A treatments.[27] The biochemical function of the *Arabidopsis* AN protein is still unknown. That the function of AN is linked to cell cycle function is suggested by three observations. First, the AN protein was shown to interact with *ZWI* in yeast two-hybrid assays suggesting that AN has a similar function to *ZWI* in cell cycle regulation.[22] In addition, the AN protein has a PEST motif close to a consensus sequence for mitotic cyclin binding, suggesting that the stability of AN protein is regulated in a cell cycle dependent manner.[22,23] Finally, the *an* mutant phenotype includes changes in leaf shape that are partly due to reduced cell numbers in the leaf width axis suggesting that cell division number is regulated by AN.[28,29]

The above data indicate that *AN* and *ZWI* can function in cell cycle regulation and suggest that also their role in branch initiation might be linked to the cell cycle. On the first glance this is difficult to understand, but the recent identification of the *siamese* (*sim*) mutant suggests a very attractive mechanistic explanation. In *sim* mutants, trichomes are multicellular and contain up to 15 cells.[30] The fact that one mutation can change the polyploid unicellular *Arabidopsis* trichomes into multicellular hairs suggests that unicellular trichomes are evolutionarily derived from multicellular trichomes. It is known from other species, for example *Verbascum*,[31] that branching of multicellular trichomes is caused by a defined cell division pattern. It is therefore speculated that the positional information underlying branch formation in *Arabidopsis* trichomes is based on the cell division machinery, such that the morphogenetic events of cytokinesis take place in the absence of actual cell division. In this scenario, *AN* and *ZWI* would be important for the control of cell divisions and as a consequence also for branching.[32]

Trichome Branching: Dosage Dependent Regulation by *STICHEL*

The strongest branching phenotype is observed in *sti* mutants; leaf trichomes are unbranched. Two sets of experiments suggest that *STI* regulates branching in a dosage dependent manner: The characterisation of a large number of *sti* alleles has shown that strong sti-alleles have fewer branches than weak sti-alleles.[33] This indicates that the number of branches is correlated with the amount of *STI* activity. Consistent with this, overexpression of STI, under the control of the strong 35S promoter from cauliflower mosaic virus, leads to extra branch formation.[33] The biochemical function of STI is still unclear. The cloning of the *STI* gene revealed that the protein contains a domain with sequence similarities to eubacterial DNA-polymerase III subunits.[33] However, as no replication defects were found in *sti* mutants, it is not very likely that *STI* functions as a DNA-polymerase subunit. Further studies are needed before biochemical function of STI in branch initiation will be understood.

Trichome Branching: Role of the Microtubule Cytoskeleton

That the microtubule cytoskeleton but not the actin cytoskeleton is important for trichome branching was demonstrated by drug inhibitor experiments. Drugs interfering with the function of actin resulted in grossly abnormal cell shapes, but did not affect trichome branching.[34,35] By contrast microtubule specific drugs resulted in reduced branching and iso-tropic cell growth.[36] The molecular analysis of several branching mutants enabled a more de-tailed analysis of microtubule function during trichome formation.

Two trichome branching genes, *FASS/ TONNEAU2* and *SPIKE1*, are likely to be involved in the spatial regulation of microtubules. The *fass/ tonneau2* mutants are characterised by spa-tially randomised cell divisions and reduced trichome branching and the organization of mi-crotubules was found to be randomised.[37,38] The recent cloning revealed that *FASS/ TON-NEAU2* encodes a protein with sequence similarity to a novel protein phosphatase 2A regulatory subunit suggesting that it regulates microtubule organisation by protein phosphorylation.[39] Mutations in the *SPIKE1* gene result in a pleiotropic phenotype including defective polarized growth of cotyledons and epidermal cells as well as reduced trichome branching. [40] The orga-nization of the microtubule cytoskeleton was also abnormal in several cell types. The *SPIKE1* gene encodes a protein with sequence similarity to CDM family adapter proteins (*C. elegans* CED-5; human DOCK180; *Drosophila* myoblast city) that are believed to integrate extracellu-lar signals and microtubule organization.[41]

In principle microtubule organization could be achieved by shifting already existing micro-tubules to new positions or by severing them and de novo polymerisation. The latter appears to be the case as mutations in the *KATANIN-P60* gene (*ectopic root hairs 3, botero1, fat root, fragile fiber2*) lead to a reduced trichome branch number phenotype.[42-44] The function of katanins is to cut existing microtubules into smaller fragments.[45] A second line of experiments supporting the idea that branch formation involves de novo synthesis of microtubules comes from the analysis of tubulin folding cofactor mutants *tfca* and *tfcc*.[46,47] TFCA and TFCC associate with α and β tubulin monomers after translation and help to form assembly competent α/β tubulin dimers.[48] Strong *tfca* and *tfcc* mutants are lethal,[49] whereas weak alleles display a range of phenotypes including a swollen and underbranched trichome phenotype. The microtubules, however, are intact and normally orientated in most cell types. Only in trichome branches that did not grow properly does the reorientation from a transverse to a longitudinal orientation fail to take place. Given the function of the *TFC* genes in creating assembly competent α/β dimers this finding suggests the mutant phenotype is due to a failure in de novo synthesis rather than the reorganisation of already existing microtubules.[46,47]

Branching in Epidermal Pavement Cells

Young protodermal pavement cells are regularly shaped. During differentiation epidermal pavement cells begin to branch in a random way, thus forming lobes leading to their final puzzle-shape appearance. A number of mutants are known in which branches are not formed and the emerging picture is that both the actin as well as the microtubule cytoskeleton are important for their formation.

Several of the mutants that affect microtubule organization and trichome branch formation also cause failure of branching in epidermal pavement cells. These include the *spike1* mutant,[40] the *katanin-P60* mutant[43] and the *angustifolia* mutant (Fig. 3).[50] Because the position of mi-crotubule bands correlates with the position of cell wall thickenings, it is believed that microtu-bules define the nonexpandable regions of the cell wall.[51,52] The regions in between remain more extensible and would give rise to a branch.

The role of actin in branch formation of pavement cells is evident from the mutant pheno-type displayed by mutants in the components of the Arp2/3 complex. The Arp2/3 complex is a conserved modulator of actin polymerisation[53,54] that plays a role in membrane protrusion and in the motility of vesicles and entero-pathogenic organisms.[55,56] The complex is able to initiate F-actin nucleation and branching of already existing actin filaments thereby generating a local fine actin meshwork.[54,57] In Arabidopis, mutations in components of the Arp2/3 com-plex exhibit unbranched pavement cells and an aberrant actin organization.[58,59] Current data indicate that the regulation of the activity of the Arp2/3 complex is to some extent conserved

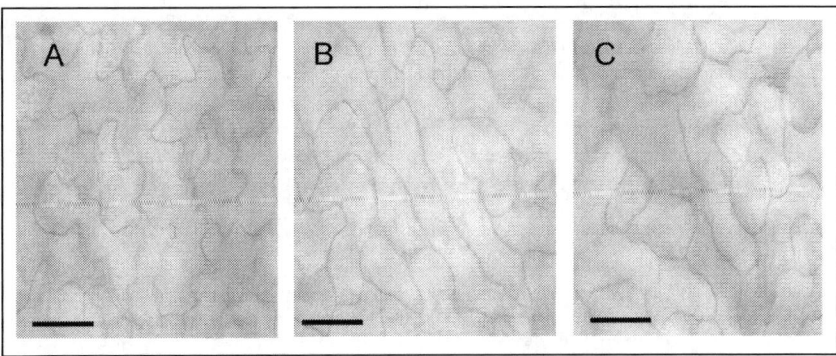

Figure 3. Pavement cells in wild type, *angustifolia* and *distorted* mutants. A) Surface view of a wild-type epidermal pavement cells. Note, that cells have many branches. B) *Angustifolia* mutant epidermal cells. Here, cells exhibit only very little branching. C) *Distorted* mutant. Cells exhibit reduced branching.

between plants and yeast and animals. One regulation pathway known in other organisms is based on small Rho-like GTPases that control the activity of the Arp2/3 complex. In plants the homologs of CDC42, Racs and Rhos are missing but a small group of Rho-like proteins called ROPs ('Rhos of plants') exist and appear to carry out a similar function.[60,61] As revealed by GFP fusions, ROP2 localizes to the extending branches of epidermal pavement cells.[55] Constitutive expression of constitutive or dominant negative form of this small GTPase interferes with lobe formation and causes aberrant accumulation of fine actin in the lobes.[62] Although not specifically shown in *Arabidopsis*, the pathway linking the regulation of the ROPs to the Arp2/3 complex appears to be conserved in plants too. From other organisms it is known that the binding of Rac to the HSPC300 complex causes its dissociation producing a HSPC300/WAVE dimer. This dimer activates ARP2/3.[63] In maize three mutants, the BRICK mutants, cause the production of nonbranched epidermal cells.[64] BRICK1 encodes a protein with sequence similarity to the human HSPC300 gene. This suggests that the whole pathway is conserved in plants though it is likely to be rather different in *Arabidopsis* as no WAVE orthologues appear to be present in the *Arabidopsis* genome.

Root Hairs

Root hair growth takes place exclusively at the very tip of the cell and the cells are therefore called tip-growing cells. Although root hairs in *Arabidopsis* normally do not form branches, two situations in which branch formation is triggered in these cells reveal principles by which branch formation is controlled in plant cells.

One situation in which root hairs are branched is when one of the above mentioned small GTPases, ROP2, is overexpressed. It is known, that several ROPs including ROP2, ROP4 and ROP6 are involved in root hair growth and that ROP2:GFP is localized to the growing tip.[65] Here they are involved in increasing the density of fine actin filaments that in turn trigger local growth of the cell. The observations that overexpression of ROP2 leads to an increase of fine actin and the initiation of new branches strongly suggests that the activation of this signalling cascade is sufficient to initiate new growth points.

Branching of root hairs can also be induced by treatments with microtubule inhibitors[66] or when the expression of α-tubulin is genetically reduced.[67] Distortions of the actin cytoskeleton, by contrast, lead to a reduction of growth but not to the initiation of new growth points. These observations indicate that microtubules function to direct growth to the tip of the cell. A more detailed analysis revealed that microtubules even hinder the formation of new branches: It is known that cytosolic Ca^{2+} is highly concentrated at the very tip or root hairs.[68]

Photoactivation of caged calcium or the local application of Calcium ionophores allows experimenters to make the Ca^{2+} influx asymmetrical which, in the tip region, leads to a reorientation of growth. Other regions of the root hair do not respond to these treatments. If, however, the same experiments are done after treatments with drugs affecting the microtubules, localized Ca^{2+} influx is sufficient to induce a branch.[68,69]

Only when more details of these three branching mechanisms are better understood will it be completely clear to what extent they share morphogenetic and regulatory mechanisms with each other and with other examples of branching by single cells.

References

1. Hülskamp M, Folkers U, Grini P. Cell morphogenesis in Arabidopsis. BioEssays 1998; 20(1):20-29.
2. Hulskamp M. How plants split hairs. Curr Biol 2000; 10(8):R308-310.
3. Marks MD, Esch J, Herman P et al. A model for cell-type determination and differentiation in plants. In: Jenkins G, Schuch W, eds. Molecular Biology of Plant Development. Cambridge: Company of Biologists Ltd., 1991:77-87.
4. Hülskamp M, Misera S, Jürgens G. Genetic dissection of trichome cell development in Arabidopsis. Cell 1994; 76:555-566.
5. Folkers U, Berger J, Hulskamp M. Cell morphogenesis of trichomes in Arabidopsis: Differential control of primary and secondary branching by branch initiation regulators and cell growth. Development 1997; 124(19):3779-3786.
6. Bouyer D, Kirik V, Hulskamp M. Cell polarity in Arabidopsis trichomes. Semin Cell Dev Biol 2001; 12(5):353-356.
7. Hulskamp M, Schnittger A. Spatial regulation of trichome formation in Arabidopsis thaliana. Semin Cell Dev Biol 1998; 9(2):213-220.
8. Szymanski DB, Lloyd AM, Marks MD. Progress in the molecular genetic analysis of trichome intiation and morphogenesis in Arabidopsis. Trends Plant Sci 2000; 5(5):53.
9. Oppenheimer D. Genetics of plant cell shape. Curr Opin Plant Biol 1998; 1:520-524.
10. Hülskamp M, Folkers U, Schnittger A. Trichome development in Arabidopsis thaliana. Int Rev Cytol 1999; 186:147-178.
11. Perazza D, Herzog M, Hülskamp M et al. Trichome cell growth in Arabidopsis thaliana can be depressed by mutations in at least five genes. Genetics 1999; 152(1):461-476.
12. Traas J, Hülskamp M, Gendreau E et al. Endoreduplication and development: Rule without dividing? Curr Opin Plant Biol 1998; 1(6):498-503.
13. Luo D, Oppenheimer DG. Genetic control of trichome branch number in Arabidopsis: The roles of the FURCA loci. Development 1999; 126(24):5547-5557.
14. Krishnakumar S, Oppenheimer DG. Extragenic suppressors of the Arabidopsis zwi-3 mutation identify new genes that function in trichome branch formation and pollen tube growth. Development 1999; 126(14):3079-3088.
15. Reddy AS, Narasimhulu SB, Safadi F et al. A plant kinesin heavy chain-like protein is a calmodulin-binding protein. Plant J 1996; 10(1):9-21.
16. Reddy ASN, Safadi F, Narasimhulu SB et al. A novel plant calmodulin-binding protein with a kinesin heavy chain motor domain. J Biol Chem 1996; 271:7052-7060.
17. Song H, Golovkin M, Reddy AS et al. In vitro motility of AtKCBP, a calmodulin-binding kinesin protein of Arabidopsis. Proc Natl Acad Sci USA 1997; 94(1):322-237.
18. Deavours BE, Reddy AS, Walker RA. Ca2+/calmodulin regulation of the Arabidopsis kinesin-like calmodulin-binding protein. Cell Motility and the Cytoskeleton 1998; 40(4):408-416.
19. Oppenheimer DG, Pollock MA, Vacik J et al. Essential role of a kinesin-like protein in Arabidopsis trichome morphogenesis. Proc Natl Acad Sci USA 1997; 94:6261-6266.
20. Bowser J, Reddy ASN. Localization of a kinesin-like calmodulin-binding protein in dividing cells of Arabidopsis and tobacco. Plant J 1997; 12(6):1429-1437.
21. Vos JW, Safadi F, Reddy AS et al. The kinesin-like calmodulin binding protein is differentially involved in cell division. Plant Cell 2000; 12(6):979-990.
22. Folkers U, Kirik V, Schobinger U et al. The cell morphogenesis gene ANGUSTIFOLIA encodes a CtBP/BARS-like protein and is involved in the control of the microtubule cytoskeleton. EMBO J 2002; 21(6):1280-1288.
23. Kim GT, Shoda K, Tsuge T et al. The ANGUSTIFOLIA gene of Arabidopsis, a plant CtBP gene, regulates leaf-cell expansion, the arrangement of cortical microtubules in leaf cells and expression of a gene involved in cell-wall formation. EMBO J 2002; 26(6):1267-1279.

24. Lippincott-Schwartz J, Yuan LC, Bonifacino JS et al. Rapid redistribution of Golgi proteins into the ER in cells treated with brefeldin A: Evidence for membrane cycling from Golgi to ER. Cell 1989; 56(5):801-813.
25. Matteis MD, Girolamo MD, Colanzi A et al. Stimulation of endogenous ADP-rebosylation by brefeldin A. Proc Natl Acad Sci USA 1994; 91:1114-1118.
26. Nibu Y, Zhang H, Levine M. Interaction of a short-range repressors with Drosophila CtBP in the embryo. Science 1998; 280:101-104.
27. Spanfo S, Silletta MG, Colanzi A et al. Molecular cloning and functional characterization of brefeldin A-ADP-ribosylated substrate. A novel protein involved in the maintenance of the Golgi structure. J Biol Chem 1999; 274(25):17705-17710.
28. Tsukaya H, Tsuge T, Uchimiya H. The cotyledon: A superior system for studies of leaf development. Planta 1994; 195:309-312.
29. Tsuge T, Tsukaya H, Uchimiya H. Two independent and polarized processes of cell elongation regulate leaf blade expansion in Arabidopsis thaliana. Development 1996; 122:1589-1600.
30. Walker JD, Oppenheimer DG, Concienne J et al. SIAMESE, a gene controlling the endoreduplication cell cycle in Arabidopsis thaliana trichomes. Development 2000; 127(18):3931-3940.
31. Uphof JCT. Plant hairs. Berlin: Gebr. Bornträger, 1962:5.
32. Schnittger A, Hulskamp M. Trichome morphogenesis: A cell-cycle perspective. Philos Trans R Soc Lond B Biol Sci 2002; 357(1422):823-826.
33. Ilgenfritz H, Bouyer D, Schnittger A et al. The Arabidopsis STICHEL gene is a regulator of trichome branch number and encodes a novel protein. Plant Physiol 2003; 131(2):643-655.
34. Szymanski DB, Marks MD, Wick SM. Organized F-actin is essential for normal trichome morphogenesis in Arabidopsis. Plant Cell 1999; 11(12):2331-2348.
35. Mathur J, Spielhofer P, Kost B et al. The actin cytoskeleton is required to elaborate and maintain spatial patterning during trichome cell morphogenesis in Arabidopsis thaliana. Development 1999; 126(24):5559-5568.
36. Mathur J, Chua NH. Microtubule stabilization leads to growth reorientation in Arabidopsis thaliana trichomes. Plant Cell 2000; 12:465-477.
37. Traas J, Bellini C, Nacry P et al. Normal differentiation patterns in plants lacking microtubular preprophase bands. Nature 1995; 375:676-677.
38. Torres-Ruiz RA, Jürgens G. Mutations in the FASS gene uncouple pattern formation and morphgenesis in Arabidopsis development. Development 1994; 120:2967-2978.
39. Camilleri C, Azimzadeh J, Pastuglia M et al. The Arabidopsis TONNEAU2 gene encodes a putative novel protein phosphatase 2A regulatory subunit essential for the control of the cortical cytoskeleton. Plant Cell 2002; 14(4):833-845.
40. Qiu JL, Jilk R, Marks MD et al. The Arabidopsis SPIKE1 gene is required for normal cell shape control and tissue development. Plant Cell 2002;14:101-118.
41. Nolan KM, Barrett K, Lu Y et al. Myoblast city, the Drosophila homolog of DOCK180/CED-5, is required in a Rac signaling pathway utilized for multiple developmental processes. Genes Dev 1998; 12(21):3337-3342.
42. Bichet A, Desnos T, Turner S et al. BOTERO1 is required for normal orientation of cortical microtubules and anisotropic cell expansion in Arabidopsis. Plant J 2001; 25(2):137-148.
43. Burk DH, Liu B, Zhong R et al. A katanin-like protein regulates normal cell wall biosynthesis and cell elongation. Plant Cell 2001; 13(4):807-827.
44. Webb M, Jouannic S, Foreman J et al. Cell specification in the Arabidopsis root epidermis requires the activity of ECTOPIC ROOT HAIR 3—a katanin-p60 protein. Development 2002; 129(1):123-131.
45. Quarmby L. Cellular Samurai: Katanin and the severing of microtubules. J Cell Sci 2000; 113(Pt 16):2821-2827.
46. Kirik V, Grini PE, Mathur J et al. The Arabidopsis TUBULIN-FOLDING COFACTOR A gene is involved in the control of the alpha/beta-tubulin monomer balance. Plant Cell 2002; 14(9):2265-2276.
47. Kirik V, Mathur J, Grini PE et al. Functional analysis of the tubulin-folding cofactor C in Arabidopsis thaliana. Curr Biol 2002; 12(17):1519-1523.
48. Lewis SA, Tian G, Cowan NJ. The α- and β-tubulin folding pathways. Trends Cell Biol 1997; 7:479-484.
49. Steinborn K, Maulbetsch C, Priester B et al. The Arabidopsis PILZ group genes encode tubulin-folding cofactor orthologs required for cell division but not cell growth. Genes Dev 2002; 16(8):959-971.

50. Tsuge T, Tsukaya H, Uchimiya H. Two independent and polarized processes of cell elongation regulate leaf blade expansion in Arabidopsis thaliana (L.) Heynh. Development 1996; 122:1589-1600.
51. Pantres E, Apostolakos P, Galatis B. Sinuous ordinary epidermal cells: Behind several patterns of waviness, a common morphogenetic mechanism. New Phytologist 1994; 127:771-780.
52. Jung G, Wernicke W. Cell shaping and microtubules in developing mesophyll of wheat (Triticum aestivum L.). Protoplasma 1990; 153:141-148.
53. Welch MD. The world according to Arp: Regulation of actin nucleation by the Arp2/3 complex. Trends Cell Biol 1999; 9:423-427.
54. Mullins RD, Heuser JA, Pollard TD. The interaction of Arp2/3 complex with actin: Nucleation, high affinity pointed end capping, and formation of branching networks of filaments. Proc Natl Acad Sci USA 1998; 95:6181-6186.
55. Welch MD, Iwamatsu A, Mitchison TJ. Actin polymerization is induced by ARP2/3 protein complex at the surface of Listeria monocytogenes. Nature 1997; 85:265-269.
56. Machesky LM, Gould KL. The ARP2/3 complex: A multifunctional actin organizer. Curr Opin Cell Biol 1999; 11:117-121.
57. Svitkina TM, Borisy GG. ARP2/3 complex and actin depolymerizing factor/cofilin in dendritic organization and treadmilling of actin filament array in lamellipodia. J Cell Biol 1999; 145:1009-1026.
58. Mathur J, Mathur N, Kernebeck B et al. Mutations in actin related proteins 2 and 3 affect cell shape development in Arabidopsis thaliana. Plant Cell 2003; in press.
59. Mathur J, Mathur N, Kirik V et al. Arabidopsis CROOKED encodes for the smallest subunit of the ARP2/3 complex and controls cell shape by region specific fine F-actin formation. Development 2003; 130(14):3137-46.
60. Li H, Wu G, Ware D et al. Arabidopsis Rho-related GTPases: Differential gene expression in pollen and polar localization in fission yeast. Plant Physiol 1998; 118:407-417.
61. Zhang Z-L, Yang Z. The Rop GTPase: An emerging signaling switch in plants. Plant Mol Biol 2000; 44:1-9.
62. Fu Y, Wu G, Yang Z. Rop GTPase-dependent dynamics of tip-localized F-actin controls tip growth in pollen tubes. J Cell Biol 2001; 152:1019-1032.
63. Eden S, Rohtagi R, Podtelejnikov AV et al. Mechanism of regulation of WAVE1-induced actin nucleation by Rac1 and Nck. Nature 2002; 418:790-793.
64. Frank MJ, Cartwright HN, Smith LG. Three Brick genes have distinct functions in a common pathway promoting polarized cell division and cell morphogenesis in the maize leaf epidermis. Development 2003; 130(4):753-762.
65. Jones MA, J-J S, Fu Y et al. The Arabidopsis Rop2 GTPase is a positive regulator of both root hair initiation and tip growth. Plant Cell 2002; 14:763-776.
66. Bibikova TN, Blancaflor EB, Gilroy S. Microtubules regulate tip growth and orientation in root hairs of Arabidopsis thaliana. Plant J 1999; 17(6):657-665.
67. Bao Y, Kost B, Chua N-H. Reduced expression of α-tubulin genes in Arabidopsis thaliana specifically affects root hair development and root gravitropism. Plant J 2001; 28:145-157.
68. Very A, Davies JM. Hyperpolarization – activated calcium channels at the tip of Arabidopsis root hairs. Proc Nat Acad Sci USA 2000; 97:9801-9806.
69. Bibikova TN, Zhigilei A, Gilroy S. Root hair growth is directed by calcium and endogenous polarity. Planta 1997; 203:495-505.

Branching in Fungal Hyphae and Fungal Tissues:
Growing Mycelia in a Desktop Computer

David Moore, Liam J. McNulty and Audrius Meskauskas

Abstract

In mycelial fungi the formation of hyphal branches is the only way in which the number of growing points can be increased. Cross walls always form at right angles to the long axis of a hypha, and nuclear division is not necessarily linked to cell division. Consequently, no matter how many nuclear divisions occur and no matter how many cross walls are formed, there will be no increase in the number of hyphal tips unless a branch arises. Evidently, for the fungi, hyphal branch formation is the equivalent of cell division in animals, plants and protists. The position of origin of a branch and its direction and rate of growth are the crucial formative events in the development of fungal tissues and organs. Kinetic analyses have shown that fungal filamentous growth can be interpreted on the basis of a regular cell cycle, and encourage the view that a mathematical description of fungal growth might be generalised into predictive simulations of tissue formation. An important point to emphasise is that all kinetic analyses published to date deal exclusively with physical influences on growth and branching kinetics (like temperature, nutrients, etc.). In this chapter we extrapolate from the kinetics so derived to deduce how the biological control events might affect the growth vector of the hyphal apex to produce the patterns of growth and branching that characterise fungal tissues and organs. This chapter presents: (i) a review of the published mathematical models that attempt to describe fungal growth and branching; (ii) a review of the cell biology of fungal growth and branching, particularly as it relates to the construction of fungal tissues; and (iii) a section in which simulated growth patterns are developed as interactive three-dimensional computer visualisations in what we call the Neighbour-Sensing model of hyphal growth. Experiments with this computer model demonstrate that geometrical form of the mycelium emerges as a consequence of the operation of specific locally-effective hyphal tip interactions. It is not necessary to impose complex spatial controls over development of the mycelium to achieve particular morphologies.

Introduction

During the life history of many fungi, hyphae differentiate from the vegetative form that ordinarily composes a mycelium and aggregate to form tissues of multihyphal structures. These may be linear organs (that emphasise parallel arrangement of hyphae):

- strands,
- rhizomorphs
- fruit body stipes,

Branching Morphogenesis, edited by Jamie A. Davies.
©2005 Eurekah.com and Springer Science+Business Media.

or globose masses (that emphasise interweaving of hyphae):

- sclerotia
- fruit bodies and other sporulating structures of the larger Ascomycota and Basidiomycota.

In microscope sections, fungal tissue appears to be composed of tightly packed cells resembling plant tissue but the hyphal (that is, tubular) nature of the components can always be demonstrated by reconstruction from serial sections or by scanning electron microscopy. Clearly, hyphal cells do not proliferate in the way that animal and plant cells do.

Plants, animals and fungi are distinct eukaryotic Kingdoms and there are fundamental differences between the three Kingdoms in the way that the morphology of multicellular structures is determined. A characteristic of animal embryology is the movement of cells and cell populations. In contrast, plant morphogenesis depends upon control of the orientation and position of the daughter cell wall, which forms at the equator of the mitotic division spindle. Fungi also have walls, like plants, but their basic structural unit, the hypha, exhibits two features which cause fungal morphogenesis to be totally different from plant morphogenesis. These are that:

- hyphae extend only at their apex, and
- cross walls form only at right angles to the long axis of the hypha.

One consequence of these "rules" is that fungal morphogenesis depends on the placement of hyphal branches. Increasing the number of growing tips by hyphal branching is the equivalent of cell proliferation in animals and plants. To proliferate, the hypha must branch, and to form an organised tissue, the position of branch emergence and its direction of growth must be controlled.

Another way in which fungal morphogenesis differs from that in other organisms is that no lateral contacts between fungal hyphae analogous to the plasmodesmata, gap junctions and cell processes that interconnect neighbouring cells in plant and animal tissues have ever been found. Their absence suggests that morphogens used to regulate development in fungi will be communicated through the extracellular environment. Since published kinetic analyses deal exclusively with external influences (like nutrient status, culture conditions, etc.) on growth and branching kinetics, this encourages our view that a mathematical description of fungal vegetative growth might be generalised into predictive simulations of tissue formation, leading to better understanding of the parameters that generate specific morphologies.

Kinetics of Mycelial Growth and Morphology

Kinetic analyses show that fungal filamentous growth can be interpreted on the basis of a regular cell cycle, and in this section we review published mathematical models that attempt to describe fungal growth and branching in the vegetative (mycelial) phase.

Measurement Methodologies

Measurements of hyphal diameter, hd, and hyphal length, hl, allow hyphal volume, hv, to be calculated, which when multiplied by the average density of the composite hyphal material, ρ, gives an estimate of biomass, X. Taking these measurements over a series of time intervals enables hyphal extension rate, E, and the rate of increase of biomass to be calculated. Currently, automated image analysis systems permit real-time analysis of these microscopic parameters,[1] and some of these analyses suggest that hyphal tips grow in pulses,[2] although this is debatable,[3] particularly because the observations use video techniques and the pixelated images generated by both analogue and digital cameras will cause pulsation artefacts.[4]

The most important macroscopic parameter is total biomass. Total hyphal length is proportional to total biomass, if hd and ρ are assumed to be constant, but measurement can be difficult. Nondestructive mass measurement is rarely feasible and in most cases separating the mycelium from the substratum is difficult (and sometimes impossible). Acuña et al[5] developed a neural network that they trained to correlate colony radius with colony biomass. However, this relationship is only relevant to circular mycelia and measurements in two dimensions. More general relationships with biomass have been suggested for particular chemical compounds, the most promising of which is ergosterol, a sterol characteristic of fungal membranes.[6,7]

Modelling Branching

A germ-tube hypha will grow in length exponentially at a rate that increases until a maximum, constant extension rate is reached. Thereafter, it increases in length linearly.[8] The primary and subsequent branches behave similarly. Trinci[9] offered a solution to the riddle of how biomass (proportional to total hyphal length) can increase exponentially when individual hyphae extend linearly by proposing that it was due to the exponential increase in tips due to branching.

Katz et al[8] studied the growth kinetics of *Aspergillus nidulans* on three different media, each with a distinct specific growth rate. From these observations they proposed a number of general relationships that are conveyed in equation (1), elucidated by Steele and Trinci:[10]

$$\overline{E} = \mu_{max} G \tag{1}$$

where $\overline{E}$ is the mean tip extension rate, μ_{max} is the maximum specific growth rate, and G is the hyphal growth unit. G is defined as the average length of a hypha supporting a growing tip according to equation (2):

$$G = \frac{L_t}{N_t} \tag{2}$$

where L_t is total mycelial length, and N_t is number of tips.

The hyphal growth unit is approximately equal to the width of the peripheral growth zone (more accurately, the volume of the hyphae within that zone), which is a ring-shaped peripheral area of the mycelium that contributes to radial expansion of the colony.[11,12] Hyphal tips growing outside this zone will only fill space within the colony. G is an indicator of branching density; Katz et al[8] postulated that a new branch is initiated when the capacity for a hypha to extend increases above $\overline{E}$, thereby regulating $G \approx 1$ unit.

Prosser and Trinci[13] described a model that successfully accounted for exponential growth and branching, constructed on the premise that tips extend by the incorporation into the tip membrane of new material that arrives packaged in vesicles.[14] This mechanism was modelled in two steps: (i) vesicles were produced in hyphal segments distal to the tip and were absorbed in tip segments; (ii) vesicles flowed from one segment to the next, towards the tip. Apical branching initiated when the concentration of vesicles in the tip exceeded the maximum rate that the apex could absorb the new material. Varying the ratio of these steps produced different flow rates and branching patterns. The model also incorporated the concept of the 'duplication cycle'.[15] This was achieved by increasing the number of nuclei in the model mycelium at a rate proportional to the rate of biomass increase. Septa were then assumed to form in growing hyphae when the volume of the apical compartment per nucleus breached a threshold level. This provided for initiation of lateral branches by assuming that vesicles accumulated behind septa to a concentration comparable to that which initiated apical branching. This model achieved good agreement with experimental data for total mycelial length, number of hyphal tips, and hyphal growth unit length in *Geotrichum candidum*.[13] In an adaptation of this model, Yang et al[16] used a stochastic element to account for the branching process, branching site and direction of branch growth being generated by probability functions. This gave rise to a much more realistic mycelial shape.

Describing Branching Patterns

Leopold[17] examined the generality of natural branching systems in trees and streams. Based on the classification system of Horton,[18] she labelled each branch of a tree or river network depending on how many tributary branches it supported. First-order branches have no tributaries; second-order branches support only first-order branches; a third-order branch supports only first and second-order branches; etc. She also measured the lengths of each branch to obtain an average value for each order of branching (the length of an n-order branch includes the length of its longest $(n-1)$-order tributary). She found that straight-line plots resulted when branch order was plotted against the logarithm of (i) the number of branches of a given order, and (ii) the average length of a branch of a given order. The gradient of these lines was interpreted as (i) the branching ratio (BR = the average number of n-order branches for each $(n+1)$-order branch); and (ii) the length ratio (LR = the average length of each n-order branch as a multiple of the average length of each $(n-1)$-order branch). Observations suggest that the values of these ratios showed little variation over a range of tree species (BR = 4.7 - 6.5; LR = 2.5 - 3.6) and river networks (BR = 3.5; LR = 2.3).

Analysing Fungal Mycelia

Gull[19] applied Leopold's analysis to the branching characteristics of mycelia of the filamentous fungus, *Thamnidium elegans*, and observed branching and length ratios of 3.8 and 4.0, respectively, for a third-order system and 2.6 and 2.7 for a fourth-order system. Though it gave no biological insight into the mechanisms of branching, Gull's work demonstrated that mycelia employ branching as a strategy for colonising the maximum area of space using the minimum total mycelial length, and indicate that the values obtained can be interpreted as a quantification of branching frequency.

Another approach to quantifying branching frequency relies on the mathematics of fractal geometry. In the box-counting method of fractal analysis a grid of boxes, each with side length ε, is placed over the pattern, and the number of boxes, N_{box}, that are intersected by the pattern is counted. If a pattern is fractal, it will be 'self-similar' at all scales. This means that a true fractal pattern has an infinite length. However, the geometry of the pattern limits the degree to which it can fill the plane. This is quantified in terms of the fractal dimension, D, according to the formula:

$$N_{box}(\varepsilon) = C\varepsilon^{-D} \tag{3}$$

where C is a constant. A straight line has a fractal dimension, $D = 1$, and a completely filled plane has $D = 2$.

By looking at higher and higher resolutions (i.e., in the limit $\varepsilon \to 0$) the repeating pattern will be revealed to cover a limited proportion of the plane. When the logarithm of N_{box} is plotted against the logarithm of $(1/\varepsilon)$, a straight line is obtained with gradient equal to D. When applied to fungal mycelia, ε is limited by the hyphal diameter microscopically and by the mycelial diameter macroscopically. Thus, mycelia are not true fractals. However, this range is sufficient to allow reasonably accurate regression analysis for D, and thus quantification of the space filling capacity, or branching frequency, of mycelia can be obtained.

Obert et al[20] applied this method to mycelia of *Ashbya gossypii*. They found that mycelia did indeed behave as fractals, and calculated a fractal dimension, $D = 1.94$. Such a high value for D indicates a mature mycelium whose centre has been almost homogenously filled by branching hyphae. For the edge of mycelia they calculated $D = 1.45$. Thus, as a mycelium develops its fractal dimension converges towards 2 when the whole mycelium is considered and towards 1.5 when only the edge is considered. Ritz and Crawford[21] and Jones et al[22] corroborated these findings.

Matsuura and Miyazima[23] used a different form of fractal analysis to quantify the 'roughness' of the edges of mycelia grown at different temperatures and on different media. Unfavourable conditions (e.g., low temperature, low nutrient concentration or stiff media) were found to produce rough edges corresponding to a lower branching frequency.

The above analyses result in a mathematical expression of the ecological description of the dual function of the fungal mycelium: that it serves to explore, and to capture resources. A rapidly growing, sparsely branched mycelium is emphasising exploration. One in which branching density increases towards homogeneity is maximising resource capture.

Generating a Circular Mycelium

This is all well and good, but the findings could apply to a mycelium of any shape, yet the fundamental morphogenetic truth about fungi is that a germinated spore on a surface (like an agar plate) will soon produce a circular colony. Testing how circularity arises requires kinetic analyses to be elaborated into simulations and this involves more demanding calculations.

The models so far described are relatively simple kinetic descriptions that, for the most part, lend themselves to manual calculation. Cohen[24] pioneered computer analysis by devising a program that was able to generate a range of branching patterns found in the natural world from a set of simple growth and branching rules. In his model, growth occurred only at the tip and branching was only initiated behind the tip. Thus, it is directly applicable to mycelial growth of most fungi. In this model, growth proceeded with respect to local density fields, calculated with reference to 36 sample points spaced 10° apart around the circumference of a circle centred on a growing tip, and quantifying the pattern density in the locality. Local density minima were key parameters that directed growth into unoccupied space. Branching probability was also made a function of local density minima. A random trial incorporated into the program decided, independently, if branch initiation should occur. Finally, the direction of both growth and branching were dependent on a 'persistence factor' that quantified to what degree they continued in the same direction in spite of gradients in the density field. The persistence factor acts rather like inertia on a moving body—changes in direction are gradual rather than instantaneous. When these rules were iterated, with the persistence factor for growth nullified (i.e., growing tips proceeded in a straight line), a circular branching pattern emerged.

Hutchinson et al[25] developed this work further by applying it directly to mycelial colonies of *Mucor hiemalis*. They determined the variability of tip growth rate, distance between branches, and branching angle throughout the colonies. They were then able to fit these data to known distribution curves with defined probability density functions. Tip growth rate was found to follow a half-normal distribution, distance between branches followed a gamma distribution, and branching angle followed a normal distribution. This formed the basis for a model in which values for the three specified variables were generated from the respective probability density functions over a series of time intervals. This model generates a circular mycelium.

This came as something of a surprise because the model includes no allowances for tropic interactions between hyphae. Yet, it seems a reasonable assumption that the readily-observed fact that growing hyphae actively avoid each other (= negative autotropism, observed by Robinson,[26,27] Trinci et al[28] and Hutchinson et al[25] among others), plays a role in determining spatial organisation in mycelia, especially colony circularity.

Indermitte et al[29] constructed models of mycelial growth with the specific intention of answering the question "How does circularization happen?" Variants of the model considered cases in which behaviour was completely random; where hyphae were forbidden from crossing other hyphae; where hyphal growth direction depended on diffusion of inhibitory substances from the mycelium into the medium; and where, in addition to diffusion of inhibitory substances, growth rate depended on the rank of the branch. All variants generated circularity. Although the method by which simulations were made is not described in the paper, the authors claim that experimentation with their model indicated that tropism increased the growth efficiency, where the latter was judged by the occupation of the surface (the ratio of biomass used to area of medium covered).

It is highly significant that a purely stochastic approach can generate realistically circular colony morphology, but it does not follow that tropisms and hyphal interactions are irrelevant to modelling hyphal growth. In real life, hyphae certainly do use autotropic behaviour (positive and negative) to control spatial organisation in particular regions of the mycelium; and where the hyphal density is high, as in fungal tissues, interactions are inevitable. However, we are not convinced that the rule "crossing of hyphae is not allowed" is realistic, though it is a crucial feature of the Indermitte et al[29] model. Even a casual glance at most fungal mycelia and tissues will reveal numerous interweaving hyphae. Even the commonly-applied informal description "hyphal mat" implies a woven texture.

Tropism and Hyphal Interactions

Edelstein[30] considered such interactions, but her approach differed from the above by considering the mechanisms operating in the mycelium as a whole rather than in discrete hyphae. She assumed that growth occurred at a constant rate throughout the mycelium. This she set at μ_{max} and so also limited her model to a tangential abstraction of the growth curve. Her model owes something to Cohen[24] in that it considers the density of the mycelium with respect to space as a key feature. Two density parameters were defined:

$p = p(x,t)$ the hyphal density per unit area
$n = n(x,t)$ the tip density per unit area

and the model was then based on two partial differential equations:

$$\frac{\partial p}{\partial t} = n\overline{E} - \delta \tag{4}$$

$$\frac{\partial n}{\partial t} = \frac{\partial n\overline{E}}{\partial x} + \sigma \tag{5}$$

where $\delta = \delta(p)$ is the rate of hyphal death, $\sigma = \sigma(n,p)$ is the rate of tip creation, and $n\overline{E}$ can be considered as tip flux.

Edelstein[30] also defined, in mathematical terms, all the hyphal interactions that affect the parameter n, and which are contained in the function σ. These included both branching mechanisms, as well as tip death and tip-tip and tip-hypha anastomoses.

She then used phase plane analysis to determine which of various combinations of hyphal interactions, expressed mathematically in the function σ, had bounded non-negative solutions of equations (4) and (5). These represented combinations that yielded spatially propagating colonies. Her results showed that when $\delta = 0$, only colonies that branched dichotomously and formed tip-hypha anastomoses could propagate. However, when $\delta > 0$, most combinations of hyphal interactions yielded propagating colonies. Thus, hyphal death was shown to be an important feature of mycelial growth; in addition to the density-dependent distribution which was an initial criterion of the model.

Ferret et al[31] adopted a similar approach to Edelstein,[30] using two partial differential equations that considered parameters defined in dimensions of density. However, they sought to apply their model to bulk cultures by adjusting mean tip extension rate ($\overline{E}$) with respect to biomass, X. This adjustment was done by collecting data that quantified how E varied when two hyphae came into close proximity with each other. This effect was incorporated into the differential equation concerned with the rate of change of biomass density (proportional to hyphal-density) so that it was more likely to be applied in regions where the density of biomass was high, and had a greater effect on regions where the density of tips was high. Thus, $\overline{E}$ decreased as the mycelium grew and biomass increased. Such an approach provides an alternative to incorporating hyphal death into the model that has the advantage of also affecting $\overline{E}$, thus limiting growth in a manner typical of batch culture, and, we speculate, perhaps also hyphal masses that contribute to fruiting bodies.

All of the published work on fungal growth kinetics has been devoted to growth of mycelium, and particularly biomass production in fermenters. It is especially important that models of mycelial growth can form the foundation for modelling pellet formation in liquid media[16,32] as this can impact on the design of bioreactors, but we do not wish to examine models primarily intended for application to bulk cultures at this time. Another approach to the subject we will not deal with here considers substrate utilisation as a means of describing and quantifying the growth process, as pioneered by Monod.[33] Edelstein and Segel[34] and Mitchell,[35] amongst others, have pursued this approach successfully. This also has obvious biotechnological significance.

Opinions

In many respects, the mycelium is the least interesting growth form. It is the 'default' growth mode of the fungal cell and any changes that occur in it are imposed by **external** forces (nutrients, environmental conditions, etc.). Of much greater biological interest is the way in which the 'default' growth mode might be altered by **internal** (that is, self-imposed) controls to generate the numerous differentiated cells that hyphae can produce and the native interactions between hyphae that cause them to cooperate and coordinate in the morphogenesis of fungal tissues.

Although some attempt has been made to extend the vesicle supply centre model of apical growth[14] into 2-dimensional and 3-dimensional models of apical growth and differentiation,[36,37] we are not aware of any kinetic analysis of fungal mycelial growth in three dimensions that might contribute to understanding fungal tissue morphogenesis. It is certain, though, that the equation $E = \mu G$ (Equation 1) is fundamental to understanding branching kinetics and that the ratio E/μ can tell us a lot about mycelial morphology, as it relates to the hyphal growth unit length, G, which can also be expressed as a volume.[38] Observation has shown that temperature increases do not affect G in some species. However, paramorphogens have been identified that do alter this ratio and hence G and morphology.[39] In our vector-based Neighbour-Sensing mathematical model, which is introduced below, the inclusion of certain tropism vectors is also able to alter this ratio by affecting the parameter E and results in a striking array of different morphologies, some of which seem to suggest a morphogenetic process that goes beyond mycelial growth and towards differentiated tissues.

In the next section, we describe briefly the sorts of tissues and hyphal interactions that must be explained eventually, and in the final section of this chapter we will describe progress in modelling key morphogenetic processes.

Construction of Fungal Tissues

Development of any multicellular structure in fungi requires modification of the normal growth pattern of a vegetative mycelium so that hyphae no longer characteristically diverge, but grow towards one another to cooperate in forming the differentiating organ.[40-42] The hyphal tip is an invasive, migratory structure. Its direction of growth after initial branch emergence must be under precise control as it determines the nature and relationships of the cells the hyphal branches will form.

Linear Organs: Strands, Rhizomorphs and Stems

Formation of parallel aggregates of hyphae (= mycelial strands and cords) is common as they provide the main translocation routes for the mycelium. They are formed in mushroom cultures to channel nutrients towards developing fruit bodies; they are also formed by mycorrhizal fungi, gathering nutrients for the host. Some fungi produce rhizomorphs, which have highly differentiated tissues and show extreme apical dominance. There is often a gradation of increasing differentiation between strand, cord (or rhizomorphs) and fruit body stipe (= stem). Linear organs arise when young branches adhere to, and grow over, an older leading hypha. From the beginning, some of the hyphae may expand to become wide-diameter but thin-walled hyphae, whilst narrow hyphal branches ('tendril' hyphae) intertwine around the inflated hyphae (Fig. 1).

Globose Structures: Sclerotia and Fruit Bodies

Sclerotia are tuber-like, with concentric zones of tissue forming an outer rind and inner medulla, with a cortex sometimes between them. They pass through a period of dormancy before utilizing accumulated reserves to 'germinate', often producing fruiting bodies immediately.

Fruiting bodies are responsible for producing and distributing spores formed following meiosis. In ascomycetes, the sexually produced ascospores are enclosed in an aggregation of hyphae termed an ascoma. Ascomata are formed from sterile hyphae surrounding the developing asci,

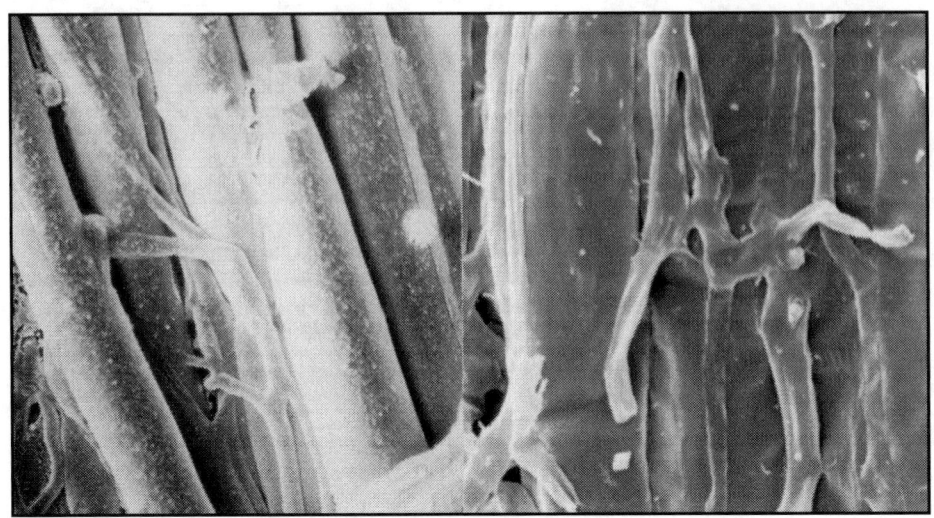

Figure 1. Two scanning electron micrographs of wide and narrow hyphae intertwined in the stem tissues of the small field mushroom, *Coprinus cinereus*. Presumably this pattern of growth is produced by positive autotropisms which ensure that the hyphae that expand to become wide diameter initially grow parallel with one another, and other tropisms that allow the narrow hyphae to grow around and intermingle with inflated hyphae.

and occur in nature in forms such as truffles and morels. The fruit-bodies of basidiomycetes, the mushrooms, toadstools, bracket fungi, puff-balls, stinkhorns, bird's nest fungi, etc., are all examples of basidiomata which bear the sexually produced basidiospores on basidia in the spore-bearing hymenial layers. These hymenia are constructed from branches of determinate growth in a precise spatial and temporal arrangement. A hyphal tip in the 'embryonic' protohymenium has a probability of about 40% of becoming a cystidium. Cystidia are large, inflated cells which are readily seen in microscope sections. When a cystidium arises, it inhibits formation of further cystidia in the same hymenium within a radius of about 30 μm. The distribution pattern of cystidia is consistent with the activator-inhibitor model that suggests that an activator autocatalyses its own synthesis, and interacts with an inhibitor that inhibits synthesis of the activator. As a result, only about 8% of the hymenial hyphal branches actually become cystidia; the rest become basidia, which proceed to karyogamy and initiate the meiotic cycle (which ends with sporulation) (Fig. 2).

Sterile packing cells, called paraphyses, then arise as branches of sub-basidial cells and insert into the hymenium (Fig. 3). About 75% of the paraphysis population is inserted before the end of meiosis, the rest insert at later stages of development. There is, therefore, a defined temporal sequence: probasidia and cystidia appear first and then paraphyses arise as branches from sub-basidial cells. Another cell type, cystesia (adhesive cells), differentiate when a cystidium grows across the gill space and contacts the opposing hymenium (Fig. 2).

Simulating the Growth Patterns of Fungal Tissues

Most models published so far simulate growth of mycelia on a single plane. However, two-dimensional space has some specific peculiarities that can affect the conclusions: forbidding crossings between hyphae in the Indermitte et al models[29] being a case in point. In a real three-dimensional world a large number of points can be connected without the need for the connection paths to cross; whereas the number of such points is limited on a flat plane. The need to cross will also have effect on models where patterning is based on a hyphal density field,

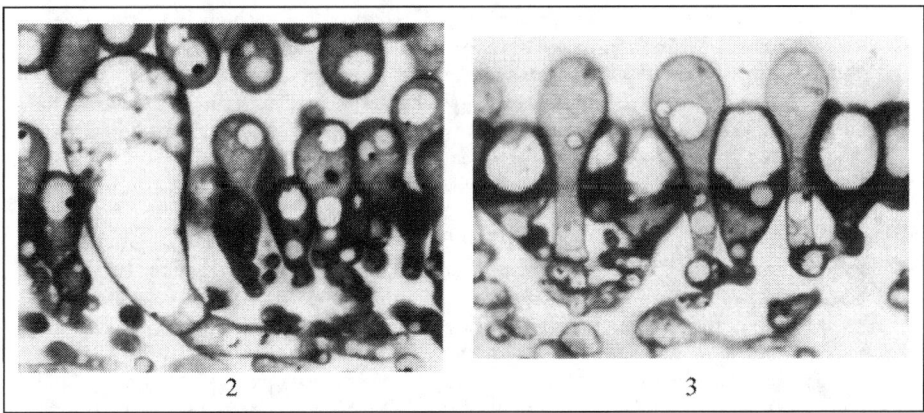

Figures 2 and 3. Light micrographs of glycolmethacrylate sections of immature hymenia of the small field mushroom, *Coprinus cinereus*. Figure 2 is the younger of the two stages shown, and the large cell in that Figure is a cystidium. Only a minority of the hyphal tips that make up the hymenium differentiate into cystidia because each cystidium establishes an inhibitory morphogenetic field around itself. Figure 3 shows that the densely-stained branches that can be seen inserting between the basidia in Figure 2, rapidly differentiate into inflated paraphyses, and in fact arise as branches from the bases of the basidia. Note also that extension growth of these hyphal tips is halted in a coordinated way so that the basidia remain as projections above the paraphyseal pavement. Overall, these images show that to construct hymenial tissue the normal divergent growth of the vegetative colony is modified to become determinate, positively autotropic, with distinctly differentiated hyphal tips of the same generation (basidia and cystidia) and distinctly different developmental fates for branches of different ranks (basidia and paraphyses).

generated by all parts of the growing mycelium, as suggested by Cohen.[24] Growth in this case is regulated by the absolute value of this field and is directed by its gradient (equivalent to negative autotropism). In two-dimensional space, turning up or down is not an option, so a tip approaching an existing hypha must go across the latter, moving against a large (possibly infinite, as the distance approaches zero) value of the density field. Cohen's original model produced polarized tree-like structures, quite different from the typical spherical fungal colonies, and while the Indermitte et al models[29] succeeded in forming circular colonies, their analysis remained in two dimensions. Consequently, knowing how the circular colony arises on a flat plane is not enough; it is crucial to understand the formation of a spherical colony in three dimensional space.

Our purpose here is to suggest a model, which we call the Neighbour-Sensing model, that, whilst being as simple as possible, is able to simulate formation of a spherical, uniformly dense fungal colony in a visualisation in three dimensional space. Following Indermitte et al[29] we gauge our success on the basis that our model successfully imitates the three branching strategies of fungal mycelia illustrated by Nils Fries in 1943.[43]

Verbal Description of the Neighbour-Sensing Model

The process of simulation is defined as a closed loop. This loop is performed for each currently existing hyphal tip of the mycelium and the algorithm:

1. Finds the number of neighbouring segments of mycelium (N). A segment is counted as neighbouring if it is closer than the given critical distance (R). In the simplest case we did not use the concept of the density field, preferring a more general formulation about the number of the neighbouring tips.
2. If $N < N_{branch}$ (the given number of neighbours required to suppress branching), there is a certain given probability (P_{branch}) that the tip will branch. If the generated random number

(0.1) is less than this probability, the new branch is created and the branching angle takes a random value. The location of the new tip initially coincides with the current tip. This stochastic branch generation model is similar overall to earlier ones[16,25,44] in which distance between branches and branching angles followed experimentally measured statistical distributions. This, however, was not required to reach the desired shape of the colony in our model. Rather, we used a uniform distribution, as did Indermitte et al.[29]

We assumed that all hyphal tips in mycelia grow at constant speed. This assumption was sufficient to get the desired shape and structure of the colony.

In the simplest version, the growth direction is defined during branching and is not altered subsequently. In other words, the initial model does not implement tropic reactions (to test the kind of morphogenesis that might arise without this component). Later versions of the model tested how implementation of the density field hypothesis would affect colony growth. The density field features were made analogous to an electrical field.[24]

Implementation of a negative autotropic reaction requires the concept of the density field, as the growth must be directed by the gradient of this field. We also implemented the suggestion[24] that the tip should change direction gradually (the so-called persistence factor; see earlier section "Generating a Circular Mycelium"). In our implementation, the growth speed remained constant and the density gradient alters only the growth direction. Otherwise, a high gradient, if formed accidentally, would cause unreliably fast growth in some parts of the mycelium.

With low values of the persistence factor, the model is able to form small linear structures. This is because, with such a parameter set, immediately after branching the hyphal density field tends to orient the new tip strictly in the opposite growth direction from the old tip. That is, the new hypha is directed to grow parallel with the old hypha but in the opposite direction. If we suppose that the hyphal density field is generated just by tips and branch points, this direction remains optimal until the tip goes sufficiently far from the branch point to start interacting with other hyphae. Changing parameters while the colony is still nearly linear can produce ellipsoidal or tubular structures.

We have experimented with a variety of extensions of the model (see illustrations in section "Conclusions", below): for example, growth being suppressed by a high number of neighbouring tips; or allowing the growing tip only to be active for a fixed time before it stops growing and branching. Such changes can result in more optimal packing of the hyphae, but are not required to form a spherical colony. Real fungal colonies are rarely uniform in structure, so the question arises whether any smaller new structures can form in a virtual colony growing in accordance with this model. We found that this could happen following abrupt changes of the model parameter set (especially R and N_{branch}).

Mathematical Description

Let, at the time $t \in Z_+$, the mycelium contain n growing hyphal tips. Let y_i and g_i be position and growth vectors, respectively, of the i-th growing hyphal tip at time $t \in Z_+$. Let Y be a set, containing other points of the mycelium that are sensed as neighbouring tips and/or branch points. Now, let

$$N_i = \sum_{j=1}^{n} \Phi\left(\left|y_i - y_j\right| - R\right) + \sum_{k \in Y} \Phi\left(\left|y_i - k\right| - R\right) \tag{6}$$

where Φ is a Heaviside function.

Let

$$v_i = \begin{bmatrix} \alpha_t & \beta_t & \sqrt{1-\alpha_t^2-\beta_t^2} \\ \beta_t & \sqrt{1-\alpha_t^2-\beta_t^2} & \beta_t \\ \sqrt{1-\alpha_t^2-\beta_t^2} & \alpha_t & \alpha_t \end{bmatrix}^{<Round(3\gamma_t+1)>} \tag{7}$$

where α_t, β_t and γ_t all form sequences of independent, uniformly distributed random variables over the range $[0...1]$.

Now compute an array b, containing all the values of i that satisfy the condition $N_i < N_{branch}$ and $\delta t < P_{branch}$. Here δt forms a sequence of independent, uniformly distributed values over the range $[0..1[$, and P_{branch} is the model parameter. Let m be the length of this array. For each $k \in [1..m]$, define:

$$y'_{n+k} = y_{b_k}, \; g'_{n+k} = v(b_k) \tag{8}$$

Finally, define $y'_i = y_i + ag_i$, $g'_i = g_i$ and $n' = n+m$ (the model parameter a determines the tip growth rate in length units per defined iteration period). Define $Y' = Y + \forall y_k : k \in b$. Now we have y', g', n' and Y' defining the state of the colony after one iteration of the model algorithm.

This basic algorithm can be extended by assuming that the tip can be active only for a fixed time (S_{max} iterations) and stops growing after its length reaches L_{max} length units. Also, it could be assumed that the growth is suppressed if $N_i > N_{growth}$. To implement these extensions, let us define the age array $S(S'_i = S_i + 1, S'_{n+k} = 0)$ and the length array $L (L'_i = L'_i + a, L'_{n+k} = 0)$. Then y'_i must be redefined as $ag_i \Phi(S_i - S_{max}) \; \Phi(L_i - L_{max}) \Phi(N_{growth} - N_i)$ and the condition for the value, i, to join the array b must be extended to $S_i < S_{max}$. In the density field version of the model, (6) must be replaced by (9):

$$N_i = \sum_{j=1}^{n(j \neq i)} \frac{1}{\left(|y_i - y_j|\right)^2} + \sum_{k=Y}^{k \neq y} \frac{1}{\left(|y_i - k|\right)^?} \tag{9}$$

Also, N_{max} changes the biological meaning to the maximal value of the density field.

Negative autotropism was implemented using Cohen's approach.[24] In this case, v_i should be replaced by:

$$v_i = norm \left[k \cdot g_i - (1-k) \cdot norm \left[\begin{array}{c} \frac{d}{dy_{(i_0)}} N_i \\ \frac{d}{dy_{(i_1)}} N_i \\ \frac{d}{dy_{(i_2)}} N_i \end{array} \right] \right] \tag{10}$$

where $norm(x) = \frac{x}{|x|}$.

Again, (6) must be replaced by (9). In (10), the parameter k is a model parameter, defining a particular coefficient of persistence, which is used to ensure that branches change direction gradually; and it operates on the previous growth vector g_i. The derivatives are computed by numeric differentiation. The function $norm(x)$ ensures that the density gradient alters the direction but not the speed of the growth.

Implementation

Both versions of the model were implemented in Java together with the simple visualiser:

$$\forall k \in (y_i \cup Y) \exists \{ y_i^{screen} = \sigma k_2, \; x_i^{screen} = \sigma(k_0 \sin \alpha + k_1 \cos \alpha) \}$$

y_i and Y being contained in a tree-like data structure. Interactive adjustment of $\sigma \in [0..\infty[$, and $\sigma \in [-\pi..\pi]$ enabled experimental observation of the growing colony and visual appreciation of its shape. To permit examination of the internal structure of the colony, the application will display a slice of chosen thickness across the colony. This complete interactive application is available for personal experimentation at this URL: http://www.world-of-fungi.org/index.htm.

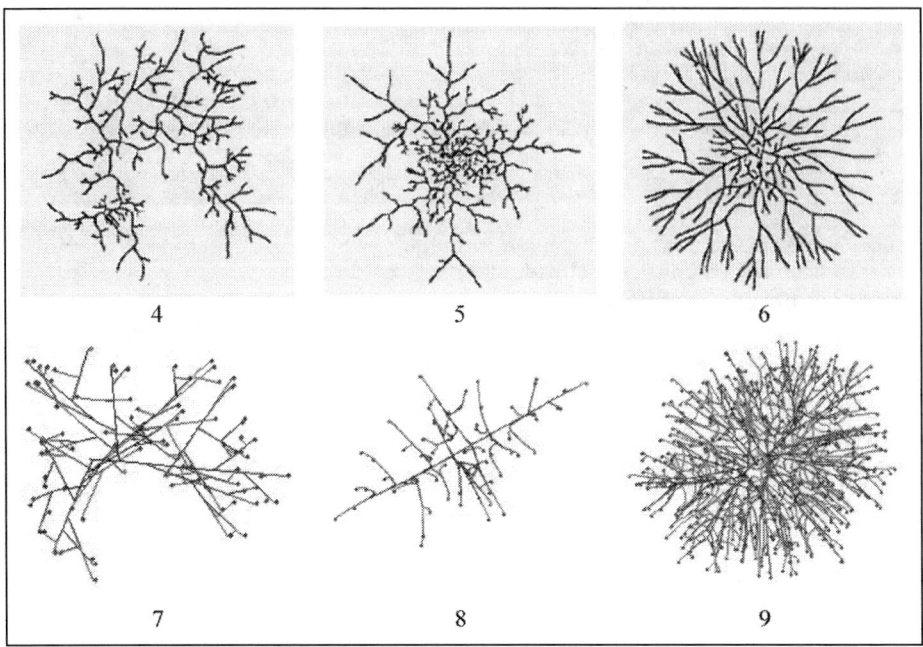

Figures 4-9. Comparison of three different colony types described by Fries[41] (Figs. 4-6) with visualisations produced by the computer model (Figs. 7-9); Figures 4 and 7 show the *Boletus* type, Figures 5 and 8 the *Amanita* type, and Figures 6 and 9 the *Tricholoma* type.

Conclusions

A random growth and branching model (i.e., one that does not include the local hyphal tip density field effect) is sufficient to form a spherical colony. The colony formed by such a model is more densely branched in the centre and sparser at the border; a feature observed in living mycelia (see earlier section "Analysing Fungal Mycelia").

Models incorporating local hyphal tip density field to affect patterning produced the most regular spherical colonies. As with the random growth models, making branching sensitive to the number of neighbouring tips forms a colony in which a near uniformly dense, essentially spherical, core is surrounded by a thin layer of slightly less dense mycelia. Using the branching types discussed by Fries[43] as our paradigm, the morphology of virtual colonies produced when branching (but not growth vector) was made sensitive to the number of neighbouring tips was closest to the so-called *Boletus* type (Figs. 4, 7).

This suggests that the *Boletus* type branching strategy does not use tropic reactions to determine patterning, nor some predefined branching algorithm (of the sort suggested by Hogeweg and Hesper[45]). Following Occam's rule that a simpler model must be preferred if it explains the experimental data equally well,[46] we conclude that hyphal tropisms are not always required to explain "circular" (= spherical) mycelia.

When our model implements the negative autotropism of hyphae, a spherical, near uniformly dense colony is also formed, but branching is still regulated by the number of neighbouring tips (not by the density field). However the structure of such a colony is different from the previously mentioned *Boletus* type, being more similar to the *Amanita rubescens* type,[43] characterised by a certain degree of differentiation between hyphae (Figs. 5, 8). First rank hyphae tending to grow away from the centre of the colony; second rank hyphae growing less

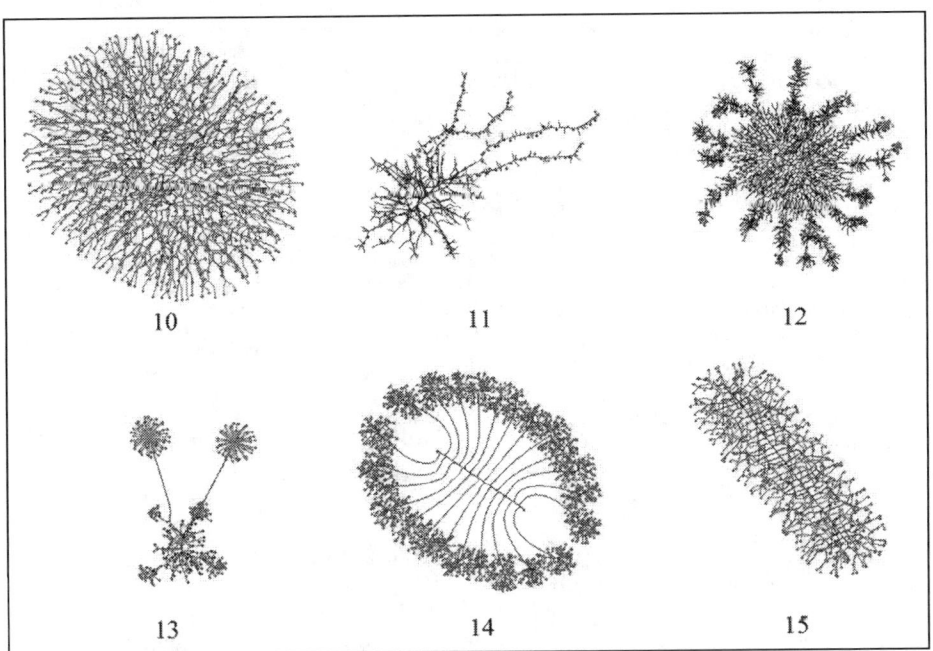

Figures 10-15. Some of the colony types obtained by varying the parameters of the model. Figure 10) *Ellipsoid*: Run for 200 time units at a growth rate of 1 length unit per time unit, with 10% negative autotropism implemented, and the density field hypothesis of branching regulation applied, where the density field threshold for branching is set at 0.1 and branching probability at 80% per time unit. Figure 11) *Mycelium with exploratory filaments*: Stage 1: Run for 100 time units at a growth rate of 1 length unit per time unit, with 10% negative autotropism implemented, and the density field hypothesis of branching regulation applied, where the density field threshold for branching is set at 0.06 and branching probability at 40% per time unit (*Tricholomas* parameter set). Stage 2: Run for 100 time units at a growth rate of 1 length unit per time unit, with 10% negative autotropism implemented, and branching and growth limited to localities where there are less than 8 and 15 neighbouring tips, respectively, in a radius of 20 length units around the growing tip, and branching occurring with a probability of 80% per time unit. Furthermore, growth and branching of tips are each stopped when the tips reach an age of 10 time units. Figure 12) *Spider*: Stage 1: Run for 200 time units at a growth rate of 1 length unit per time unit, with 10% negative autotropism implemented, and the density field hypothesis of branching regulation applied, where the density field threshold for branching is set at 0.1 and branching probability at 80% per time unit. Stage 2: Run for 100 time units at a growth rate of 1 length unit per time unit, with 10% negative autotropism implemented, and branching and growth limited to localities where there are less than 50 and 80 neighbouring tips, respectively, in a radius of 50 length units around the growing tip, and branching occurring with a probability of 80% per time unit. Furthermore, growth and branching of tips are each stopped when the tips reach an age of 20 time units. Figure 13) *Parent and daughter mycelia*: Stage 1: Run for 200 time units at a growth rate of 1 length unit per time unit, with 10% negative autotropism implemented, and the density field hypothesis of branching regulation applied, where the density field threshold for branching is set at 0.06 and branching probability at 40% per time unit (*Tricholomas* parameter set). Stage 2: Run for 100 time units at a growth rate of 1 length unit per time unit, with 10% negative autotropism implemented, and growth limited to localities where there are less than 135 neighbouring tips in a radius of 100 length units around the growing tip, and branching occurring with a probability of 0.1% per time unit. Stage 3: Run for 100 time units at a growth rate of 0.5 length unit per time unit, with 10% negative autotropism implemented, and branching and growth limited to localities where there are less than 100 and 150 neighbouring tips, respectively, in a radius of 100 length units around the growing tip, and branching occurring with a probability of 80% per time unit. The figure legend is continued on the next page.

Figures 10-15, continued. Figure 14) *Doughnut*: Stage 1: Run for 90 time units at a growth rate of 1 length unit per time unit, with no negative autotropism implemented, and the density field hypothesis of branching regulation applied, where the density field threshold for branching is set at 0.1 and branching probability at 80% per time unit. Stage 2: Run for 110 time units at a growth rate of 1 length unit per time unit, with no negative autotropism implemented, and the density field hypothesis of branching regulation applied, where the density field threshold for branching is set at 0.005 and branching probability at 20% per time unit. Furthermore, branching is stopped when the tips reach an age of 50 time units. Stage 3: Run for 30 time units at a growth rate of 1 length unit per time unit, with no negative autotropism implemented, and the density field hypothesis of branching regulation applied, where the density field threshold for branching is set at 0.5 and branching probability at 80% per time unit. Furthermore, growth is stopped when the tips reach an age of 25 time units. Figure 15) *Rod*: Run for 1000 time units at a growth rate of 1 length unit per time unit, with no negative autotropism implemented, and the density field hypothesis of branching regulation applied, where the density field threshold for branching is set at 0.005 and branching probability at 80% per time unit.

regularly, and filling the remaining space. In the early stages of development such a colony is more star-like than spherical. We wish to emphasise that this remarkable differentiation of hyphae emerges in the visualisation even though the program does not include routines implementing differences in hyphal behaviour. In the mathematical model, all virtual hyphae are driven by the same algorithm. By altering the persistence factor, it is possible to generate the whole range of intermediate forms between *Boletus* and *Amanita* types.

Finally, when both autotropic reaction and branching are regulated by the hyphal density field, a spherical, uniformly dense colony is also formed. However, the structure is different again, such a colony being similar to the *Tricholoma* type illustrated by Fries[43](Figs. 6 and 9). This type has the appearance of a dichotomous branching pattern, but it is not a true dichotomy. Rather the new branch, being very close, generates a strong density field that turns the older tip. In the previous model a tip nearby has no stronger effect than a more distant tip as long as they are both closer than R.

Hence the *Amanita rubescens* and *Tricholoma* branching strategies may be based on a negative autotropic reaction of the growing hyphae while the *Boletus* strategy may be based on the absence of such a reaction, relying only on density-dependent branching. Differences between *Amanita* and *Tricholoma* in the way that the growing tip senses its neighbours may be obscured in life. In *Amanita* and *Boletus* types, the tip may sense the number of other tips in its immediate surroundings. In the *Tricholoma* type, the tip may sense all other parts of the mycelium, but the local segments have the greatest impact.

Our models show that the broadly different types of branching observed in the fungal mycelium are likely to be based on differential expression of relatively simple control mechanisms. We presume that the "rules" governing branch patterning (that is, the mechanisms causing the patterning) are likely to change in the life of a mycelium, as both intracellular and extracellular conditions alter. We have imitated some of these changes by making alterations to particular model parameters during the course of a simulation. Some of the results are illustrated in Figures 10-15, and they show that the Neighbour-Sensing model is capable of generating a range of morphologies in its virtual mycelia which are reminiscent of fungal tissues. These experiments make it evident that it is not necessary to impose complex spatial controls over development of the mycelium to achieve particular geometrical forms. Rather, geometrical form of the mycelium emerges as a consequence of the operation of specific locally-effective hyphal tip interactions. We hope that further experimentation with the model will enable us to predict how tissue branching patterns are established in real life.

Acknowledgement

LJMcN thanks the British Mycological Society for the award of a Bursary that enabled his contribution to this chapter.

References

1. Adams HL, Thomas CR. The use of image analysis for morphological measurements on filamentous organisms. Biotechnol Bioeng 1988; 32:707-712.
2. Money NP. The pulse of the machine - reevaluating tip-growth methodology. New Phytologist 2001; 151:553-555.
3. Jackson SL. Do hyphae pulse as they grow? New Phytologist 2001; 151:556-560.
4. Hammad F, Watling R, Moore D. Artifacts in video measurements cause growth curves to advance in steps. J Microbiol Meth 1993; 18:113-117.
5. Acuña G, Giral R, Agosin E et al. A neural network estimator for total biomass of filamentous fungi growing on two dimensional solid substrate. Biotech Tech 1998; 17(7):515-519.
6. Desgranges C, Vergoignan C, Georges M et al. Biomass estimation in solid state fermentation. I. Manual biochemical methods. Appl Microbiol Biotechnol 1991; 35(2):200-205.
7. Desgranges C, Vergoignan C, Georges M et al. Biomass estimation in solid state fermentation. II. On-line measurements. Appl Microbiol Biotechnol 1991; 35(2):206-209.
8. Katz D, Goldstein D, Rosenberger RF. Model for branch initiation in Aspergillus nidulans based on measurement of growth parameters. J Bacteriology 1972; 109:1097-1100.
9. Trinci APJ. A study of the kinetics of hyphal extension and branch initiation of fungal mycelia. J Gen Microbiol 1974; 81:225-236.
10. Steele GC, Trinci APJ. The extension zone of mycelial hyphae. New Phytologist 1975; 75:583-587.
11. Pirt SJ. A kinetic study of the mode of growth surface colonies of bacteria and fungi. J Gen Microbiol 1967; 47:181-197.
12. Trinci APJ. Influence of the peripheral growth zone on the radial growth rate of fungal colonies. J Gen Microbiol 1971; 67:325-344.
13. Prosser JI, Trinci APJ. A model for hyphal growth and branching. J Gen Microbiol 1979; 111:153-164.
14. Bartnicki-Garcia S. Fundamental aspects of hyphal morphogenesis. Symposia of the Society of General Microbiology 1973; 23:245-267.
15. Trinci APJ. The duplication cycle and branching in fungi. In: Burnett JH, Trinci APJ, eds. Fungal Walls and Hyphal Growth. Cambridge, UK: Cambridge University Press; 1979:319-358.
16. Yang H, King R, Reichl U et al. Mathematical model for apical growth, septation, and branching of mycelial microorganisms. Biotechnol Bioeng 1992;39:49-58.
17. Leopold LB. Trees and streams: The efficiency of branching patterns. J Theor Biol 1971; 31:339-354.
18. Horton RE. Erosional development of streams and their drainage basins: Hydrophysical approach to quantitative morphometry. Bull Geographical Soc Amer 1945; 56:275-370.
19. Gull K. Mycelium branch patterns of Thamnidium elegans. Transactions of the British Mycological Society 1975; 64:321-324.
20. Obert M, Pfeifer P, Sernetz M. Microbial growth patterns described by fractal geometry. J Bacteriol 1990; 172:1180-1185.
21. Ritz K, Crawford J. Quantification of the fractal nature of colonies of Trichoderma viride. Mycol Res 1990; 94:1138-1152.
22. Jones CL, Lonergan GT, Mainwaring DE. A rapid method for the fractal analysis of fungal colony growth using image processing. Binary 1993; 5:171-180.
23. Matsuura S, Miyazima S. Colony of the fungus Aspergillus oryzae and self-affine fractal geometry of growth fronts. Fractals 1993; 1:11-19.
24. Cohen D. Computer simulation of biological pattern generation processes. Nature 1967; 216:246-248.
25. Hutchinson SA, Sharma P, Clarke KR et al. Control of hyphal orientation in colonies of Mucor hiemalis. Transactions of the British Mycological Society 1980; 75:177-191.
26. Robinson PM. Chemotropism in fungi. Transactions of the British Mycological Society 1973; 61:303-313.
27. Robinson PM. Autotropism in fungal spores and hyphae. Botanical Rev 1973; 39:367-384.
28. Trinci APJ, Saunders PT, Gosrani R et al. Spiral growth of mycelial and reproductive hyphae. Transactions of the British Mycological Society 1979; 73:283-292.
29. Indermitte C, Liebling TM, Clémençon H. Culture analysis and external interaction models of mycelial growth. Bull Mathematical Biol 1994; 56(4):633-664.
30. Edelstein L. The propagation of fungal colonies: A model for tissue growth. J Theor Biol 1982; 98:679-701.
31. Ferret E, Siméon JH, Molin P et al. Macroscopic growth of filamentous fungi on solid substrate explained by a microscopic approach. Biotechnol Bioeng 1999;65(5):512-522.
32. Koch KL. The kinetics of mycelial growth. J General Microbiol 1975; 89:209-216.

33. Monod J. The growth of bacterial cultures. Ann Rev Microbiol 1949; 3:371-394.
34. Edelstein L, Segel LA. Growth and metabolism in mycelial fungi. J Theor Biol 1982; 104:187-210.
35. Mitchell DA, Do DD, Greenfield PF. A semimechanistic mathematical model for growth of Rhizopus oligosporus in a model solid-state fermentation system. Biotechnol Bioeng 1991; 38(4):353-362.
36. Bartnicki-Garcia S, Hergert F, Gierz G. Computer simulation of fungal morphogenesis and the mathematical basis of hyphal (tip) growth. Protoplasma 1989; 153:46-57.
37. Gierz G, Bartnicki-Garcia S. A three-dimensional model of fungal morphogenesis based on the vesicle supply center concept. J Theor Biol 2001; 208(2):151-164.
38. Trinci APJ. Regulation of hyphal branching and hyphal orientation. In: Jennings DH, Rayner ADM, eds. Ecology and Physiology of the Fungal Mycelium. Cambridge, UK: Cambridge University Press; 1984:23-52.
39. Trinci APJ, Wiebe MG, Robson GD. The mycelium as an integrated entity. In: Wessels JGH, Meinhardt F, eds. Growth, Differentiation and Sexuality. (The Mycota, vol. 1). Berlin, Heidelberg: Springer-Verlag; 1994:175-193.
40. Moore D. Tissue Formation. In: Gow NAR, Gadd GM, eds. The Growing Fungus. London: Chapman and Hall; 1994:423-465.
41. Chiu SW, Moore D. Patterns in Fungal Development. Cambridge, U.K.: Cambridge University Press; 1996.
42. Moore D. Fungal Morphogenesis. New York: Cambridge University Press; 1998.
43. Fries N. Untersuchungen über Sporenkeimung und Mycelentwicklung bodenbewohneneder Hymenomyceten. Symbolae Botanicae Upsaliensis 1943; 6(4):633-664.
44. Kotov V, Reshetnikov SV. A stochastic model for early mycelial growth. Mycol Res 1990; 94:577-586.
45. Hogeweg P, Hesper B. A model study on biomorphological description. Pattern Recognition 1974; 6:165-179.
46. Witten IH, Frank E. Data Mining: Practical Machine Learning Tools and Techniques with Java Implementations. San Francisco: Morgan Kaufmann Publishers; 1999.

CHAPTER 5

Branching in Colonial Hydroids

Igor A. Kosevich

Abstract

Cnidarians are primitive multi-cellular animals whose body is constructed of two epithelial layers and whose gastric cavity has only one opening. Most cnidarians are colonial. Colonial hydroids with their branched body can be regarded as a model for the whole phylum and are the most- studied cnidarian group with respect to developmental biology. Their colonies are constructed by repetition of limited number of developmental modules. The new modules are formed in the course of activity of terminal elements—growing tips of stolons and shoots. The growing tips of cnidarians, in contrast to those of plants, lack cell proliferation and drive morphogenesis instead by laying down and shaping the outer skeleton and formation of new colony elements. Cell multiplication takes place proximally to the growing tips. Branching in colonial hydroids happens due to the emergence of the new growing tip within the existing structures or by subdivision of the growing tip into several rudiments. Marcomorphogenetic events associated with different variants of branching are described, and the problems of pattern control are discussed in brief. Less is known about genetic basis of branching control.

Introduction

Cnidarians are generally considered to be the basic primitive group of multi-cellular organisms. The main feature of their general body plan is a two-layer body in a form of a blind sack with only one mouth opening; the body is composed of two tissue layers, ectoderm and endoderm, separated by extracellular matrix—the mesoglea. One of the most distinctive features is the presence of the nematocytes—epithelial cells containing sting capsules (cnidae or nematocysts) that are used for defence, capture of prey and temporary attachment. Cnidarians remain at the epithelial level of organisation—they have no real tissues or organs. The ectoderm and endoderm are composed of several cell types, namely epithelia-muscular cells with contractile processes at the base, several types of gland cells, nerve cells, nematocytes, and multipotent interstitial cells (i-cells). The whole diversity of cells types is maintained by the presence of three independent and self-supporting cell lineages—ectodermal epithelia-muscular cells, endodermal epithelia-muscular cells and i-cells that give rise to the nerve cells, different gland cells, nematocytes and germ cells.[1-4] It is believed that these cell lineages are determined during early stages of embryogenesis[5-7] and show no ability for reciprocal trans-differentiation under normal conditions.[1,8-11]

The phylum Cnidaria is composed of four classes: Anthozoa, Hydrozoa, Scyphozoa and Cubozoa.[12,13] With respect to the question of branching morphogenesis, I will discuss the representatives of the class Hydrozoa, which have received most attention from developmental biologists. This group of cnidarians is characterised by metagenetic life-cycle: the larva undergoes metamorphosis into the polyp stage (mostly sessile and attached) and this stage sheds the motile planktonic medusae.[14] Polyps multiply asexually through different variants of budding,

---

Branching Morphogenesis, edited by Jamie A. Davies.

while medusae generally reproduce sexually. The ability of polyps to produce buds was the basis for the development of colonial (or modular) organisation within the polypoid stage of cnidarians and hydroids in particular.[15]

Organisation of Hydroid Colony

The main parts of hydroid colony are the creeping hydrorhiza and the hydranths, or shoots, that protrude into surrounding water (Fig. 1). The hydrorhiza is composed of a net of the tube-like stolons. Hydranths are either located directly on the stolons (sessile hydranths) or have a pedicel. The shoots have a different structural organisation: they may have a stem and lateral branches of successive orders, and may bear numerous hydranths. Modular organisation of the organism implies that the its body is constructed by the repetition of the limited number of definite elements (modules). In the case of colonial hydroids, these modules are: stolon internodes, shoot internodes, hydranths, and growing tips of stolons and shoots. Commonly, the stolon internode is a section of the stolon tube between two adjacent bases of the sessile polyps or shoots (Fig. 2A). The organisation of the shoots is more complex in most cases. The simplest variant is repetition of almost identical shoot internodes (Fig. 2A,B). The branches and the shoot stem in that case are organised similarly. In more highly-integrated shoots, the internodes within the stem and branches may differ and are frequently complicated by formation of secondary (complex) internodes (Fig. 2C,D).

Schematically, a hydroid colony may be imagined as a system of branching tubes with hydranths at one end and growing tips at the others. The nongrowing terminus of the shoot either is occupied by the hydranth or has no specific structure. The nongrowing end of the stolon is a blind end of the tube without any specific structure either. The hydranths are organised more or less as a solitary polyp *Hydra* with one exception—they lack the foot structure and are connected to the tube of the colonial body tissue—the coenosarc. The coenosarc is a two-layer tube with practically unvarying organisation along its length. From the outside the coenosarc is covered with the outer rigid skeleton—the chitinious perisarc. The perisarc is used for tight attachment to the substrate along the stolons, gives some protection against predators, and provides mechanical support for soft tissue for development of the elevated structures of the shoots (Marfenin, Kosevich, in press).[16]

The presence of the hard skeleton (perisarc) and branching points along the colony limit the mode of elongation of the colony. Growth of the colony can be achieved only by the extension of tubes at their termini. The terminal part of the stolon tube is occupied by a

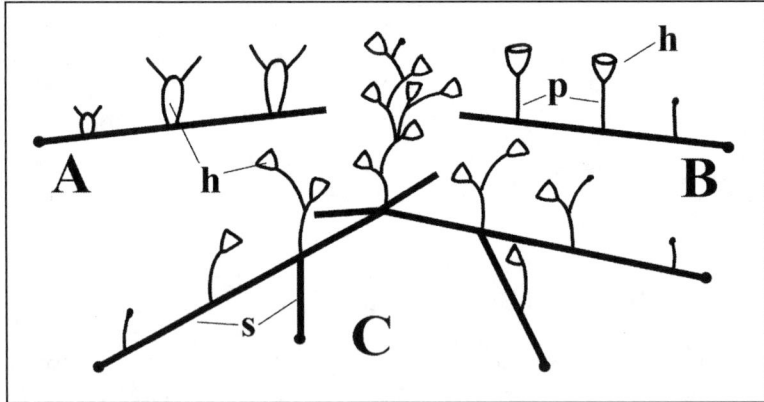

Figure 1. Scheme of the hydroid colony organisation. A) Stolonal colony with sessile hydranths. B) Stolonal colony with hydranth with pedicels. C) Colony with sympodial shoots. s—stolons, h—hydranths, p—hydranth pedicels.

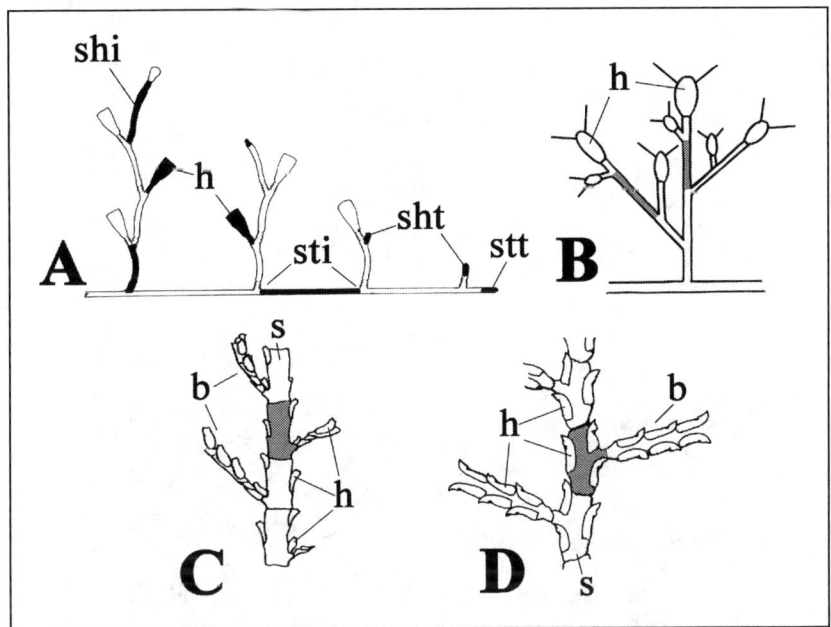

Figure 2. Main elements of hydroid colonies. A) Colony with sympodial shoots. B) Colony with monopodial shoots with terminal hydranth. C,D) Parts of highly-integrated shoots with complex shoot internodes. b—shoot branch, h—hydranths, s—stem of the shoot, shi—shoot internode, sht—shoot growing tip, sti—stolon internode, stt—stolon growing tip.

growing stolon tip. The termini of the shoots are occupied either by shoot growing tips (Thecate hydroids) or by terminal hydranths (Athecate hydroids). The growing tip in colonial hydroids is a morphogenetic element whose job is to shape new colonial elements by laying down new portions of perisarc and to move ahead by repetitive growth pulsations—the series of elongation-contractions of the growing tip—with a periodicity of several minutes.[17-26] Morphologically, the growing tip differs from the rest of the coenosarc: its tissue has permanent contact with the perisarc tube and the cells of the growing tip have a characteristic organisation. The soft tissue extends within the part of the stolon or shoot between the growing tip and the

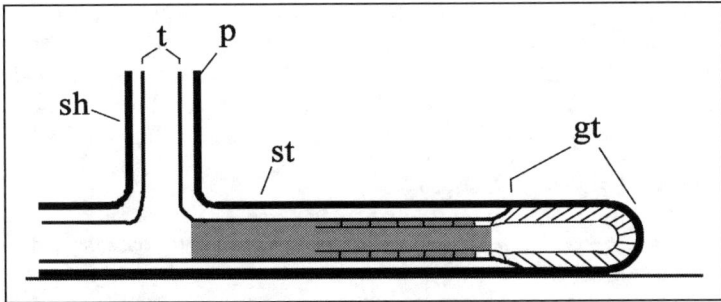

Figure 3. Scheme of the terminal part of the colony stolon showing the relative organisation of the outer skeleton and soft tissue. Only one tissue layer is marked. gt—growing tip, p—perisarc (outer skeleton), sh—shoot base, st—stolon, t—tissue tube. The region of the tissue extension is shadowed.

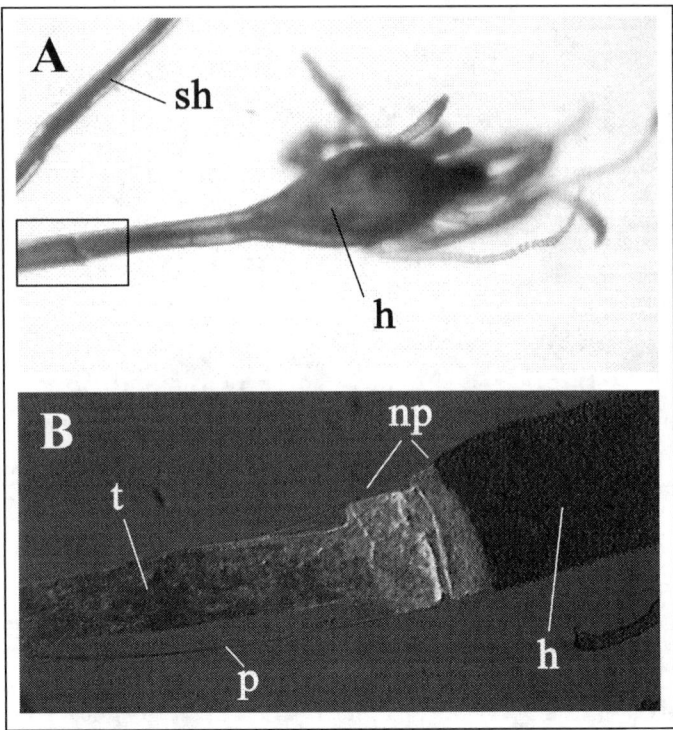

Figure 4. Place of the tissue and skeleton extension in shoots with terminal hydranths. A) Photo of the terminal hydranth with part of the shoot. B) Magnified view of the hydranth base marked by the rectangle in A—white fluorescence corresponds to the newly laid perisarc (staining with Calcofluor White). h— hydranth, np—newly laid perisarc (skeleton), p—old perisarc, sh—part of the shoot, t—soft tissue (coenosarc).

last branching point (Fig. 3).[27-29] In those shoots where the termini are occupied by the hydranth, the soft tissue and new perisarc are added just under the hydranth's base (Fig. 4).

The material for elongation of the coenosarc tube comes from more proximal regions of the colony.[30-36] The growing tip completely lacks cell divisions and has relatively permanent cell composition. Proliferation has been observed in cells just behind the growing tip and proliferation is distributed more or less evenly, at least along the nearest 3-5 internodes.[37] Direct observation of the ectoderm revealed that single cells and entire tissue sheets move towards the growing tip. The speed of such migrations decreases with distance from the growing tip coming to nought within the third or forth internode. But within the most distal uncompleted internode just behind the tip, the speed of ectoderm cells' migration can even be higher than the movement of the tip itself.[33]

Branching in Colonial Hydroids

Each node within a stolon or a shoot is a branching point. This node can be formed either as a result of the appearance of a new growing tip upon the already formed structure, or in a form of the subdivision of the growing tips into two or more parts during the course of their growth. The first case is characteristic for stolon branching in most species, for the growth and branching in sympodial shoots and for other types of shoots with irregular mode of branching (Fig. 5A,B). Subdivision of the tip into several parts (rudiments) is an attribute of a monopodial type of the shoot growth with a regular mode of branching (Fig. 5C).

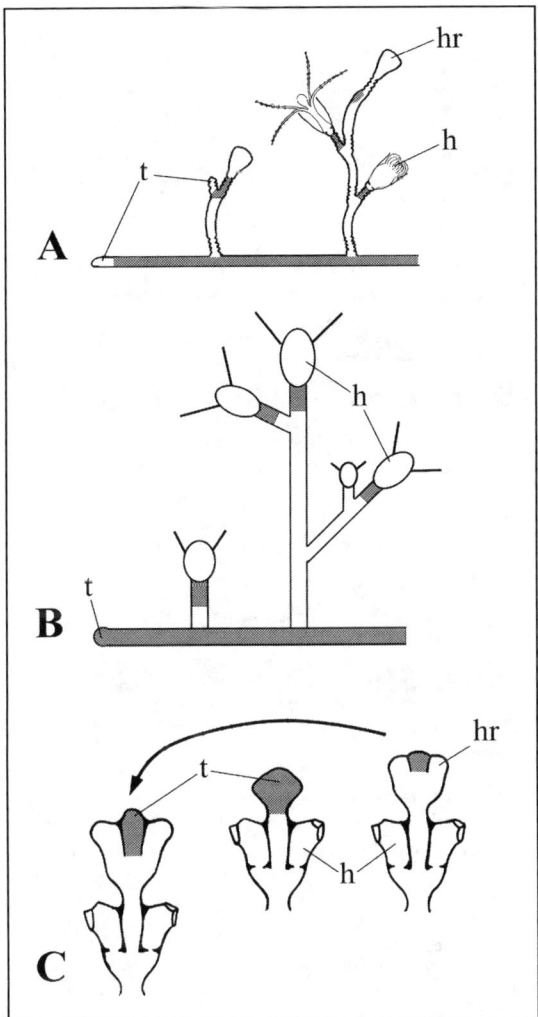

Figure 5. Zones of branching within different types of colonies. A) Colony with sympodial shoots and irregular branching. B) Colony with monopodial shoots, terminal hydranth and irregular branching. C) Terminal part of the monopodial shoot with terminal growing tip and regular branching. h—hydranths, hr—hydranth rudiment, t—growing tips. Zones of branching are shadowed.

Emergence of a New Growing Tip

The branching in stolons and shoots starts with the appearance of a new growing tip. In stolons and sympodial shoots, the new tip emerges from the coenosarc tissue which is similar in composition and is characterised by flattened ectodermal and loosely organised endodermal cells. The first visible changes are associated with formation of a plate of columnar ectodermal cells at the point of branching. It is very likely that this reorganisation of the ectoderm is linked with the simultaneous reorganisation of underlying endoderm cells, including their vacuolisation. Later, the plate starts to pulsate and forms out-folds (Fig. 6). During the course of the initial steps of growth, the new tip reaches its final dimensions and form and gradually gains the highest speed of its growth.

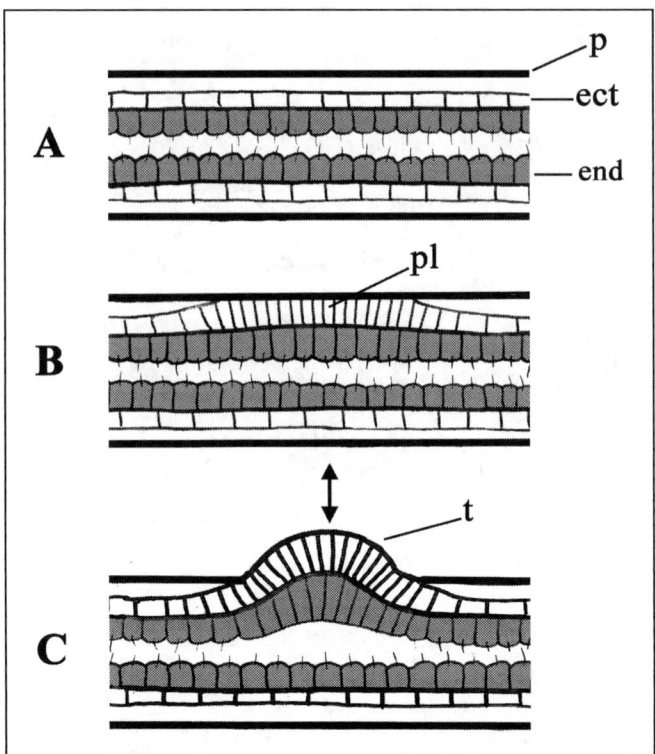

Figure 6. Scheme of the new growing tip emergence upon the stolon. ect—ectoderm layer, end—endoderm layer of coenosarc, p—perisarc (skeleton), pl—plate of ectodermal cells, t—new growing tip. Arrows indicate direction of growth pulsations of the tip.

Emergence of a new tip requires that the existing perisarc tube must 'open'. Unfortunately the biochemical mechanism of this process is not known. No chitinase activity capable of digesting the chitinious matrix of the perisarc has been detected during new tip emergence (personal observations). From the outside view, it seems that in stolons and at least some of the species with monopodial shoots, the existing perisarc at the point of the new tip emergence 'melts' over the surface of the new tip. This process may be similar to the growth and budding of fungal cell walls: new portions of the polymers are added to the existing ones that constitute the matrix of the cell wall and this causes extension the surface of the cell wall which itself is not elastic.[38-43] If this is the case in such cnidaria, there would be no need for rupture of the old perisarc.

In the sympodial shoots of *Laomedea flexuosa* Hincks (Campanulariidae, Thecathora), the perisarc 'opening' is achieved in a different manner, but the basic biochemical machinery may be the same. At the very first moment of the tip emergence, when the ectodermal plate is just forming, the circular set of ectodermal cells start to release amorphous chitin, the precursor of the perisarc matrix. The release has been visualised by staining with Calcofluor White (Fluorescent Brightener 28)(Sigma)(Fig. 7), which stains various carbohydrate fibrils, including amorphous chitin.[44] Later, the entire apical surface of the growing tip releases the amorphous chitin and after the new tip has emerged one can see that the plate of the old perisarc is pushed out like the lid of a tin (Fig. 8) and the growing tip itself is covered by new perisarc.

The model proposed for the mechanism of the tip growth due to pulsations[18,45] implies that the growing tip has a mechanical support from the hard (already hardened and practically

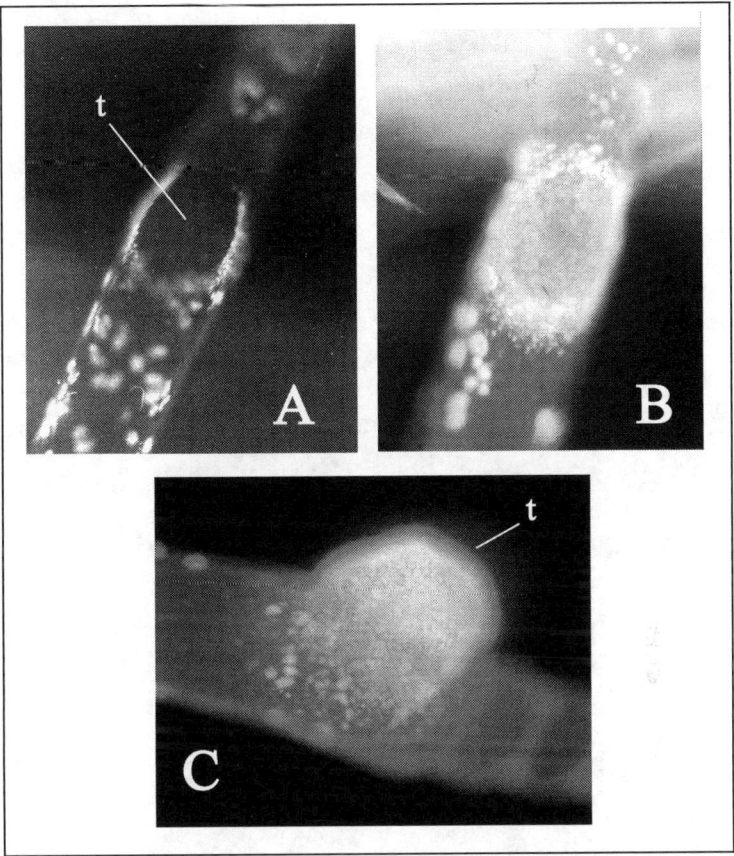

Figure 7. Staining for the amorphous chitin with Calcofluor White during the initial moments of the new tip emergence upon the shoot internode in *Laomedea flexuosa*. White fluorescence corresponds to the places of the amorphous chitin release. A) Initial moment of the tip appearance—formation of the ectodermal plate (see Fig. 6B). B) About an hour later—new tip is formed but had not yet opened the old perisarc. C) The new growing tip get out of the old perisarc. t—new growing tip.

not stretchable) perisarc that surrounds its circumference. In the course of growth, new soft perisarc is released by the tip tissue at its spherical apex. As soon as it is left behind by the advancing apex, the perisarc quickly hardens. So the growing tip forms the perisarc tube by itself and simultaneously uses it as a mechanical support for forward movement by growth pulsations.

Initially, the new tip on the sympodial shoot internode is supported by the old perisarc on only one side, the outer surface of the internode. That is why, after the onset of the pulsation and before the 'opening' of the maternal perisarc, the new tip has to form additional perisarc wall from the inside of the existing perisarc tube (Fig. 9A). This way, it obtains sufficient mechanical support along its circumference.

Subdivision of an Existing Tip

The same is true for morphogenesis by subdivision of the growing tip into several parts (rudiments). For example, the growing tip of the monopodial shoot in *Dynamena pumila* L. (Sertulariidae, Thecaphora) has a morphogenetic cycle that includes the formation of a pair of

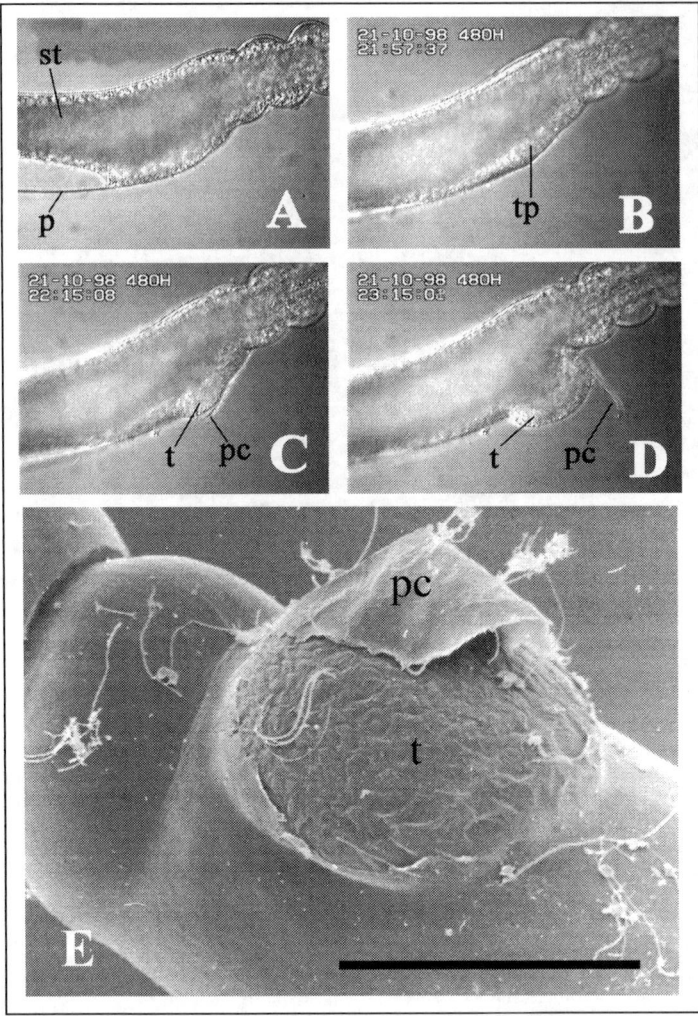

Figure 8. New growing tip emergence upon the shoot internode in *Laomedea flexuosa*. A-D) Sequence of events during new tip emergence (video microscopy - dissecting microscope). E) Scanning electron micrograph of the newly emerged tip (corresponds to D). Scale bar—100 mkm. p—perisarc, pc—perisarc covering, st—soft tissue, t—new growing tip, tp—ectoderm plate at the beginning of the tip organisation.

oppositely- located hydranths with a growing tip between them. In the course of this cycle, the tip starts as a practically spherical bulb that later becomes oval in the plane of the shoot due to greater growth in the dimension of this plane. The apical surface of this growing tip then divides into three parts; lateral ones, which will form the hydranths, and the central one will produce the growing tip for the next cycle of shoot growth.[18,46,47] Subdivision of the entire growing tip is accompanied by formation of additional perisarc walls between the rudiments. These walls are formed not only along newly-grown lengths of the rudiment, but also by in-growth of the perisarc as the apical tissue divides (Fig. 9B). Such additional walls support the development of the rudiments without decreasing the speed of growth, and perhaps play certain role in determining the fate of the rudiment.[46]

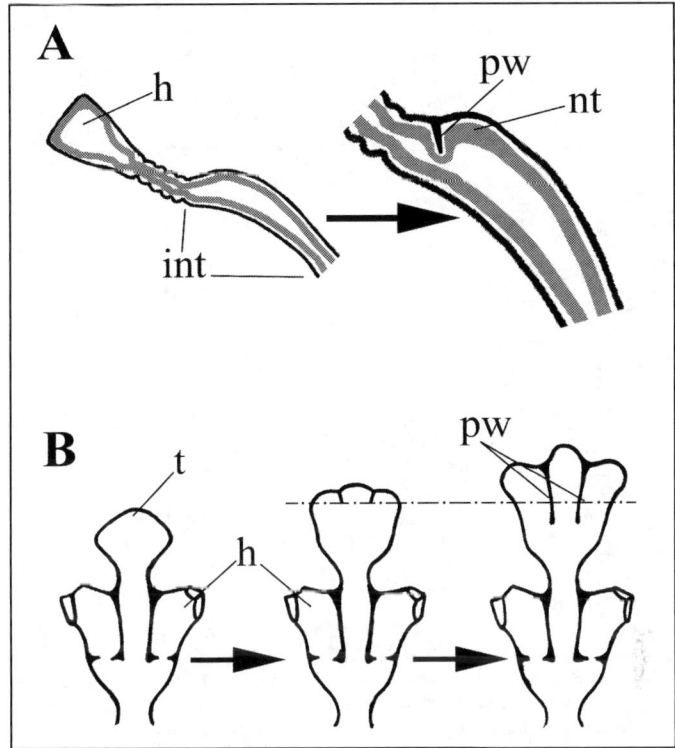

Figure 9. Formation of the additional perisarc walls from inside of the existing perisarc tube during new tip(s) emergence. A) Scheme of the additional perisarc plate formation in sympodial shoots. A grey line shows the tissue. Perisarc is shown in black. B) Formation of the perisarc walls by ingrowth during the subdivision of the growing tip into 3 parts. Only perisarc is shown. Dashed line shows the primary level of the tip subdivision. h—hydranth, int—shoot internode, nt—new tip, pw—additional perisare wall, t—growing tip.

Interaction of a New Tip with Adjacent Structures

In the majority of colonial hydroid species studied to date, the emergence of new growing tips is spatially connected to existing growing tips or to the bases of the hydranths (hydranth pedicels). The tip of a new shoot appears on the stolon just at the proximal part of the stolon tip.[48,49] The new tip of the sympodial shoot (e.g., in *Laomedea flexuosa, Gonothyraea loveni, Obelia longissima*) emerges at the border of the smooth part and the hydranth pedicel.[48,50-52] The lateral branches of the shoots in highly-organised species of the Sertulariidae family begin as a part of a morphogenetic cycle of the shoot growing tip in which the tip subdivides into several rudiments (Marfenin, Kosevich, in press). That means that the condition of the surrounding tissue is not homogenous along the circumference of the tip base. From the proximal side (along the axis of the maternal internode) the tissue is more stretchable in comparison with the tissue layer distal to the new tip base. Morphogenesis could be regulated simply by the extensibility of this tissue layer. New tips mechanically interact with one another, competing for the tissue and, because of the synchronous pulsations that take place just after subdivision, these daughter tips pull the same small portion of tissue connecting them. This may cause new tips to bend towards the existing one (Fig. 10A).[18] Later on, the distance between the daughter tips increases and they practically cease interacting mechanically.

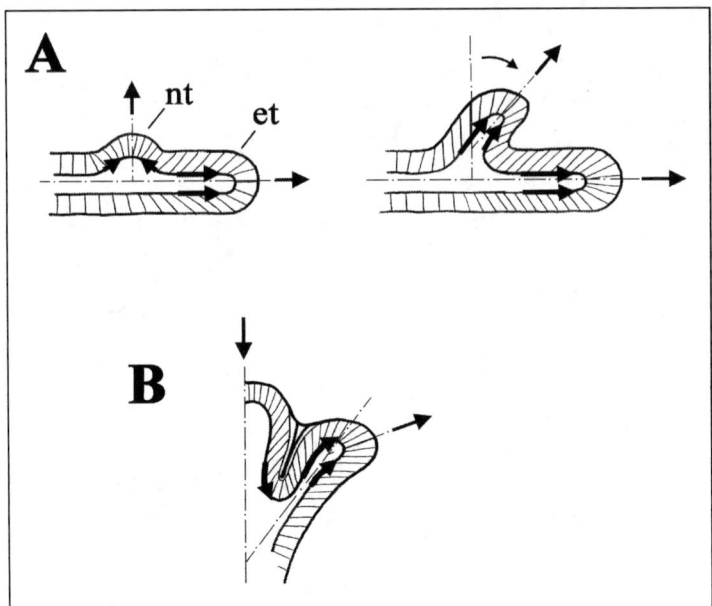

Figure 10. Schematic explanation of the bending of the growing tip due to interaction with adjacent tip. A) Initial bending towards the adjacent tip due to synchronous pulsations. B) Outward bending of the tip in the case of asynchronous pulsations. et—existing growing tip, nt—new growing tip. Arrows inside the rudiment indicate the direction and magnitude of tissue movement during growth pulsations; arrows outside shows the direction of simultaneous growth pulsations. Dashed lines indicate primary axes of the tip growth.

In certain situations there could be additional mechanical interaction between adjacent tips. For example, during the morphogenetic cycle of *D. pumila*, hydranth rudiments are forced to bend towards the central rudiment initially. After several growth pulsations, however, the parameters of pulsations change and instead of being synchronous they gradually switch so that the central rudiment pulsates in antiphase with the lateral rudiments. This means that as the central rudiment retracts and the tissue on its sides shifts disto-proximally, the lateral tips move forward pulling the tissue behind them. The tissue between the rudiments is therefore practically pushed in the direction of the hydranth's tip by the central retracting rudiment. As a result the tissue on this side moves forwards more than on the opposite side and causes the tip itself to bend (Fig. 10B).[53] This might explain why the orifices of the hydrothecae in complex shoots with 'sunken' hydrothecae are always directed outwards, away from the axis of the shoot.

The mechanical interaction between adjacent tips explains why the axis at the base of a shoot is always bent towards the stolon tip. This is never seen in the primary shoots that de-velop from the settled planula larvae of frustules (small stolon-like pieces of the coenosarc separated from the colony for asexual reproduction). In primary shoots, the new growing tip is the only one and emerges in the centre of the structure, so the tissue state is symmetrical at the point of the tip emergence.

Branching Control

The question of the branching morphogenesis in colonial hydroids is inseparable from the problem of pattern formation: what controls the distance between the adjacent structures (hy-dranths, shoots, branches)—the length of the internode, and how the fate of the new tip is determined?

Necessary Conditions

An important condition for new tip initiation is that there must be sufficient 'excess' production of new cells over and above that required by the colony for replacement of spent cells and maintenance of growth of existing tips. If the quantity of new cells exceeds these needs, then there will be sufficient for the initiation of a new tip. The presence of such condition is obvious but can be illustrated by the ratio between the number of growing tips and the length of the coenosarc (where the cell divisions take place) in the colony under different nutrition levels. For *Gonothyraea loveni*, *Obelia longissima* and *Dynamena pumila*, even under most favourable conditions, the value of this ratio never exceeds 0.3. With decrease of nutrition the ratio diminishes.[49,54,55] Under starvation the branching and growth of the tips within the colony stops,[56] although the cell proliferation can still take place, as in *Hydra*,[57,58] to replace spent cells.

Control of Branch Spacing

We will discuss the determination of branching points within the stolon and shoots separately.

Branching of a Stolon

Emergence of the lateral stolon tips is the least regular branching process, at least for the majority of colonial hydroids. It strongly depends on the nutrition of the colony. In most athecate species, there is no exact spatial preference for the appearance of a new stolon tip. Generally, the new stolon tips appear in peripheral parts of the colony near the base of the sessile hydranths or shoots. When nutrition increases, however, or when there is a lack of free substrate, new stolon tips can emerge in the old part of the colony too.[59,60] There appears to be only one rule: a lateral stolon branch will never be formed very close to the apex of existing tip. The smallest distance differs between species but approximately is about 200-300 micrometers. This could be explained by the inhibitory effect from the existing tip according to the predominant model of local activation and lateral inhibition.[61-67]

When the general arrangement of a colony is more regular, the stolon branching is too. This regularity is demonstrated by the appearance of points within the stolon at which branching is more probable. The simplest rule is that lateral stolons emerge close to the base of the shoots; but other positions remain possible (e.g *Laomedea flexuosa*, *Gonothyraea loveni*, *Obelia geniculata* (Campanulariidae)—species with smooth tube-form perisarc of the stolons). It is difficult to explain such predominance. One possibility is that it is somehow is connected with the peristaltic waves of the coenosarc contractions that provide the gastro-vascular flow within the colony.[59,68-74] At the base of the shoots the oppositely directed peristaltic waves could meet to produce a "standing wave" and therefore cause prolonged pressing of the coenosarc over the perisarc from inside. This may initiate the emergence of a new tip. The possible role of mechanical pressure upon the initiation of the new tip is supported experimentally.[64,65] Another possibility is that tip initiation is driven by an accumulation of 'free' cell material which arises from migration of cells into the stolon from the shoot.

In highly-integrated species (e.g., certain species of Campanulariidae, Sertulariidae, Plumulariidae families) the branching points are 'preformed' during the growth of existing stolon. Each stolon internode (segment between adjacent shoots) ends with formation of the wide plate at the base of the next shoot. In the simplest case, this consists simply of a widening of the stolon but in many species it becomes plate-like and the inner space is subdivided into regular pockets by perisarc partitions (Fig. 11). These pockets are the potential points of initiation of new stolon branches. Within one shoot base not all pockets will be used, which ones perhaps being determined by chance. In these species, stolon branching is restricted strictly to bases of the shoots.

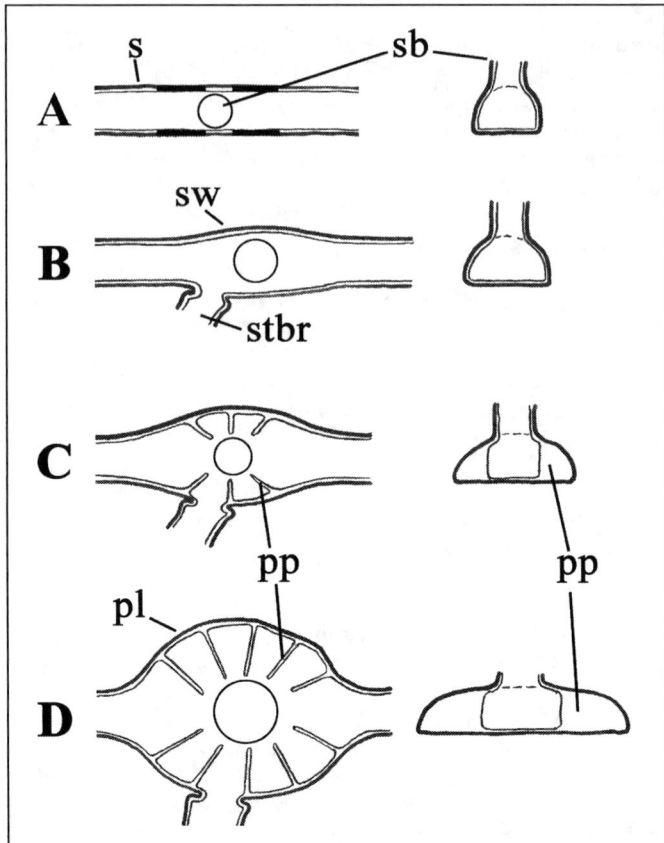

Figure 11. Schematic sketch of the variants of stolon shape at shoot base in different species of colonial hydroids. Left column—view from above, left column—cross section through shoot base. Only the perisarc is shown. A) *Laomedea flexuosa* (Campanulariidae). Thicker line indicate the region of predominant stolon branching. B) *Obelia longissima* (Campanulariidae). C) *Dynamena pumila* (Sertulariidae). D) *Sertularia mirabilis* (Sertulariidae). pl—stolon plate at the shoot base, pp—perisarc partitions, s—stolon, sb—shoot base, stbr—stolon branch, sw—stolon widening.

Appearance of the New Hydranths or Shoots on the Stolon

New hydranths or shoots appear in a regular way during stolon growth. The emergence of the new tip on the upper side of the stolon takes place close to the stolon tip. Sometimes it appears that the stolon tip buds off the new hydranth tip on its upper side (Fig. 12A). In some simple colonies of athecate species, however, the stolon tip raises itself up from the substrate and transforms into the hydranth bud and, as it does so, a new stolon tip emerges from the point of bending up to continue the growth of the stolon (Fig. 12B-E). In these species, a stolon clearly has its own morphogenetic cycle that starts with tip emergence and ends with formation of a hydranth/shoot tip. In most of the colonial species, this sequence is secondarily modified and the stolon tip has the appearance of a permanent element.

The distance between two adjacent hydranth/shoots is strictly controlled in most of colonial hydroids. It can vary under different nutrition conditions and can also be altered artificially by surgical operations.[48,75] In the predominant hypothesis about how spacing works—a reaction-diffusion model which includes local activation and lateral inhibition in different

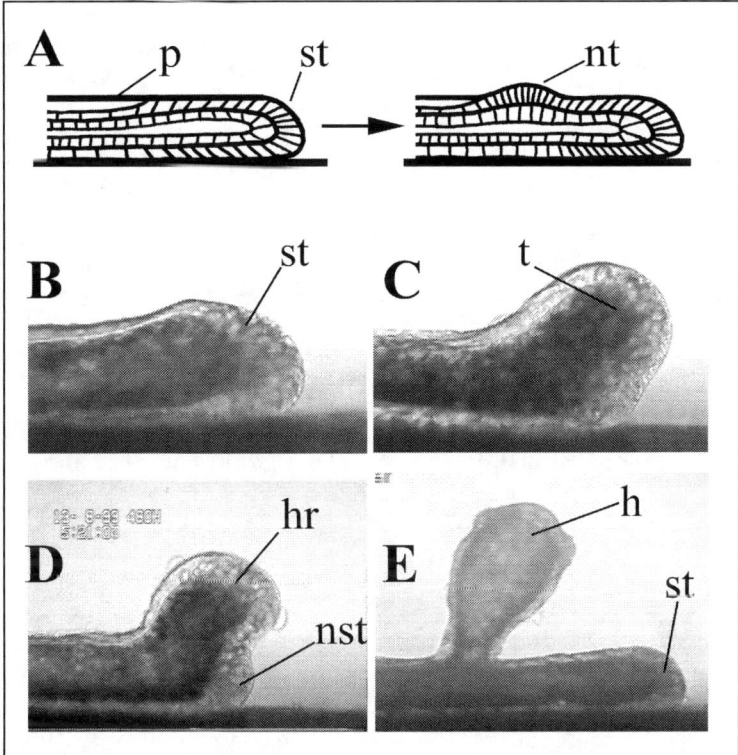

Figure 12. Appearance of the new hydranth/shoot tip on the stolon. A) Scheme of the emergence of the new tip upon upper side of the existing stolon growing tip. B) Transformation of the stolon growing tip into the hydranth rudiment (video microscopy - dissecting microscope). h—hydranth, hr—hydranth rudiment, nst—new (next) stolon tip, nt—new tip, p—perisarc, st—stolon tip, t—former stolon tip.

modifications[46,61-63,76-84]—the distance is controlled by some inhibitory effect from the existing hydranth/shoot that diminishes with distance from that shoot. When the concentration of inhibitor has fallen below some threshold level, a new hydranth/shoot tip can be initiated on the stolon. The main problem with this model is that no molecules responsible for it have been identified.

One proposal is that the changes in the value of ROX potential could play a decisive role.[85-89] At first glance the results of experimental perturbations and measuring of potential in colonial hydroid *Hydractinia* support this hypothesis, but it is difficult to separate the effect and the result. As most chemicals affect numerous targets the question remains: does the ROX state alter the colony proportions, or it is just the result of the altered colony composition?

There are models other than the reaction-diffusion one. One feature that could play a role in determining the distances between adjacent hydranths/shoots might be the cell density immediately behind the stolon tip. As has been shown by several different approaches[28,33] the speed of cell movement towards the stolon tip is higher than the speed of the tip growth itself. This can only mean that the density of cells must rise behind the tip, and it has been suggested that this increase provides both the signal and the raw materials for new tip initiation. The problem with this model is that the distance between the hydranth/shoots remains approximately constant regardless of the nutrition of the colony and, even under starvation conditions when the stolon itself ceases its growth, this distance is not affected.

Branching of the Shoots

Branching of shoots includes at least three processes: regular appearance of growing tips that continue elongation of sympodial shoot; subdivision of the tips into separate rudiments (with different fates in shoots with monopodial growth); and emergence of the lateral branches over the shoot stem. In many cases, elongation of the shoot is complicated by the general complexity of shoot morphogenesis (Marfenin, Kosevich, in press), and it becomes difficult to separate these processes. But the main rules seem to be the same, so we will examine several examples.

Emergence of the Next Tip in Sympodial Shoots

In all cases, a new tip appears after the maternal shoot internode has been formed. The completion of the internode is defined by formation of the hydranth. In some groups (e.g., the Campanulariidae, Campanulinidae families), the hydranths have an annulated pedicel, the distal portion of the internode below the hydrotheca. In others, the hydranth lacks such a pedicel and the internode perisarc gradually turns into the hydrotheca. In all cases, the new growing tip emerges close to the base of the hydranth (Fig. 13): in species with a hydranth pedicel this occurs at the border of the smooth part of the internode and the pedicel.

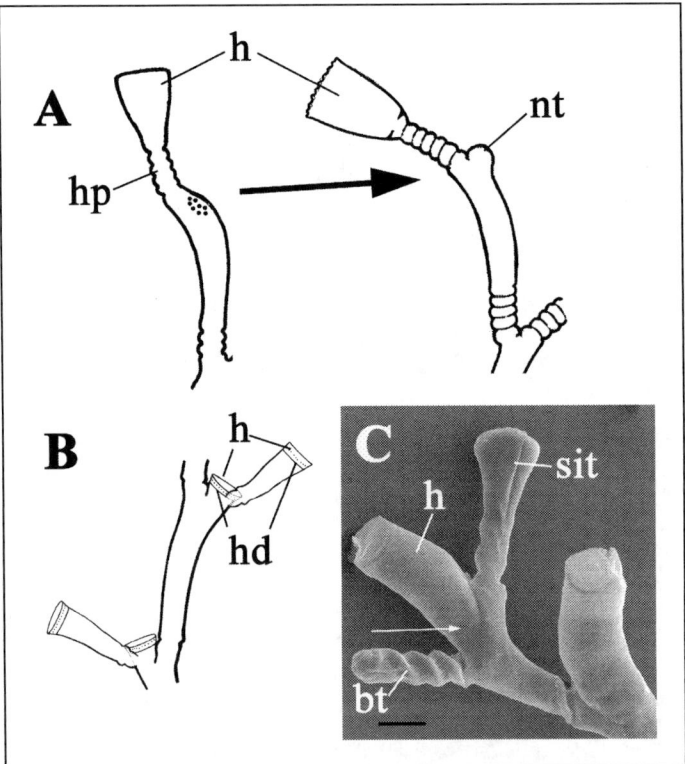

Figure 13. Places of the next shoot growing tip emergence in different hydroids with sympodial shoots. A) gg.*Obelia, Campanularia, Gonothyraea, Laomedea, Calicella*, etc. Dots indicate the place of the next tip appearance. B) g.*Halecium*. C) g.Sertularia (Scanning electron micrograph. Arrow indicates the level of the hydranth diaphragm. Scale bar—100 μm). bt—branch tip, h—hydranth, hd—hydranth diaphragm, hp – hydranth pedicel, nt—next tip, sit—shoot internode tip.

At least two hypothetical explanations can be proposed to account for this pattern. One is based on the hypothesis of positional information.[90-93] Briefly, it proposes that during the course of internode formation, the positional value of the tip tissue gradually increases from a basal value and causes the transition from development of one part of the internode to development of the other. Once the positional value reaches its highest possible value, the growing tip initiates the hydranth formation.[46] The next tip of the shoot has an innate tendency to be form within the tissue with highest positional value, but it is opposed in this by an inhibitory signal emerging from the hydranth. These two opposing tendencies result in the next tip being initiated at the border between the hydranth pedicel and the rest of coenosarc tissue.

The other hypothesis is mechanical rather than biochemical. It is based on the observation that the tissue and cells actively migrate towards the growing tip, and the coenosarc tube shows peristaltic-like waves of contraction and expansion. As the shoot internode develops, the tip moves forward to shape new parts of the perisarc and directs and uses cell material for formation of the new coenosarc. When the hydranth bud at the distal terminus of the internode reaches its final dimensions it ceases to consume cell material and therefore results in an accumulation of cell material still being delivered. If the conditions are favourable, the dense accumulation of cells forms a new tip. The initial stimulus that determines the general location of tip emergence therefore comes from the asymmetry in mechanical forces within the tissue layer during interaction between the coenosarc and perisarc. Periodically the coenosarc is pressed over the perisarc from inside, and the border between the hydranth pedicel and the rest of the perisarc is the most curved region of the internode. This curvature will cause local mechanical stress, and fixes the precise place of tip emergence (Fig. 13A).[94] Experimental alteration of the position of maximal curvature of the perisarc results in emergence of a new tip at the new position of maximal curvature, providing strong support for this hypothesis.[52]

Subdivision of a Shoot Tip

The best hypothesis for growth, morphogenesis and subdivision of the tip into several rudiments in monopodial shoots with terminally located tips, was proposed by L.Beloussov.[95,96] The central idea of this hypothesis is that the transition from one form to the other in thecate hydroids is based of the shifts of the region of the maximal active stretching of the rudiment (tip) surface in basicoapical direction within one growth pulsation. These shifts cause symmetric or asymmetric narrowing or widening of the tip. In the case of successive widening the tip would subdivide into several parts according to mechanical properties of such structures. The forms of the rudiments predicted from this theory fit well with the main types of branching actually seen in thecate hydroids (Fig. 14). The relative activities of the ectoderm and endoderm in the tip are considered to be the main mechanism of such shifts,[97] and the 'physical' properties of the tissue layers—quasi-elasticity and mechanical cell-cell interaction—are used as the main varying parameters of the model.[96]

An additional condition of the model is that the successive changes in the shape of the developing tip have to be fixed by the perisarc and changes in tip form are possible only during growth pulsation.[98] If the border between the already-hardened and still-soft perisarc (which is released on the apical surface of the growing tip) shifts closer to the tip apex, the tip becomes narrower. If the border shifts away from the tip instead, the tip expands in width. Spherical tips become intrinsically unstable as the tips enlarge,[18] leading to the splitting of the tip into several rudiments. Alternatively it is possible that asymmetry in the local rate of hardening of the perisarc would provoke subdivision of the tip.

Emergence of the Lateral Branches

There are two main variants of branching in colonial hydroids: (1) the process of branching is not regular and the branches appear on an already-formed stem; (2) the branching is regular and the next branch tip is formed by subdivision of the stem tip in the course of shoot internode development. In the latter case, the cyclic morphogenesis of the shoot becomes

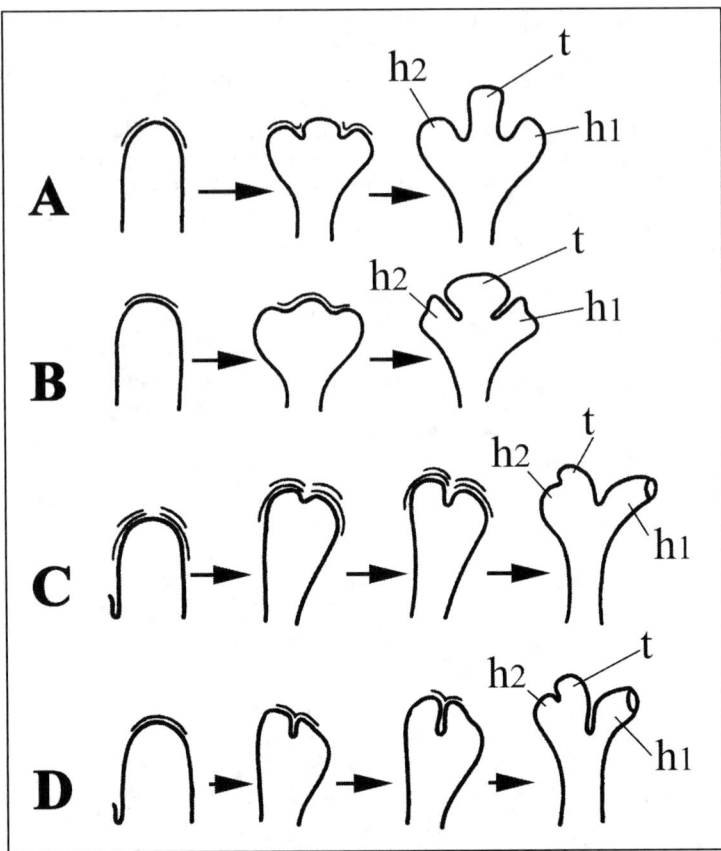

Figure 14. Schematic illustration of the different fates of the tip growth and development in the case of the basicoapical shifts of the regions of the maximal active stretching of the tip surface (shown by second contour) during growth pulsations. A,B) Symmetric tips (rudiments). C,D) Asymmetric tips (rudiments). h1—first hydranth, h2—second hydranth, t—growing tip. (Modified after Belousov, 1975.)

more complicated by inclusion of branch tip formation into the growth cycle: the secondary morphogenetic cycle now includes formation of several internodes and one branch (Marfenin, Kosevich, in press) (Fig. 15).

If the branch is started later, the main rules will perhaps be the same as those for the appearance of a new growing tip in sympodial shoots. The branching point is close to the base of the hydranth and the relative state of the soft tissue and the outer skeleton provide necessary conditions for initiation of the new tip (Fig. 16A). There are still many unanswered questions. For example, in most species with a compact shoot stem and no regular branching, the bases of branches are localised on one side of the stem rather than being symmetrical with respect to the axis of the shoot (Fig. 16B). Nothing is known about why.

Control of Developmental Fate

When a tip subdivides into rudiments that give rise to different structures, such as hydranth, lateral branch, stem etc., some system must act to regulate the developmental fate of each. The models that seem most reasonable are based on the variations of the hypothesis of

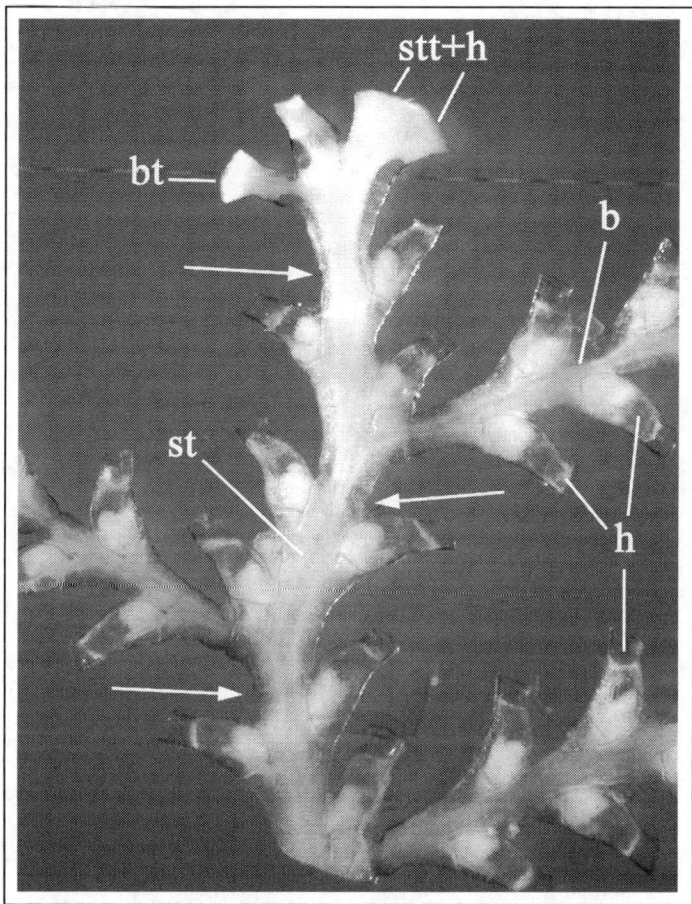

Figure 15. Microphotograph of the terminal part of *Abietinaria abietina* (Sertulariidae) shoot illustrating complex shoot internodes that include obligatory lateral branch formation. b—branch, bt—branch growing tip, h—hydranth, st—shoot stem, stt+h—shoot tip in the course of subdivision into the shoot stem tip and hydranth rudiment. Arrows indicate the boarders of the internodes (marked by light furrows).

local activation and lateral inhibition.[46,80,90,99,100] These models describe the determination of the rudiment fate on the bases of distance control, and are founded on the interaction of three mutually dependent 'players'; one activator and two inhibitors. This models fit most of the observed data and experimental results on branching processes in cnidaria and in plants and they set out an agenda for experimental identification of their molecular components. The models do have various problems, however, in the case of certain highly-integrated species of colonial hydroids and will require improvements or introduction of additional parameters.

The Genetic Basis of Branching in Cnidarians

The genetic and molecular basis for branching in cnidarians remains unclear, mainly because cnidarian genomes have not yet been studied in detail. In *Hydra,* the Wnt signalling pathway has been shown to control formation of head structures.[101-103] The budding that results in the organisation of a second axis and head structures in *Hydra* can be compared with

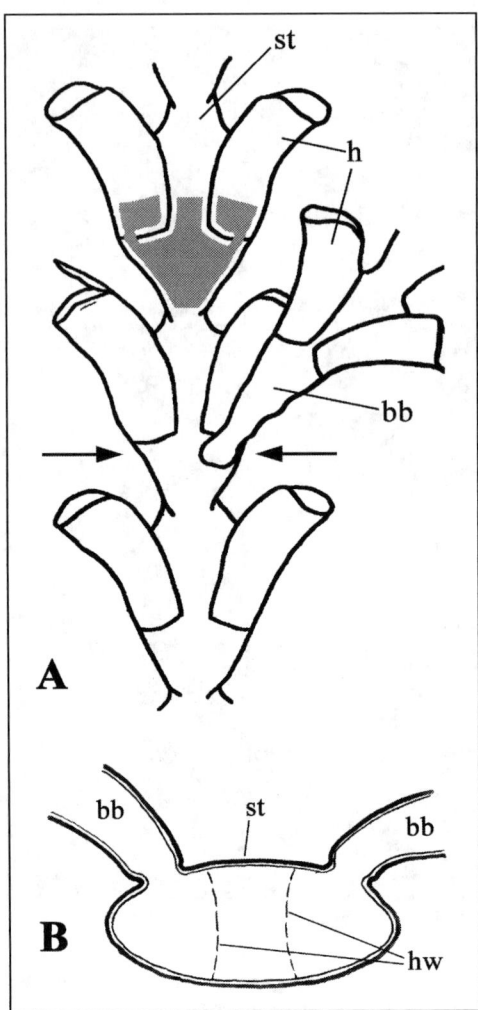

Figure 16. Places of the lateral branches formation upon shoot with irregular branching. A) Part of the shoot of *Diphasia fallax* with the base of the lateral branch. Shadowed area shows relative position of the soft tissue within the skeleton. B) Cross-section through the shoot stem at the level of the branches bases displaying their asymmetric position. bb—branch base, h—hydranth, hw—relative position of the inner walls of the upper hydrothecae, st—shoot stem.

branching in colonial forms. Some conserved genes are expressed at early stages of bud formation.[102-110] In *Hydractinia* the gene *budhead*, a *fork head* homologue, seems to be involved in the earliest stages of the polyp head determination during larva metamorphosis.[111] Some genetic information is now being obtained for the fate control between stolon and shoots (or hydranths); in *Hydractinia,* the gene *Cn-ems*, an *empty-spiracle* homologue, is expressed in the head region of gastrozooids (feeding polyps) and not in blastostyles (reproductive polyps); *Cnox-2* expression differs between polyp types and between polyps and stolons, implying possible specificity in expression during development of different elements within the colony.[111-113] Nevertheless, the really important genetic questions remains open: what is the primary signal that starts the whole sequence of events leading to the initiation of the new tip (new axis) formation? What determines the fate of a new axis? The expression patterns of all of the genes studied to date are simply consequences, and not causes, of these unknown regulatory events.

References

1. Fujisawa T, Sugiyama T. Genetic analysis of developmental mechanisms in Hydra. IV. Characterization of a nematocyst-deficient strain. J Cell Sci 1978; 30:175-85.
2. Bode H, Dunne J, Heimfeld S et al. Transdifferentiation occurs continously in adult Hydra. In: Yamada Science Foundation, ed. Current Topics in Developmental Biology. Japan: Academic Press, 1986:20:257-80.
3. Smid I, Tardent P. Migration of I-cells from ectoderm to endoderm in Hydra attenuata Pall (Cnidaria, Hydrozoa) and their subsequent differentiation. Dev Biol 1984; 106:469-77.
4. Teragawa CK, Bode HR. Migrating interstitial cells differentiate into neurons in hydra. Dev Biol 1995; 171:286-93.
5. Van de Vyver G. Etude du developpment embryonnaire des hydraires athecates (Gymnoblastiques) a gonophores. 2. Formes a planula. Arch Biol (Paris) 1967; 78:451-518.
6. Bodo F, Bouillon J. Etude histologique du developpment embryonnaire de queiques Hydromeduses de Roscoff: Phialidium hemisphaericum (L.), Obelia sp Peron et Lesueur, Sarsia eximia (Allman), Podocoryne carnea (Sars), Gonionemus vertens Agassiz. Cah Biol 1968; 9:69-104.
7. Lenhoff HM. Our link with the Trambleys - Abraham (1710-1784), MAurice (1874-1942) and Jean-Gustave (1903-1977). In: Tardent P, Tardent R, eds. Developmental and cellular biology of coelenterates. Amsterdam: Elsevier/N. Holland Biomed Press, 1980:xvii-xxiv.
8. Burnett AL. A model of growth and cell differentiation in Hydra. Amer Naturalist 1966; 100:165-89.
9. Marcum BA, Campbell RD. Developmental roles of epithelial and interstitial cell lineages in hydra: Aanalysis of chimeras. J Cell Sci 1978; 32:233-47.
10. Rose PG, Burnett AL. The origin of secretory cells in Cordylophora caspia during regeneration. Wilhelm Roux's Arch EntwMech Org 1970; 165:192-216.
11. Rose PG, Burnett AL. The origin of the mucous cells in Hydra viridis. II. Mid-gastric regeneration and budding. Wilhelm Roux's Arch EntwMech Org 1970; 165:177-91.
12. Werner B. New investigations on systematics and evolution of the class Scyphozoa and the phylum Cnidaria. Publs Seto Mar Biol Lab 1973; 20:35-61.
13. Werner B. Die neue Cnidariaklasse Cubozoa. Verh Dtsch Zool Ges 1976; 230.
14. Werner B. Life cycles of the Cnidaria. In: Tardent P, Tardent R, eds. Development and Cellular Biology of Coelenterates. Elsevier / North-Holland: Elsevier / North-Holland Biomedical Press 1980:3-10.
15. Beklemishev VN. Principles of comparative anatomy of Invertebrates. Edinburgh and University of Chicago Press: Oliver & Boyd Ltd 1970; 1-490.
16. Marfenin NN. Evolution of colonial organisation in hydroid order Leptolida. Transactions of Paleontological Institute, Ac Sci USSR 1987; 222:4-19.
17. Beloussov LV. Growth and morphogenesis of some marine hydrozoa according to histological data and time-lapse studies. Publ Seto Mar Biol Lab 1973; 20:315-66.
18. Beloussov LV, Dorfman YaG. On the mechanics of growth and morphogenesis in hydroid polyps. Amer Zool 1974; 14:719-34.
19. Beloussov LV, Badenko LA, Labas JuA. Growth rhythms and species-specific shape in Thecaphora hydroids. In: Tardent P, Tardent R, eds. Developmental and Cellular Biology of Coelenterates. Amsterdam: Elsevier/North-Holland Biomedical Press, 1980:175-8.
20. Beloussov LV, Kazakova NI, Labas JuA. Growth pulsations in hydroid polyps: Kinematics, biological role, and cytophysiology. In: Rensing L, ed. Oscillations and Morphogenesis. New York: Marcel Dekker Inc., 1993:183-93.
21. Beloussov LV, Labas JuA, Badenko LA. Growth pulsations and rudiment shapes in hydroid polyps. Zhurnal Obshchej Biologii 1984; 45:796-806.
22. Crowell S. Morphogenetic events associated with stolon elongation in colonial hydroids. Amer Zool 1974; 14:665-72.
23. Hale LJ. Contractility and hydroplasmic movements in the hydroid Clytia johnstoni. Quarterly J Microsc Sci 1960; 101:339-50.
24. Wyttenbach ChR. The dynamics of stolon elongation in the hydroid, Campanularia flexuosa. J Exp Zool 1968; 167:333-52.
25. Wyttenbach ChR, Crowell S, Suddith RL. The cyclic elongation of stolons and uprights in the hydroid, Campanularia. Biol Bull 1965; 129:429

26. Wyttenbach ChR, Crowell S, Suddith RL. Variations in the mode of stolon growth among different genera of colonial hydroids, and their evolutionary implications. J Morphol 1973; 139:363-75.
27. Braverman M. Studies on hydroid differentiation. VII. The hydrozoan stolon. J Morphol 1971; 135:131-52.
28. Hale LJ. Cell movement, cell division and growth in the hydroid Clytia johnstoni. J Embryol Exp Morph 1964; 12:517-38.
29. Hale LJ. The pattern of growth of Clytia johnstoni. J Embryol Exp Morphol 1973; 29:283-309.
30. Braverman M. Studies on hydroid differentiation IV. Cell movements in Podocoryne carnea hydranths. Growth 1969; 33:99-111.
31. Braverman M. The cellular basis of hydroid morphogenesis. Publ Seto Mar Biol Lab 1973; 20:221-56.
32. Crowell S, Wyttenbaàæ ChR, Suddith RL. Evidence against the concept of growth zones in hydroids. Biol Bull 1965; 129:403
33. Kossevitch IA. Cell migration during growth of hydroid colony. Zhurnal Obshchej Biologii 1999; 60:91-8.
34. Suddith RL. Cell proliferation in the terminal regions of the internodes and stolons of the colonial hydroid Campanularia flexuosa. Amer Zool 1974; 14:745-55.
35. Wyttenbach ChR. Sites of mitotic activity in the colonial hydroid, Campanularia flexuosa. Anat Rec 1965; 151:483
36. Wyttenbach ChR. Cell movements associated with terminal growth in colonial hydroids. Amer Zool 1974; 14:699-717.
37. Marfenin NN, Burykin YuB, Ostroumova TV. Organismal regulation of the balanced growthy in hydroid colony Gonothyraea loveni (Allm.). Zhurnal Obshchej Biologii 1999; 60:80-90.
38. Gooday GW. An autoradiographic study of hyphal growth of some fungi. J Gen Microbiology 1971; 67:125-33.
39. Mulisch M. Chitin in Protistan organisms. Distribution, synthesis and deposition. Europ J Protistol 1993; 29:1-18.
40. Katz D, Rosenberger RF. Hyphal wall synthesis in Aspergillus nidulans: Effect of protein synthesis inhibition and osmotic shock on chitin insertion and morphogenesis. J Bacteriol 1971; 108:184-90.
41. Robertson NF. The growth process in fungi. A Rev Phytopath 1968; 6:115-136.
42. Stratford M. Another brick in the wall? Recent developments concerning the yeast cell envelope. Yeast 1994; 10:1741-52.
43. Wessels JGH, Sietsma JY, Sonnenberg ASM. Wall synthesis and assembly during hyphal morphogenesis in Schizophyllum commune. J Gen Microbiology 1983; 129:1607-1616.
44. Compere P. Cytochemical labelling of chitin. In: Giraud-Guille MM, ed. Chitin in Life Sciences. Lyon, France: Andre Publisher, 1996:66-87.
45. Zaraisky AG, Beloussov LV, Labas JuA et al. Studies of cellular mechanisms of growth pulsations in hydroid polyps. Russian Journal of Developmental Biology 1984; 15:163-169.
46. Berking S, Hesse M, Herrmann K. A shoot meristem-like organ in animals; monopodial and sympodial growth in Hydrozoa. Int J Dev Biol 2002; 46:301-308.
47. Marfenin NN. The phenomenon of coloniality. Moscow: Moscow State Univ Publisher, 1993:1-239.
48. Kosevich IA. Regulation of formation of the elements of the hydroid polyps colony. Russian Journal of Developmental Biology 1996; 27:95-101.
49. Marfenin NN, Kosevich IA. Colonial morphology of the hydroid Obelia loveni (Allm.)(Campanulariidae). Vestnik Moskovskogo Universiteta Biologia 1984; 2:37-45.
50. Kosevich IA. Development of stolon's and stem's internodes in hydroid genera Obelia (Campanulariidae). Vestnik Moskovskogo Universiteta Biologia 1990; 3:26-32.
51. Kosevich IA. Regulation of the "giant" shoot structure in the colonial hydroid Obelia longissima (Campanulariidae). Russian Journal of Developmental Biology 1991; 22:204-210.
52. Kossevitch IA. Role of the skeleton in determination of the branching points in hydroid colonies. Zhurnal Obshchej Biologii 2002; 63:40-49.
53. Belousov LV, Dorfman YaG. Mechanisms of growth and morphogenesis in hydroid polyps by the data of time lapse microcinematography. Russian Journal of Developmental Biology 1974; 5:437-445.
54. Kosevich IA, Marfenin NN. Colonial morphology of tyhe hydroid Obelia longissima (Pallas, 1766)(Campanulariidae). Vestnik Moskovskogo Universiteta Biologia 1986; 3:44-52.
55. Marfenin NN. The hydroid colony as an organism: Regulation of growth in the entire colony. Proceedings of the 6th International Conference on Coelenterate biology 1995; 315-320.
56. Leontovich AA, Marfenin NN. Connection of major intercolonial processes at branching in colonial hydroids. Zhurnal Obshchej Biologii 1990; 51:353-362.

57. Bode HR. Activity of Hydra cells in vitro and in regenerating cell reaggregates. Amer Zool 1974; 14:543-550.

58. Bosch TC, David CN. Growth regulation in Hydra: Relationship between epithelial cell cycle length and growth rate. Dev Biol 1984; 104:161-171.

59. Blackstone NW. Gastrovascular flow and colony development in two colonial hydroids. Biol Bull 1996; 190:56-68.

60. Blackstone NW. Dose-response relationships for experimental heterochrony in a colonial hydroid. Biol Bull 1997; 193:47-61.

61. Gierer A, Meinhardt H. A theory of biological pattern formation. Kybernetik 1972; 12:30-39.

62. Meinhardt H, Gierer A. Applications of a theory of biological pattern formation based on lateral inhibition. J Cell Sci 1974; 15:321-346.

63. Meinhardt H. Models of biological pattern formation: Common mechanism in plant and animal development. Int J Dev Biol 1996; 40:123-134.

64. Muller WA, Plickert G. Quantitative analysis of an inhibitory gradient field in the hydrozoan stolon. Wilhelm Roux's Arch. 1982; 191:56-63.

65. Plickert G. Mechanically induced stolon branching in Eirene viridula (Thecata, Campanulinidae). In: Tardent P, Nardent R, eds. Developmental and cellular biology of Coelenterates. Amsterdam: Elsevier/North-Holland Biomedical Press, 1980:185-190.

66. Plickert G. Low-molecular-wight factor from colonial hydroids affect pattern formation. Wilhelm Roux's Archiv Dev Biol 1987; 248-256.

67. Plickert G, Heringer A, Hiller B. Analysis of spacing in a periodic pattern. Dev Biol 1987; 120:399-411.

68. Dudgeon SR, Buss LW. Growth with the flow: On the maintenance and malleability of colony form in the hydroid Hydractinia. Amer Naturalist 1996; 147:667-91.

69. Dudgeon SR, Wagner A, Vaisnys RJ et al. Dynamics of gastrovascular circulation: Clues to understanding colony integration and morphogenesis in hydrozoans. In: Grassle JP, Kelsey A, Oates E, Snelgrove PV, eds. Twenty Third Benthic Ecology Meeting. New Brunswick, Nj, USA: Rutgers the State Univ Inst Marine Coastal Sciences, 1995:5.

70. Dudgeon S, Wagner A, Vaisnys JR et al. Dynamics of gastrovascular circulation in the hydrozoan Podocoryne carnea: The one-polyp case. Biol Bull 1999; 196:1-17.

71. Fulton C. Rhythmic movements in Cordylophora. J cell comp Physiol 1963; 61:39-52.

72. Karlsen AG, Marfenin NN. Hydroplasm movements in the colony of hydroids, Dynamena pumila L. and some other species taken as examples. Zhurnal Obshchej Biologii 1984; 45:670-680.

73. Marfenin NN. The functioning of the pulsatory-peristaltic type transport system in colonial hydroids. Zhurnal Obshchej Biologii 1985; 46:153-164.

74. Winkle Van DH, Blackstone NW. Video microscopical measures of gastrovascular flow in colonial hydroids. Invertebrate Biology 1997; 116:6-16.

75. Marfenin NN. Study of the integration of the colony of Dynamena pumila (L.) using quantitative morphologival criteria. Zhurnal Obshchej Biologii 1977; 38:409-422.

76. Berking S. Metamorphosis of Hydractinia echinata. Insights into pattern formation in Hydroids. Roux's Arch Dev Biol 1984; 193:370-378.

77. Berking S. Hydrozoa metamorphosis and pattern formation. Curr Top Dev Biol 1998; 38:81-131.

78. Berking S. A model for budding in hydra: Pattern formation in concentric rings. J Theor Biol 2003; 222:37-52.

79. Leontovich AA. Regularities in spatial distribution of hydranth and stolons in a hydroid, Cordylophora inkermanica (Hydrozoa, Clavidae). Zhurnal Obshchej Biologii 1991; 52:534-550.

80. Meinhardt H. A model for pattern formation of hypostome, tentacles, and foot in hydra: How to form structures close to each other, how to form them at a distance. Dev Biol 1993; 157:321-333.

81. Meinhardt H. Organizer and axes formation as a self-organizing process. Int J Dev Biol 2001; 45:177-188.

82. Meinhardt H, Gierer A. Pattern formation by local self-activation and lateral inhibition. Bio Essays 2000; 22:753-760.

83. Pfeifer R, Berking S. Control of formation of the two types of polyps in Thecocodium quadratum (Hydrozoa, Cnidaria). Int J Dev Biol 1995; 39:395-400.

84. Walther M, Ulrich R, Kroiher M et al. Metamorphosis and pattern formation in Hydractinia echinata, a colonial hydroid. Int J Dev Biol 1996; 40:313-322.

85. Blackstone NW. Morphological, physiological and metabolic comparisons between runner-like and sheet-like inbred lines of a colonial hydroid. J Exp Biol 1998; 201:2821-2831.

86. Blackstone NW. Redox control in development and evolution: Evidence from colonial hydroids. J Exp Biol 1999; 202(24):3541-53.

87. Blackstone NW. Redox control and the evolution of multicellularity. BioEssays. 2000; 22:947-953.

88. Blackstone NW. Redox state, reactive oxygen species and adaptive growth in colonial hydroids. J Exp Biol 2001; 204:1845-1853.

89. Blackstone NW. Redox signaling in the growth and development of colonial hydroids. J Exp Biol 2003; 206:651-658.

90. Wolpert L. Positional information and pattern formation. Philos Trans R Soc Lond B Biol Sci 1981; 295:441-4450.

91. Wolpert L. The evolutionary origin of development: Cycles, patterning, privilege and continuity. Development 1994; Supplement:79-84.

92. Wolpert L. One hundred years of positional information. Trends Genet 1996; 12:359-364.

93. Kerszberg M, Wolpert L. The origin of metazoa and the egg: A role for cell death. J Theor Biol 1998; 193:535-537.

94. Beloussov LV. Patterns of mechanical stresses and formation of the body plans in animal embryos. Verh Dtsch Zool Ges 1996; 89:219-229.

95. Belousov LV. Possible ontogenetic mechanisms governing formation of principal morphogenetic types of thecaphoran hydroids. Zhurnal Obshchej Biologii 1975; 36:203-211.

96. Beloussov LV, Grabovsky VI. A Geometro-mechanical model for pulsatile morphogenesis. Comput Methods Biomech Biomed Engin 2003; 6:53-63.

97. Beloussov LV. Basic morphogenetic processes in Hydrozoa and their evolutionary implications: An exercise in rational taxonomy. In: Williams RB, Cornelius PFS, Hughes RG, Robson EA, eds. Coelenterate Biology: Recent Research On Cnidaria And Ctenophora. Dortrecht: Kluwer Acad Publ., 1991; 61-67.

98. Kossevitch IA, Herrmann K, Berking S. Shaping of colony elements in Laomedea flexuosa Hinks (Hydrozoa, Thecaphora) includes a temporal and spatial control of skeleton hardening. Biol Bull 2001; 201:417-423.

99. Wolpert L. Positional information revisited. Development 1989; (Supplement):3-12.

100. Wolpert L, Stein WD. Positional information and pattern formation. In: Maqlacinski GM, Bryant SV, eds. Pattern formation. A primer in developmental biology. New-York: Macmillan Publishing Company, A Division of Macmillan Inc., 1984:3-21.

101. Hobmayer B, Rentzsch F, Kuhn K et al. WNT signalling molecules act in axis formation in the diploblastic metazoan Hydra. Nature 2000; 407:186-189.

102. Gauchat D, Kreger S, Holstein T et al. prdl-a, a gene marker for hydra apical differentiation related to triploblastic paired-like head-specific genes. Development 1998; 125:1637-1645.

103. Technau U, Cramer VL, Rentzsch F et al. Parameters of self-organization in Hydra aggregates. Proc Natl Acad Sci USA 2000; 97:12127-12131.

104. Broun M, Sokol S, Bode HR. Cngsc, a homologue of goosecoid, participates in the patterning of the head, and is expressed in the organizer region of Hydra. Development 1999; 26:5245-5254.

105. Gauchat D, Mazet F, Berney C et al. Evolution of Antp-class genes and differential expression of Hydra Hox/paraHox genes in anterior patterning. Proc Natl Acad Sci USA 2000; 97:4493-4498.

106. Hermans-Borgmeyer I, Schinke B, Schaller HC et al. Isolation of a marker for head-specific cell differentiation in hydra. Differentiation 1996; 61:95-101.

107. Martinez DE, Dirksen ML, Bode PM et al. Budhead, a fork head/HNF-3 homologue, is expressed during axis formation and head specification in hydra. Dev Biol 1997; 192:523-536.

108. Technau U, Bode HR. HyBra1. a Brachyury homologue, acts during head formation in Hydra. Development 1999; 126:999-1010.

109. Shenk MA, Bode HR, Steele RE. Expression of Cnox-2, a HOM/HOX homeobox gene in hydra, is correlated with axial pattern formation. Development 1993; 117:657-667.

110. Shenk MA, Gee L, Steele RE et al. Expression of Cnox-2, a HOM/HOX gene, is suppressed during head formation in Hydra. Dev Biol 1993; 160:108-118.

111. Mokady O, Dick MH, Lackschewitz D et al. Over one-half billion years of head conservation? Expression of an ems class gene in Hydractinia symbiolongicarpus (Cnidaria: Hydrozoa). Proc Natl Acad Sci USA 1998; 95:3673-368.

112. Cartwright P, Buss LW. Colony integration and the expression of the Hox gene, Cnox-2, in Hydractinia symbiolongicarpus (Cnidaria: Hydrozoa). J Exp Zool 1999; 285:57-62.

113. Cartwright P, Bowsher J, Buss LW. Expression of a Hox gene, Cnox-2, and the division of labor in a colonial hydroid. Proc Natl Acad Sci USA 1999; 96:2183-2186.

CHAPTER 6

How Is the Branching of Animal Blood Vessels Implemented?

Sybill Patan

The Blood Circulatory System: Tree Analogy versus Network Concept

For centuries, the cardiovascular system of animals has been described as a branching tree with the heart in its very centre.[1] Although this description dates back to Galen[1] (c. 130-200 A.D.), and structural similarities with trees are obvious, it has to be emphasized that, unlike a tree, the animal blood circulation does not contain blind ending branches. The discoveries of Harvey (1578-1657) and Malpighi (1628-1694) demonstrated that the blood circulation, especially when studied at light microscopic dimensions, forms a "closed system" of circuits in which every single blood vessel is continuous with another one. Arterial vessels leave the heart, branch, and connect through a capillary network to corresponding veins that drain the blood back to the heart. It is therefore more appropriate to understand the circulatory system as a network of conductive units of varying sizes.

Given the structural similarities with branching trees, it has been suggested that the growth of the cardiovascular system follows a mode in which tiny tubules in the size range of capillaries "sprout" from existing "mother" vessels. These sprouts, which form new vascular branches, then grow in length until they meet another vessel to which they connect. This process allows for the establishment of flow between both preexisting vessels.[1-7] It is, however, still not clear how the growth of sprouts is directed towards one another so that their fusion is possible. The sprouting mechanism can thus only be of physiological benefit within a network in which the distribution of blood vessels is already quite dense. If the sprout deviates only a few degrees from its closest vessel, the chances for a successful connection are minimal. In addition, since the critical step for the elongation of sprouts is mitosis of endothelial cells, it will take time until a sprout invading an avascular region of tissue meets another vessel to establish the circulation. Only after the onset of the blood perfusion will the nutritional supply of the previously avascular compartment begin. Does an alternative mechanism exist to allow faster and more efficient expansion of the vascular network? To answer this question, we have to review current concepts of vascular morphogenesis in general.

Mechanisms of Blood Vessel Formation, Network Growth, and Remodeling; What Is Their Relationship to the Branching Process?

The cellular mechanisms that are responsible for the establishment, growth, and functional organization (remodeling) of the circulatory system in animals (a) cause the de novo formation of a network pattern and (b) are suited to expand and adapt this network to match the physiological needs of growing tissues and organs. In the embryo, a simple uniform network of tubes of similar sizes is formed initially by the mechanism of vasculogenesis. Endothelial- and blood cells differentiate in situ from mesodermal precursor cells in a pattern of islands (termed "blood

Branching Morphogenesis, edited by Jamie A. Davies.

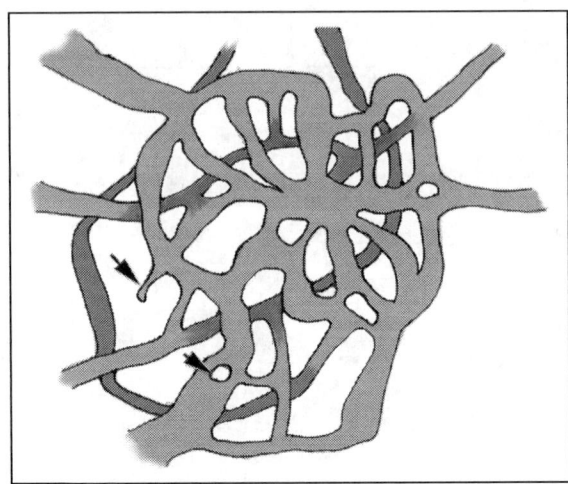

Figure 1. Sprouting- versus intussusceptive concept of vascular morphogenesis. Network expansion by insertion of a new blind-ending tubular branch (sprout, long arrow) inside a branching point (tissue between branches) by out-migration of endothelial cells from the wall of a "mother" vessel. Sprouts thus randomly divide branching points inside the network. Network expansion by insertion of a new pillar (branching point) inside an existing branch, which divides the latter into two segments (branches) according to intussusceptive microvascular growth (short arrow). The pillar has a constant distinct ultra-structure ("pillar core") to guarantee for the stabilization of the network. Branching points within the network and pillars, white areas; blood vessel lumens, grey areas.

islands"). The endothelial cells line these blood islands (the 3-dimensional organization of which is one of connected tubular branches), while the free blood cells fill their lumens. The latter will begin to circulate after the heart has started to function.[8-16]

Because primitive vascular networks grow denser and become more extended to reach so-far nonvascularized tissues and organs during subsequent development, new tubular branches have to be added to the system. This process is referred to as angiogenesis. The existence of angiogenesis implies that new branching points between these tubular segments must be established. Angiogenesis by endothelial sprouting involves migration of single endothelial cells to leave the wall of the "mother" vessel, endothelial proliferation to replace the migrating cells, and finally assembly of the latter to form new tubes, which consequently "branch out" of existing ones.[2,17] Branching points thus refer to tissue remnants between the growing and forming sprouts. On the other hand, network expansion and remodeling can also be achieved by the insertion of new branching points within the tubular branches of the network. The analysis and reconstruction of serial sections has revealed that these branching points correspond to tissue columns and possess a specific ultra-structure that is important for the stabilization of the network.[18-24] This process permits growth "within itself", i.e., it follows an intussusceptive* mechanism. One important difference between the sprouting- and the intussusceptive concept of angiogenesis is therefore the formation of branches versus branching points to accomplish vascular network growth and remodeling (Fig. 1).[21-22]

Simultaneously with its growth and expansion, the network has to remodel to achieve an adaptation to the physiological needs of the tissue that it supports. Network remodeling means the differential growth of the uniform branches formed during vasculogenesis to establish a network that contains segments of different sizes (the process is also termed "pruning"[17]) and

---

* Intussusception refers to "growth by deposition of new particles among the existing particles" (Webster's Encyclopedic Unabridged Dictionary of the English Language, New York, Portland House, 1989).

exhibits a nonuniform distribution of the latter. The addition and deletion of segments (branches) is another important characteristic of the remodeling process and is largely achieved by intussusceptive growth.[21-23] The importance of network remodeling is emphasized by the fact that the main intra-embryonic blood vessels, as the dorsal aortae and the cardinal veins, arise from plexuses established by vasculogenesis.[24] Remodeling based on intussusception occurs while vasculogenesis is still proceeding. This has been confirmed by the discovery of tiny tissue columns, the landmarks of intussusceptive growth, in the yolk sac circulation of the chicken embryo at an extremely early stage, on day E4.0, during vasculogenesis (Patan, unpublished data).

What Are the Cellular Mechanisms That Lead to the Formation of Branching Points According to Intussusceptive Microvascular Growth?

Intussusceptive microvascular growth (IMG) is a mechanism of network formation, growth and remodeling that forms an alternative concept to the sprouting model of angiogenesis. It is based on the idea that the vascular system expands by dividing the blood vessel lumen into subordinate segments (branches) through the insertion of columns of tissue, termed interstitial- or intervascular tissue structures (ITSs, diameter >2.5 µm) and tissue pillars or posts (diameter <2.5 µm).[18-23,25-32] These "tissue columns" correspond to branching points of the network, since they split the blood circulation in divergent directions. Using video microscopy we have demonstrated the existence of tissue pillars, i.e., tiny branching points (diameter often <1 µm), and their formation in vivo.[18-20,30] The analysis and reconstruction of sequential serial sections provide insight about the three dimensional network architecture as well as the pillar morphology and reveal the existence of their precursor stages. The latter technique, together with in vivo video microscopy, permits the determination of the cellular mechanisms that lead to pillar- or branching point formation and has confirmed the wide spread existence of IMG.[18-20,30-33]

New pillars or branching points can form in varying ways. Several different mechanisms have been uniformly detected in the embryo,[23,19-20] as well as in pathological states in the adult organism, as in tissue repair,[32] in cancer,[30-31] during the recanalization of thrombotic lesions,[31-32] and after myocardial infarctions (Patan et al, in prep.). Others, such as segmentation and apposition (see below) are preferentially related to adult angiogenesis and are based on the existence of intra-and extra-vascular fibrin deposits (Patan et al, in prep.).[31-32]

Pillar Formation by Folding of the Blood Vessel Wall

Pillars can separate from the tips of tissue folds that project inside the vessel lumen. Folding is initiated by retraction of the endothelial layer of the vessel wall into the adjacent tissue around the region of the future fold. The fold is lined by endothelial cells of the lateral vessel wall and contains peri-endothelial cell extensions and collagen fibers within its center. A pillar core forms within the intra-luminal tip of the fold during fold elongation. The fold does not contain organized pillar cores at its initial stages of formation. A pillar core is composed of a collagen fiber bundle that can be ensheathed by peri-endothelial cell extensions. The pillar core can also be viewed as the core of the future branching point. In a next step, the intraluminal tip of the fold that contains the pillar core separates to give rise to a free tissue pillar. The latter remains, however, connected to the fold at its bottom and its top. Pillar separation occurs based on thinning of the fold adjacent to the pillar core to form a thin cytoplasmic bridge composed of the extensions of one endothelial cell. Fusion of opposite cell membranes within this extension causes the formation of a transcellular hole that separates the pillar from its fold. This is followed by endothelial cell rearrangement to increase the distance between the pillar and the fold (Fig. 2A).[19,21-23,30-33]

Pillar Formation by Splitting of Intervascular Walls or Larger ITSs and Pillars

Intervascular walls that separate two adjacent vessels frequently contain pillar cores, which permit their splitting. This is based on thinning and merging of the endothelial layer around this core and cell membrane fusion to form two transcellular holes that separate the pillar at both of its sides. This process causes fusion of neighboring vessels (Fig. 2B, a-b).

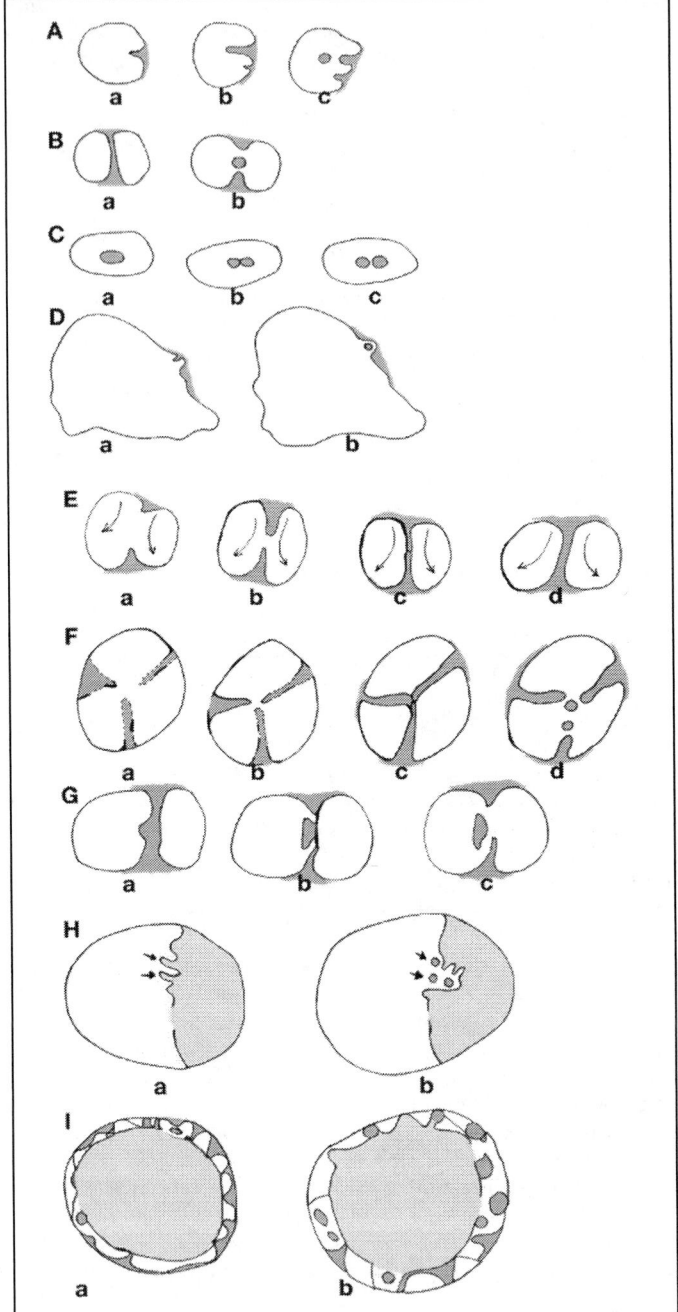

Figure 2. Different cellular mechanisms of intussusceptive microvascular growth. The vessels are depicted in cross-section, pillars (or branching points) appear thus as round spots, i.e., "islands" within the vessel lumens. The figure legend is continued on the next page.

Figure 2, continued. A) Pillar (branching point) formation based on folding of the lateral vessel wall. a: Folding is initiated by concentric retraction of the endothelial layer of the vessel wall into the adjacent tissue around the region of the future fold. b: Fold elongation. Folding can simultaneously occur within another region of the vessel wall. c: The intraluminal tip of the fold that contains the pillar core separates to give rise to a free tissue pillar surrounded by the circulation. B) Splitting of a wall that separates two vessels (intervascular wall). a: An intervascular wall can split, if it contains a pillar core. b: This process separates a free pillar and results in fusion of both neighbor vessels. B, b-a: Splitting of an intervascular wall can be reversed based on connection of pillars and folds and their insertion at opposite vessel walls. C) a-c: Following a related mechanism as illustrated in B a-b, a larger pillar or ITS can divide to give rise to two or more smaller pillars. C, c-a: The same mechanism shown in C a-c can be reversed causing the connection of two pillars or ITSs. D) a-b: A similar process as described in A causes the in-situ formation of a small elementary loop within the lumen of a larger vein. Based on this mechanism and remodelling of the elementary loop by division of its central ITS (similar to C, a-c) large systems of compound loops can form. E) Formation of the "flow pillar" a: Within a branching point of blood flow two slender tissue folds project into the lumen from opposite vessel walls, b: elongate and c: attach to each other, if the flow conditions remain un-altered. d: The connection of both opposite folds follows to form a stable tissue pillar. F) a-b: Segmentation involves the deposition of fibrin cords in the vessel lumen. Migration of single endothelial cells along these cords. c. This causes the transformation of the cords to tiny folds and their connection in a spoke-like pattern. d. Further remodelling allows the separation of tiny pillars or ITSs from these folds. G) a: Apposition leads to the formation of a larger ITS via folding of the lateral vessel wall. b: The separating ITS becomes surrounded by a larger part of the vessel lumen termed "outpouch" and spatially related to the wall of an adjacent vessel. c: The process concludes with separation of the central ITS and fusion of the outpouch with the adjacent vessel thus forming a collateral vascular segment that connects both preexisting vessels. H) A large thrombus occludes part of the lumen of a vein. a: Endothelial migration from the lateral venous wall covers the luminal surface of the thrombus and is followed by invasion of fibroblasts. This allows for formation of tiny folds from the endothelial cover layer. b: The folds give rise to pillars and ITSs, which reexpands the venous lumen inside the thrombus. I) A large thrombus occludes the entire lumen of an artery. a-b: Fibroblasts from the media layer of the arterial wall migrate into the intima which subsequently thickens to form two-to three cell layers including thrombus matrix composed of fibrin and coagulated blood. Folding, pillar and ITS-formation and fusion processes transform this circular layer into a ring of communicating vessels. Black lines = endothelial cells, grey areas = extracellular matrix containing also peri-endothelial cells (interstitial tissue), white spaces = vessel lumen.

The critical requirement for the division of ITSs or pillars is the existence of two pillar cores in one ITS or pillar. The endothelial cell extensions will then merge around these cores, followed by thinning and cell membrane fusion to form a transcellular hole that divides the pillar or ITS. This leads simultaneously to the formation of two smaller ITSs or pillars and a new vascular segment between both. The pillars and ITSs remain, however, connected at their bottoms and at their tops (Fig. 2C, a-c).[20,21-23,30-32]

Connection of Adjacent Pillars

The process of pillar or ITS division can be reversed, causing pillar connection with simultaneous deletion of the vascular segment located between them (Fig. 2C, c-a). This means that branching points between vascular segments can fuse to give rise to larger ones. It can also result in the separation of segments to form distinct vessels through the establishment of intervascular walls (Fig. 2B, b-a). The latter mechanism is of special importance in the pathological state, especially in tumors, where it can lead to fragmentation of vessel loops with the subsequent formation of blind ending tubes and vessel regression (S. Patan, in prep.).[30-32,21-22]

"In-Situ Loop Formation"

This mechanism of IMG has been detected in the mouse embryo, but also in the adult organism during tissue repair, in tumors and after myocardial infarctions, especially in humans. Vessel loops are formed in a one step process within the walls of larger veins. The endothelial layer of the vessel wall retreats towards the surrounding tissue to separate a free tissue pillar, which is encircled by a vessel loop (Fig. 2D). In-situ loop formation is thus responsible for the establishment of a new vessel loop composed of two segments that connect it to the vessel from which it derived. The separation of the tissue pillar or ITS (branching point) forming the center of the loop is a concentric process that occurs simultaneously around the entire circumference of the pillar (and results in fold formation as an initial step, compare to Fig. 2A). This guarantees that the onset of loop perfusion takes place in time to benefit the physiological requirements of the supported tissue. Further remodeling and growth of this "elementary" loop based on splitting of its central ITS (see Fig. 2C) causes the formation of loop systems and ultimately transforms a part of the large vessel into a new vascular network of connected loops.[31-32,21-23]

Pillar Formation by Connection of Opposing Intraluminal Tissue Folds

This mechanism is directly related to blood flow conditions and has been documented using in vivo microscopy and the analysis of tissue serial sections in the chicken chorio-allantoic membrane and in tumors. In those segments of the vascular network in which flow is diverted into different directions, intraluminal tissue folds derived from opposite vessel walls have a chance to connect and merge to form a new branching point. This is followed by fusion of the adjacent cell membranes of the endothelial cells that line the tips of the folds to form two trans-cellular holes. Matrix elements will invade these holes to stably connect both folds, thus establishing a tiny pillar. This process of fold connection is facilitated by the fact that the opposing folds are located precisely between both "branching" streams of flow and are subsequently pushed into the center of the lumen by the latter. If the flow conditions remain stable and do not change for a longer period of time (2-3 hours), both folds will connect. If, however, the flow directions change and cross over the region where the folds are located, connection cannot occur and the branching point will not be established (Fig. 2E). The latter case has been especially observed in tumors (S. Patan, in prep.).[30,21-22]

Segmentation

In contrast to most other mechanisms of IMG that cause the formation of new branching points, segmentation has so far been detected only during adult neovascularization. It is prominent in tissue repair and in tumors and is responsible for the expansion of a previous network towards regions of new vessel formation. In addition, it is dominant in the remodeling process of large veins and arteries after myocardial infarctions to give rise to multiple smaller segments. A prerequisite for segmentation is the deposition of fibrin in the vessel lumen; the fibrin deposits form a scaffold along which endothelial cells of the vessel wall migrate and which they line. This causes the formation of thin intraluminal folds that connect to each other in the center of the lumen in a spoke like pattern. Gradually the fibrin in the center of these folds is replaced by collagen fibers, which have a stabilization effect and facilitate the further separation of pillars and ITSs (Fig. 2F). In a pathological variation of this mechanism, the folds form valve-like structures corresponding to blind-ending pockets that are not patent and are consequently without perfusion (Patan et al, in prep.).[31-32,34]

Apposition

Apposition is a special mechanism of IMG and has so far been detected only in mouse and human myocardial infarctions. It is a variation of in-situ loop formation. However, in contrast to in-situ loop formation, apposition results in the establishment of a larger segment parallel to the vessel from which it is derived. We term this new segment an "outpouch". Outpouches form by retraction of the endothelial layer around an extra-vascular fibrin deposit which forms the core of the future ITS or branching point. The latter becomes surrounded by the expanding outpouch to form a free tissue column inside the newly established lumen. The fibrin will be replaced by collagen fibers and other matrix elements during this process. Corresponding to the architecture of the smaller loops, the outpouch has a very thin wall composed of a single endothelial layer that facilitates the connection to an adjacent vessel. This occurs by attachment of the endothelial layer that lines the outpouch to the wall of the adjacent vessel to create a common intervascular wall. Splitting of this wall follows the sequence described above (compare to Fig. 2B). Apposition has thus two important characteristics not involved in in–situ loop formation; the establishment of a new larger vascular segment, the outpouch, and its fusion with a preexisting neighboring vessel (Fig. 2G). We concluded that the outpouch might thus be connected to the "collaterals" that have been observed to connect two preexisting vessels and are specific for newly established vessels after myocardial infarctions. Apposition can occur on the venous, as well as on the arterial side of the circulation (Patan et al, in prep.).

Recanalization of Thrombotic Lesions

Recanaliztion of thrombotic lesions that are located in large arteries and veins has been detected in a model of tissue repair and in experimentally grown tumors utilizing the same model system.[31-32] In the case of venous recanalization, the thrombus filled the vessel lumen only partially. Therefore the endothelial layer of the lateral vessel wall was able to grow over its free intraluminal surface, followed by in-migration of underlying fibroblasts. This process probably attached the thrombus to the vessel wall and stabilized it by transformation of its fibrin-rich matrix into collagen fibres. In a second step, the endothelial layer at the thrombus surface retreated toward its center to form folds, as well as ITS- and pillar cores composed of fibrin and collagen fibers (Fig. 2H, a). This was followed by separation of ITSs and pillars from these folds and subsequent establishment of vascular segments inside the thrombus matrix (Fig. 2H, b). The latter remodeled to give rise to a new vascular network, which allowed for reperfusion of the region of the former stenosis.[32]

Arterial recanalization followed a different mechanism of IMG, since the thrombus had usually filled the vessel lumen entirely. The endothelial cells of the intima layer of the vessel wall largely retained their original position as a circular lining. At some places, however, the intima ruptured, allowing for blood deposits to undermine it and for subsequent invasion of fibroblasts from the underlying media layer. This caused the intima to thicken and form two- to three cell layers. In places, intimal endothelial cells aligned themselves around the fibrin deposits, coagulated blood and collagen fibers that were ensheathed by fibroblast extensions. This process gave rise to pillar cores and intervascular walls (Fig. 2I, a-b). The latter subsequently formed future branching points between vessel segments and thus transformed the circular layer of cells and matrix into a ring of communicating vessels. Blood flow was subsequently reestablished in the former area of occlusion.[31]

What Are the Advantages of IMG versus Sprouting in the Expansion and Remodeling of the Circulatory Network?

There are several major advantages of adding new segments to the network by dividing vessel lumens based on the insertion of new branching points compared to the addition of new sprouts (branches). First, dividing the lumen based on insertion of a column of tissue can form daughter segments of different sizes, depending on the size and position of the core of the pillar or ITS. In contrast, a newly developing sprout is always very small, about the size of a capillary. Secondly, the tissue pillar or ITS forms while the vessel is still being perfused. The sprout, however, needs to grow, elongate, and fuse with another vessel before its perfusion is established. This implies that inserting tissue columns to modify the network pattern has an immediate physiological benefit. Finally, vessel wall maturation provides the "naked" sprout with a layer of peri-endothelial support cells. Since lumen division based on IMG often occurs in vessels that already possess a fully- matured wall, this step is frequently not necessary.

Pathological Variations of the Cellular Mechanisms of IMG

In addition to the physiological mechanisms of IMG that are listed above, intussusception can follow pathological variations in the adult organism, in tumor angiogenesis and after myocardial infarctions, but apparently not in tissue repair.[31-32] Many of these pathological variations lead to the formation of blind-ending segments. The latter cause the fragmentation of the network and thus impair the oxygenation of the tissue. This explains the hypoxia that is a characteristic of tumors and the post-infarction myocardium. High and persistent levels of hypoxia contribute to the perpetuation of insufficient angiogenesis in these pathological conditions, which can be deduced from the continuous release of angiogenic growth factors, such as VEGF, FGF-2 and hypoxia inducible factor α (HIF-α).[31-32,35,21-22] Their local concentration inside tumors is assumed to be higher than the one of secreted inhibitors, so that angiogenesis dominates.[36]

Two different pathological features of IMG have been detected in tumors and to some extent also during neovascularization process after myocardial infarctions:

1. Intravital microscopy of the circulation of colon adenocarcinomas transplanted to the skin of nude mice demonstrated that pillars in tumors can form more frequently, but are continuously and rapidly remodeled, subsequently reversing pillar formation through its subsequent deletion. This makes the process of vascularization highly ineffective and probably even increases the oxygen consumption of the local tissue. It also causes the permanent alteration of blood flow on a time-scale of minutes, which is known as "intermittent blood flow" and is a characteristic of the tumor circulation.[30]

2. Pathological variations of the mechanisms of IMG detected in tumors and after myocardial infaction interfere with the formation of regular folds, pillars and ITSs and thus impair the deposition of normal branching points inside the network. A common feature of these pathological variations is the formation of blind-ending tubes. One example of pathological IMG, especially of in-situ loop formation, is the late separation of the central ITS or branching point in the center of the loop when the latter has already reached large dimensions. Another pathological mechanism refers to the secondary disconnection of loop segments, which interferes with the patency of the loop or loop system.[31] Disturbed mechanisms of IMG also include the formation of intervascular walls in tumors that occlude vessel segments, even resulting in vessel regression (Patan et al, in prep.).[30]

What Causes the Existence of Pathological Mechanisms of IMG and Subsequently Disturbed Branching Point Formation?

Since the pattern of gene expression in tumor endothelial cells is very similar to that in wound healing,[37] it can be assumed that endothelial cells are not responsible for the disturbed mechanisms of IMG. Thus the best available explanation is the existence of nonendothelial, fibroblast-like cells that are incorporated into vessel walls during IMG to replace endothelial cells. Under tissue repair conditions these cells proliferate strongly and express endothelial-specific markers, such as platelet-endothelial adhesion molecule (PECAM/CD31), which indicates that they might differentiate to form endothelial cells.[32] In tumors these fibroblast-like cells are less common.[31] The vessel-wall-invading tumor cells are obviously substituting for these fibroblast-like cells to form "mosaic vessels", in which the endothelial layer is composed of both endothelial-derived and tumor-derived cells.[38-40] Although tumor cells mimic the endothelial cell morphology and thus hide in the vessel wall, they exhibit a pattern of gene expression that is different to the one of endothelial cells that does not include endothelial specific markers (S. Patan, in prep.). This will ultimately interfere with the normal mechanisms of IMG. Therefore it cannot be expected that mosaic vessels give rise to regular tissue folds, pillars, and ITSs.

How Is the Formation of Branching Points at the Molecular Level Regulated?

The currently known pro-angiogenic molecules have been characterized by their effect on endothelial cell proliferation, migration and tube formation. The most prominent ones of these growth factors are fibroblast growth factor 1 and 2 (FGF-1, acidic and FGF-2, basic), platelet-derived growth factor (PDGF), hepatocyte growth factor (HGF, scatter factor), vascular endothelial growth factor/vascular permeability factor (VEGF-A, VEGF-B,

VEGF-C), transforming growth factor alpha (TGF-α), and interleukin 8 (IL-8).[41-44] Additionally, an important role in angiogenesis has been established for the Tie/Angiopoietin and the Eph-B/ephrin-B system of tyrosine kinase receptors and their ligands.[23,33,45-50]

It has been shown that the heparin-binding isoform of VEGF-A is especially responsible for vascular branching. Transgenic mice that solely expressed the VEGF-A isoform, which lacks heparin binding (VEGF 120/120), exhibited larger vessels with more endothelial cells but fewer vessel branches as compared to wild-type littermates.[51]

Concerning the formation of pillars at branching points of the circulation (see Fig. 2E), it is known that the alteration of shear stress profiles can cause the expression of angiogenic growth- and transcription factors by endothelial cells; one example is platelet-derived growth factor B (PDGF-B). The receptors for these growth factors are up-regulated in peri-vascular cells of the vessel wall.[52-54] Endothelial cells that are exposed to large shear stress gradients in vitro respond with increased cell division and motility in the vicinity of flow separation.[54] It can be expected that the shear stress profile is high in these areas of flow divergence, while it is low within the branching point itself. Furthermore it has been recently shown that VEGF receptor 2 (VEGFR-2, Flk-1) can be activated by fluid shear stress in a ligand-independent manner, which causes further downstream-activation of endothelial nitric oxide synthase (eNOS).[55] Since eNOS produces nitric oxide (NO) as a response to shear stress [56-57] and NO induces neovascularization,[58-60] a direct link between shear stress and its influence on vascular morphogenesis through a growth factor- receptor independent signaling is established.

Recently, it has been demonstrated that the Angiopoietin-1 (Ang-1)/Tie-2 growth factor receptor system regulates embryonic IMG.[23,33,45-47,50] The ligand, Angiopoietin-1 (Ang-1), is expressed by peri-vascular fibroblasts and mesenchymal cells in the embryo[23] and by pericytes in wound healing.[61] Endothelial cells in the vessel wall express the Tie-2 tyrosine kinase receptor in a paracrine manner.[23] In the embryo, the Ang-1/Tie-2 system is responsible, in cooperation with VEGF, for the recruitment of peri-endothelial cells to the endothelial layer. It also facilitates the interaction between endothelial cells and the extra-cellular matrix to promote cell stretching and motility. The Ang-1/Tie-2 system is critical for both functions, as has been demonstrated in Ang-1- and Tie-2 deficient embryonic mice.[23,33] As is obvious from the analysis and reconstruction of tissue folds, pillars, and ITSs, all cellular mechanisms of IMG depicted in Figure 2 can occur with contribution of peri-endothelial cells to form stable pillar cores. This means that the establishment of regular tissue folds and stable pillars and ITSs is dependent on Ang-1/Tie-2. This fact has been confirmed by the analysis of mice deficient of either the ligand or the receptor. Ang-1[-/-] or Tie-2[-/-] mice exhibited defects in vascular network growth and remodeling resulting in an abnormal network pattern and death of the embryo.[23,33]

In this respect it is important to note that a recent analysis of embryonic mice deficient of the Tie-1 tyrosine kinase receptor confirmed its role as an inhibitor of IMG. Tie-1 (the ligand for which is still unknown) appears to negatively regulate endothelial cell motility (through interaction with the extracellular matrix) as well as the ability of these cells to form transcellular holes. Consequently, the comparison to wild type embryos demonstrated that Tie-1 deficient mice exhibited increased vessel numbers and "hyperactive" endothelial cells. The latter appeared overstretched, and possessed a large number of filopodia and transcellular holes.[23,50]

FGF-2 facilitates regular neo-vascularization after myocardial infarctions by interfering with pathological mechanisms of IMG. In experimental myocardial infarctions in adult mice (in which the left coronary artery is ligated), the one-time intramyocardial injection of FGF-2 close to the suture increases additionally the number of normal vessels in the transitional zone between healthy myocardium and scar tissue (Patan et al, in prep.).

On contrary the Thromboxane receptor (TP) that has a function in blood coagulation is responsible for a strong inhibition of neo-vascularization after experimental infarctions. As demonstrated by the analysis of TP-deficient transgenic mice, the absence of TP causes an extremely strong and long lasting vascularization response in the scar tissue based on increased pillar formation (Patan et al, in prep.).

Interestingly, the analysis of transgenic mice deficient of another tyrosine kinase receptor family, the Eph/B receptors and their ligands, the ephrins, has demonstrated functions in embryonic angiogenesis that correspond to the one of the Ang-1/Tie-2 system.[48-49] In this case ligand and receptor are attached to the cells and are differentially expressed on arterial (the ligand ephrin-B2) and venous (the receptor Eph-B4) endothelial cells.[48] Ephrin-B2 is interestingly also expressed on peri-vascular cells.[49,62] Ligand and receptor that engage in bi-directional signaling are subsequently also responsible for the coordinated interaction of arterial- and venous endothelial cells.[62-63]

The Notch gene family encodes large transmembrane receptors that are involved in intercellular signaling. Notch1 and Notch4 are expressed in endothelial cells in the embryonic vasculature.[64] Notch1[-/-] and Notch1/Notch4 double mutant embryos displayed severe defects of embryonic vascular remodeling corresponding to ones detected in Ang-1-, Tie-2-, ephrin-B2 deficient embryos.[65]

The analysis of transgenic mice deficient of connexin 45 (Cx45) demonstrates a similar phenotype concerning embryonic angiogenesis. Not surprisingly, Cx45, a gap junction protein, is involved in the communication between endothelial-, as well as endothelial- and peri-endothelial cells.[66]

RECK, a membrane anchored glycoprotein, was found to be essential for mouse vascular remodeling in the embryo. It inhibits three matrix metalloproteinases (MMPs: MMP-9, MMP-2, and MT1-MMP)[67] and is widely expressed in mesenchymal and vascular smooth muscle cells. RECK deficiency caused increased proteolysis of collagen 1 and defects in the basal lamina with subsequent failure of remodeling of the primitive vascular plexus to form a mature one.[68] Since collagen fibers (collagen type 1) are an indispensable component of every pillar- or ITS core, the important role of RECK for the regulation of the balance between synthesis and proteolysis of collagen can be estimated. This guarantees the sufficient formation of branching points.

In an attempt to investigate the role of the cytoskeleton and Rho GTPases in the process of branching morphogenesis, small GTPase Rac was found to be responsible for matrix induced changes in endothelial cell morphology, while p21-activated kinase, an effector of Rac, was required for cell motility.[69] Correspondingly WAVE2, a protein preferentially expressed in endothelial cells during embryogenesis is responsible for Rac-induced membrane ruffling as it is necessary for changes of cell morphology. Embryos lacking WAVE2 exhibit an increased number of transcellular holes corresponding to Tie-1[-/-] embryos. In contrast to Tie-1 deficient embryos, WAVE2[-/-] mice possess fewer endothelial filopodia and reduced vessel numbers compared to wildtype littermates.[70]

Future work should identify the interaction and precise function of these molecular regulators, which are transformed into a sequence of cues to regulate each cellular mechanism of branching morphogenesis.

References

1. Galen, De foetuum formatione. Harris CRS. The heart and the vascular system in ancient Greek medicine. Oxford: Clarendon Press, 1973.
2. Fülleborn F. Beiträge zur Entwicklung der Allantois der Vögel. Inaug Diss. Berlin: Francke 1895.
3. Danchakoff V. The position of the respiratory vascular net in the allantois of the chick. Am J Anat 1917; 21:407-420.
4. Clark ER. Studies on the growth of blood vessels, by observation of living tadpoles and by experiments on chicken embryos. Anat Rec 1915; 9:17-68.
5. Clark ER. Studies on the growth of blood-vessels in the tail of the frog larva – by observation and experiment on the living animal. Am J Anat 1918; 23:37-88.
6. Clark ER, Clark EL. Microscopic observations on the growth of blood capillaries in the living mammal. Am J Anat 1939; 64:251-299.
7. Ausprunk D, Folkman J. Migration and proliferation of endothelial cells in preformed and newly formed blood vessels during tumor angiogenesis. Microvasc Res 1977; 14 53-65.
8. Sabin FR. Origin and development of the primitive vessels of the chick and of the pig. Contrib Embryol Carnegie Inst Publ Wash 1917; 6:61-124.

9. Sabin FR. Studies on the origin of blood-vessels and of red blood-corpuscles as seen in the living blastoderm of chicks during the second day of incubation. Contrib Embryol Carnegie Inst Publ Wash 1920; 36:213-259.

10. Pardanaud L, Altmann C, Kitos P et al. Vasculogenesis in the early quail blastodisc as studied with a monoclonal antibody recognizing endothelial cells. Development 1987; 100:339-349.

11. Pardanaud L, Yassine F, Dieterlen-Lièvre F. Relationship between vasculogenesis, angiogenesis and hematopoiesis during avian ontogeny. Development 1989; 105:473-485.

12. Noden DM. The formation of avian embryonic blood vessels. Am Rev Respir Dis 1989; 140:1097-1103.

13. Poole TJ, Coffin JD. Vasculogenesis and angiogenesis: Two distinct morphogenetic mechanisms establish the embryonic vascular pattern. J Exp Zool 1989; 251:224-231.

14. Risau W. Vasculogenesis, angiogenesis and endothelial cell differentiation during embryonic development. In: Feinberg RN, Sherer GK, Auerbach R, eds. The development of the vascular system. Issues Biomed 14. Basel: Karger 1991:58-68.

15. Poole TJ, Coffin D. Morphogenetic mechanisms in avian vascular development. In: Feinberg RN, Sherer GK, Auerbach R, eds. The development of the vascular system. Issues Biomed 14, Basel: Karger 1991:25-36.

16. Risau W, Flamme I. Vasculogenesis. Annu Rev Cell Dev Biol 1995; 11:73-91.

17. Risau W. Mechanisms of angiogenesis. Nature 1997; 386:671-674.

18. Patan S, Haenni B, Burri PH. Evidence for intussusceptive capillary growth in the chicken chorio-allantoic membrane (CAM). Anat Embryol 1993; 187:121-130.

19. Patan S, Heanni B, Burri PH. Implementation of intussusceptive microvascular growth in the chicken chorio-allantoic membrane (CAM): 1. Pillar formation by folding of the capillary wall. Microvasc Res 1996; 51:80-98.

20. Patan S, Haenni B, Burri PH. Implementation of intussusceptive microvascular growth in the chicken chorio-allantoic membrane (CAM): 2. Pillar formation by capillary fusion. Microvasc Res 1997; 53:33-52.

21. Patan, S. Vasculogenesis and angiogenesis as mechanisms of vascular network formation, growth and remodeling. J Neuro-Oncol 2000; 50:1-15.

22. Patan S. Vasculogenesis and angiogenesis. In: Black P, Kirsch M, eds. Angiogenesis in brain tumors. Cancer Treatment and Research, Boston: Kluwer Academic Publishers, 2004:3-32.

23. Patan S. TIE1 and TIE2 receptor tyrosine kinases inversely regulate embryonic angiogenesis by the mechanism of intussusceptive microvascular growth. Microvasc Res 1998; 56:1-21.

24. Gilbert SG. Pictorial human embryology. Seattle and London: University of Washington Press, 1989.

25. Short RHD. Alveolar epithelium in relation to growth of the lung. Philos Trans R Soc London Ser B 1950; 235:35-87.

26. Caduff JH, Fischer LC, Burri PH. Scanning electron microscope study of the developing microvasculature in the postnatal rat lung. Anat Rec 1986; 216:154-164.

27. Burri PH, Tarek MR. A novel mechanism of capillary growth in the rat pulmonary microcirculation. Anat Rec 1990; 228:35-45.

28. Van Groningen JP, Wenink ACG, Testers LHM. Myocardial capillaries: Increase in number by splitting of existing vessels. Anat Embryol 1991; 184:65-70.

29. Patan S, Alvarez MJ, Schittny JC et al. Intussusceptive microvascular growth: Common alternative to endothelial sprouting. Arch Histol Cytol 1992; 55(Suppl):65-75.

30. Patan S, Munn LL, Jain RK. Intussusceptive microvascular growth in a human colon adenocarcinoma xenograft: A novel mechanism of tumor angiogenesis. Microvasc Res 1996; 51:260-272.

31. Patan S, Tanda S, Roberge S et al. Vascular morphogenesis and remodeling in a human tumor xenograft. Blood vessel formation and growth after ovariectomy and tumor implantation. Circ Res 2001; 89:732-739.

32. Patan S, Munn LL, Tanda S et al. Vascular morphogenesis and remodeling in a model of tissue repair. Blood vessel formation and growth in the ovarian pedicle after ovariectomy. Circ Res 2001; 89:723-731.

33. Suri C, Jones PF, Patan S et al. Requisite role of Angiopoietin-1, a ligand for the TIE2 receptor during embryonic angiogenesis. Cell 1996; 87:1171-1180.

34. Nagy JA, Morgan ES, Herzberg KT et al. Pathogenesis of ascites tumor growth: Angiogenesis, vascular remodeling, and stroma formation in the peritoneal lining. Cancer Res 1995; 55:376-85.

35. Dvorak HF. Tumors: Wounds that do not heal. Similarities between tumor stroma generation and wound healing. N Engl J Med 1986; 315:1650-1659.

36. Folkman J. Angiogenesis in cancer, vascular, rheumatoid and other disease. Nature Med 1995; 1:27-31.

37. St. Croix B, Rago C, Velculescu V et al. Genes expressed in human tumor endothelium. Science 2000; 289:1197-1202.
38. Warren BA, Shubik P. The growth of the blood supply to melanoma transplants in the hamster cheek pouch chamber. Lab Invest 1966; 15:464-478.
39. Hammersen F, Osterkamp-Baust U, Endrich B. Ein Beitrag zum Feinbau terminaler Strombahnen und ihrer Entstehung in bösartigen Tumoren. Mikrozirk Forsch Klin 1983; 2:15-51.
40. Hammersen F, Endrich B, Messmer K. The fine structure of tumor blood vessels. I. Participation of nonendothelial cells in tumor angiogenesis. Int J Microcirc Clin Exp 1985; 4:31-43.
41. Ware JA, Simons M. Angiogenesis in ischemic heart disease. Nature Med 1997; 3:158-164.
42. Fernandez B, Buehler A, Wolfram S et al. Transgenic myocardial overexpression of fibroblast growth factor-1 increases coronary artery density and branching. Circ Res 2000; 87:176-178.
43. Xin X, Yang S, Ingle G et al. Hepatocyte growth factor enhances vascular endothelial growth factor induced angiogenesis in vitro and in vivo. Am J Pathol 2001; 158:1111-1120.
44. Hellstrom M, Gerhardt H, Kalen M et al. Lack of pericytes leads to endothelial hyperplasia and abnormal vascular morphogenesis. J Cell Biol 2001; 153:543-553.
45. Sato TN, Quin Y, Kozak CA et al. tie-1 and tie-2 define another class of putative receptor tyrosine kinase genes expressed in early embryonic vascular system. Proc Natl Acad Sci USA 1993; 90:9355-9358.
46. Dumont DJ, Gradwohl G, Fong GH et al. Dominant-negative and targeted null mutations in the endothelial receptor tyrosine kinase, tek, reveal a critical role in vasculogenesis of the embryo. Genes Dev 1994; 8:1897-1909.
47. Davis S, Aldrich TH, Jones PF et al. Isolation of angiopoietin-1, a ligand for the angiogenic TIE2 receptor, by secretion-trap expression cloning. Cell 1996; 87:1161-1169.
48. Wang HU, Chen CF, Anderson DJ. Molecular distinction and angiogenic interaction of embryonic arteries and veins revealed by ephrin-B2 and its receptor Eph-B4. Cell 1998; 93:741-753.
49. Adams RH, Wilkinson GA, Weiss C et al. Roles of ephrin-B ligands and EphB receptors in cardiovascular development: Demarcation of arterial/venous domains, vascular morphogenesis, and sprouting angiogenesis. Genes Dev 1999; 3:295-306.
50. Marron MB, Hughes DP, Edge MD et al. Evidence for heterotypic interaction between the receptor tyrosine kinases Tie-1 and Tie-2. J Biol Chem 2000; 275:39741-39746.
51. Ruhrberg C, Gerhardt H, Golding M et al. Spatially restricted patterning cues provided by haparin-binding VEGF-A control blood vessel branching morphogenesis. Genes Dev 2002; 16:2684-2698.
52. Tardy Y, Resnick N, Nagel T et al. Shear stress gradients remodel endothelial monolayers in vitro via a cell proliferation-migration-loss cycle. Arterioscler Thromb Vasc Biol 1997; 17:3102-3106.
53. Sumpio BE, Du W, Galagher G et al. Regulation of PDGF-B in endothelial cells exposed to cyclic strain. Arterioscler Thromb Vasc Biol 1998; 18:349-355.
54. Nagel T, Resnick N, Dewey Jr CF et al. Vascular endothelial cells respond to spatial gradients in fluid shear stress by enhanced activation of transcription factors. Arterioscler Thromb Vasc Biol 1999; 19:1825-1834.
55. Jin Z-G, Ueba H, Tanimoto T et al. Ligand-independent activation of vascular endothelial growth factor receptor 2 by fluid shear stress regulates activation of endothelial nitric oxide synthase. Circ Res 2003; 93:354-363.
56. Boo YC, Jo H. Flow-dependent regulation of endothgelial nitric oxide synthase: Role of protein kinases. Am J Physiol Cell Physiol 2003; 285:C4999-C508.
57. Rizzo V, Morton C, DePola N et al. Recruitment of endothelial caveolae into mechanotransduction pathways by flow-conditioning in vitro. Am J Physiol Heart Circ Physiol 2003; 285:H1720-H1729.
58. Kawasaki K, Smith Jr RS, Hsieh CM et al. Activation of the phosphatidylinositol 3-kinase/protein kinase Akt pathway mediates nitric oxide-induced endothelial cell migration and angiogenesis. Mol Cell Biol 2003; 23:5726-5737.
59. Schwentker A, Billiar TR. Nitric oxide and wound repair. Surg Clin North Am 2003: 83:521-530.
60. Lin Z, Chen S, Ye C et al. Nitric oxide synthase expression in human bladder cancer and its relation to angiogenesis. Urol Res 2003; 31:232-235.
61. Sundberg C, Kowanetz M, Brown LF et al. Stable expression of angiopoietin-1 and other markers by cultured pericytes: Phenotypic similarities to a subpopulation of cells in maturing vessels during later stages of angiogenesis in vivo. Lab Invest 2002; 82:387-401.
62. Gerety SS, Anderson DJ. Cardiovascular ephrin B2 function is essential for embryonic angiogenesis. Development 2002; 129:1397-1410.
63. Davis S, Gale NW, Aldrich TH et al. Ligands for EPH-related receptor tyrosine kinases that require. membrane attachment or clustering for activity. Science 1994; 266:816-819.

64. Franco del Amo F, Smith DE, Swiatek PJ et al. Expression of Motch, a mouse homologue of Drosophila Notch, suggests an important role in early postimplantation mouse development. Development 1992; 115:737-745.
65. Krebs LT, Xue Y, Norton CR et al. Notch signaling is essential for vascular morphogenesis in mice. Genes Dev 2000; 14:1343-1352.
66. Krüger O, Plum A, Kim J-S et al. Defective vascular development in connexin 45 deficient mice. Development 2000; 127:4179-4193.
67. Chang C, Werb Z. The many faces of metalloproteases: Cell growth, invasion, angiogenesis and metastasis. Trends Cell Biol 2001; 11:S37-S43.
68. Oh J, Takahashi R, Kondo S et al. The membrane-anchored MMP inhibitor RECK is a key regulator of extracellular matrix integrity and angiogenesis. Cell 2001; 107:789-800.
69. Connolly JO, Simpson N, Hewlett L et al. Rac regulates endothelial morphogenesis and capillary assembly. Mol Biol Cell 2002; 13:2474-2485.
70. Yamazaki D, Suetsugu S, Miki H et al. WAVE2 is required for directed cell migration and cardio-vascular development. Nature 2003; 424:452-456.

CHAPTER 7

Extracellular Matrix Remodeling in Mammary Gland Branching Morphogenesis and Breast Cancer:
The Double-Edged Sword

Eva A. Turley and Mina J. Bissell

Introduction

That differentiation and malignancy are different faces of the same coin is now almost a cliché.[1-4] Although widely accepted as fact, exactly what are the points of similarity and differences that contribute to normal morphogenesis on the one hand and to neoplastic progression on the other? How can mechanisms that permit, guide and determine differentiation also contribute to malignancy? More specifically, what are the molecules that guide normal morphogenesis yet contribute to neoplastic transformation and progression? These processes probably involve arrays of genetic programs. For the purpose of this review, we will focus on the roles of several genes that appear to fill these contradictory functions.

Breast tissue morphogenesis is unusual in mammals in that it occurs in the adult and is coupled to the periodicity of mammalian reproductive and pregnancy cycles (Fig. 1).[2-8] During the estrous cycle, and to a greater extent during pregnancy and following parturition, breast tissue branching morphogenetic programs dominate over those promoting differentiation or apoptosis and culminate in the expansion of ducts resulting from regulated growth, migration and invasion into the fat pad. Branching morphogenesis is followed in pregnancy by differentiation of ductal cells into lobular alveolar epithelium that produces milk after parturition. This in turn is followed by extensive extracellular matrix remodeling that occurs concomitantly with tissue involution resulting from the regulated apoptosis of lobular alveolar breast epithelium in concert with the proliferation of adipocytes that occurs after weaning. These morphogenetic/differentiation/involution cycles are repeated throughout the reproductive life span of female mammals. The cyclically regenerative capability of normal adult breast tissue is clearly substantial and is currently thought to result from both hierarchies of stem/progenitor cells within the luminal epithelial cell population and myoepithelial cells lining the ducts. These cell types have marked regenerative abilities and can, for instance, give rise to entire mammary trees when they are transplanted into cleared mammary fat pads.[9-12] Progenitor cell types must also survive involution to participate in the future expansion of ducts during subsequent estrous cycles and pregnancies. We will address potential mechanisms that make them resistant to apoptosis-inducing properties associated with a remodeling ("reactive") stroma, and since evidence suggests that breast stem/progenitor cell populations with regenerative capabilities may give rise to highly aggressive tumorigenic cell subsets within breast tumors,[5,9,10,13] how these properties might be utilized by transformed cells to survive and to proliferate.[10,14-19]

Branching Morphogenesis, edited by Jamie A. Davies.
©2005 Eurekah.com and Springer Science+Business Media.

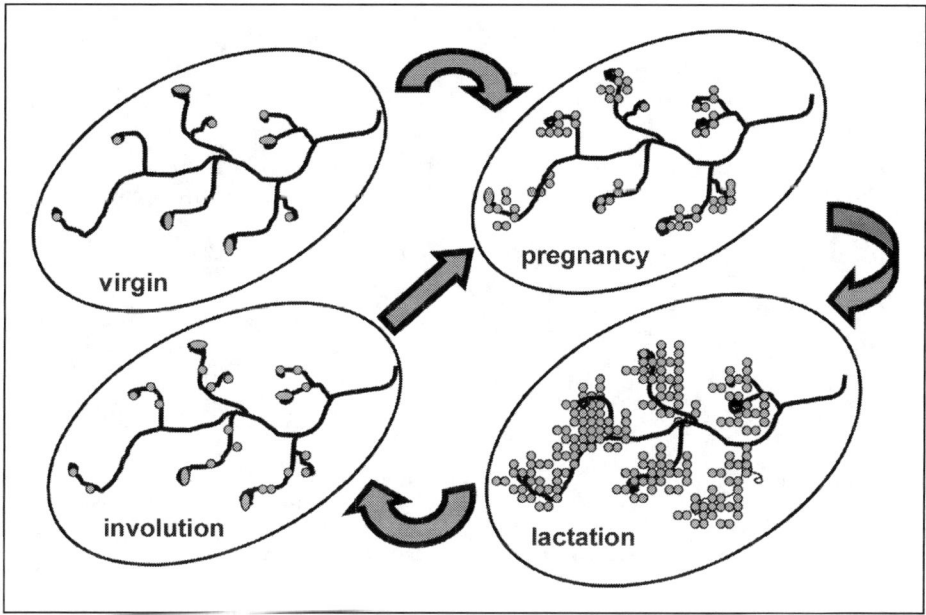

Figure 1. The branching morphogenesis cycles of the mammary gland. Diagrams show the extent of branching that occurs in virgin, pregnant, lactating and involuting mammary glands. In subsequent pregnancies, branching of the mammary glands increases from the involuted state to that typical for pregnancy and lactation, returning to the involuted state again following weaning.

The cyclical morphogenesis of breast tissue results from paracrine morpho-regulatory programs that occur in the context of a clear division of labor amongst the cell types that participate in this program and which include three distinct phenotypes: luminal epithelium, myoepithelium and mesenchyme (Fig. 2). These cell types coordinate branching by two distinct processes: bifurcation of end buds and side branching from the primary ducts.[20,21] Morphogenetic programs are not only spatially regulated but are also temporally confined.[22,23] For example, mesenchymal or stromal cells produce morphogens such as bFGF, HGF, epimorphin, MMPs and proteoglycans that control the expression of gene clusters in luminal epithelial cells involved in coordinating tubule formation and branching.[4,24-27] In addition to their role in cyclical regeneration of the mammary tree, myoepithelial cells provide contractile, inductive and proliferation/tumor suppressive functions in the normal mammary gland.[9-12,28,29] Adipocytes support mammary gland morphogenesis although their inductive and metabolic role(s), while present,[29] have not been as well dissected as those of myoepithelial and stromal fibroblasts.[20,29-31] Disruption of the morpho-regulatory programs that regulate branching results in apoptosis of ductal epithelium that is destined to become lobular alveolar cells.[3,32,33] Selective pressure on such normal cells is therefore towards achieving differentiation. It follows that at least one essential event in neoplastic conversion has to be a resistance to apoptosis, such as must be exhibited by stems cells during involution or must be achieved by transformed epithelial cells.

The Stroma, Branching Morphogenesis, and Breast Cancer

A considerable literature indicates that stromal extracellular matrix components are major players in determining resistance to apoptosis, and regulation of gene sets that control cell proliferation, migration, and invasion during both branching morphogenesis and tumorigenesis. We define this conundrum of a double-edged sword as follows: Those ECM components

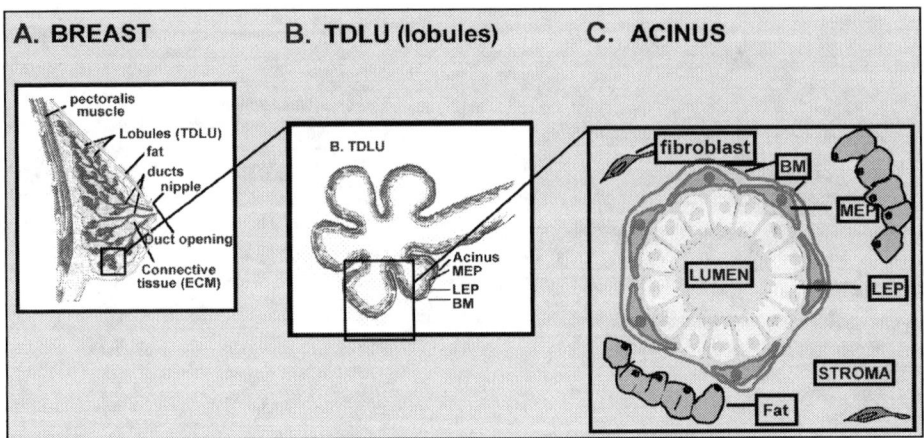

Figure 2. Structure of the human breast. The diagram shows a cross-section of a typical mammary end bud of a mature virgin gland that contains luminal breast epithelial cells, and myo-epithelial cells that are surrounded by a basal lamina. The end bud is incased by stromal fibroblasts and by fat cells. through a mammary duct that includes the different cell lineages that make up the differentiated mammary gland.

that permit cyclical breast morphogenesis and ultimately enable our survival as a species may also predispose us to the risk of breast cancer. The identification of key genes involved in, or associated with ECM remodeling, that are commonly expressed during normal morphogenesis and during neoplastic conversion, can reasonably be expected to yield important markers and possible therapeutic targets for detecting and suppressing breast cancer progression. A number of factors have been identified that commonly regulate breast branching morphogenesis and cancer initiation/progression. Many of these have been well reviewed, and some very recently.[2,3,20,34-36] This review therefore summarizes our current knowledge of how two distinct groups of stromal factors, proteoglycans/glycosaminoglycans (GAG) and metalloproteinases (MMPs), contribute to mammary branching morphogenesis on the one hand and neoplastic conversion/progression on the other hand. These particular stromal factors are functionally and physically interconnected, and they provide an excellent example of how morpho-regulatory stromal factors can interact to coordinate collectively signaling pathways in epithelium necessary for branching morphogenesis, and how changes in the regulation of these associations sustain neoplastic properties of breast epithelium.

Proteoglycans/GAGs As Regulators of Branching Morphogenesis

Hyaluronan

Here, we will review in detail the roles of hyaluronan (HA), a stromal glycosaminoglycan (GAG), and CD44, an HA receptor that is also expressed as a proteoglycan in branching morphogenesis. CD44 also performs docking functions for growth factors such as erbB4 and MMPs such as MMP-9 (gelatinase B), MMP-7 (matrilysin), and MMP-14 (MT1-MMP).[37-40] CD44 thus links signaling pathways regulated by HA/proteoglycans to those regulated by MMPs and growth factors. This integration is essential for efficient branching morphogenesis and appears to be involved also in neoplastic transformation/progression.[37,39,41] Both HA[42-44] and MMPs, such as MMP-3 (stromelysin-1) and MMP-7,[20,45-49] exert effects on mammary stromal tissue that may promote a reliance of breast cells on CD44-mediated signaling for resistance to apoptosis,[50] and that may also predispose to transformation or offer a growth advantage once cells are transformed.[15] Therefore, this group of molecules provides an excellent paradigm for examining the assumptions underlying our concept of a double-edged sword.

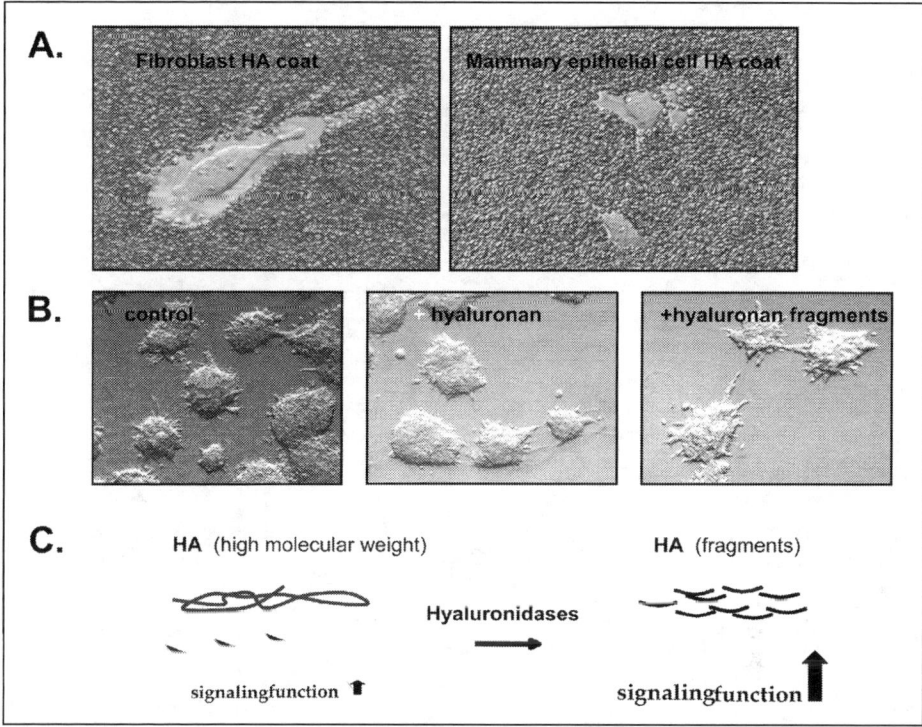

Figure 3. Hyaluronan and breast epithelial cells (EPH4) in vitro. A) Coats of hyaluronan (HA) surround fibroblasts and mammary epithlelial cells (EPH4) and are visualized with particle exclusion. HA is normally produced primarily by the stroma in quiescent mammary glands[68] and normal fibroblasts produce more HA than normal mammary epithelial cells. Magnification X200. Mammary epithelial cell aggregates (EPH4) produce branching structures in response to EGF alone (control) but branching is enhanced when HA fragments (+hyaluronan fragments) are combined with EGF. High molecular weight HA (+ hyaluronan) does not affect EGF-mediated branching. Magnification X120 HA is "activated" to a signaling mode when fragmented. The studies shown in (B) and other reports (see text) suggest that fragmented HA is more active than high molecular weight HA in promoting cell signaling that results, for example, in branching in vitro (B)

GAGs are a class of polysaccharides that typically comprise unbranched chains of repeating units of glucuronic acid and amino sugars. GAG chains rarely exist as free polysaccharides but are covalently attached to proteins forming proteoglycans such as syndecan, perlecan and versican.[51-53] HA, which is composed of disaccharide units of B-glucuronic acid and N-acetyl-glucosamine, is unique in this family of polysaccharides because it lacks sulfated residues, it is rarely covalently attached to proteins (therefore rarely exists as a proteoglycan), and can exceed 10^6 Daltons in mass.[43] Additionally, HA is uniquely synthesized at the plasma membrane by one of three synthase isoforms (HAS 1, 2, 3)[54-58] in contrast to other GAGs which are synthesized in the golgi apparatus.[59,60] A growing nascent HA chain is extruded through the plasma membrane by as yet unknown mechanisms but possibly one that involves oligomerization of the synthase itself.[55,57] HA is retained in the extracellular matrix (ECM) by binding to proteins such as versican and aggrecan, and HA associates with and coats both stromal and epithelial cells by binding to specific cell surface receptors such as CD44, RHAMM, LYVE-1/CSRSBP-1 and layillin;[37,43] this can be visualized in vitro using particle exclusion assays (Fig. 3A), Several recent reviews summarize the mechanisms by which HA binds to

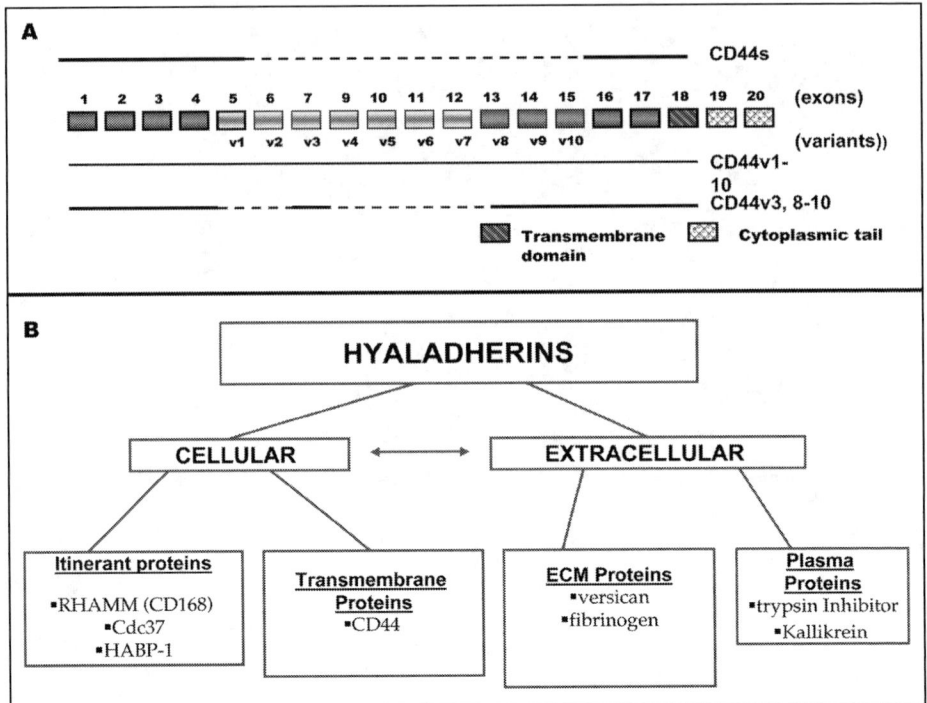

Figure 4. CD44 variants and other hyaladherins expressed in mammary glands. A) Diagram of the splice variants of CD44 that have been documented to be expressed in the mammary glands (see text for references). B) Diagram of hyaluronan receptors/binding proteins (hyaladherins) that have been reported to be expressed in mammary glands.

specific cell receptors such as CD44 and RHAMM, and document details of the signaling cascades regulated by these interactions.[40,46,47] Briefly, the association of HA with cell receptors such as CD44 and RHAMM activates signaling cascades through Src, PI3 kinase, Ras and Erk[36,61-65] pathways that are known to regulate epithelial to mesenchymal transition (EMT), migration/invasion and resistance to anchorage- dependent apoptosis.[66,67] A current summary of HA receptors and of extracellular binding proteins that have been linked to either breast branching morphogenesis or to breast cancer initiation/progression is shown in Figure 4.

HA is produced in and primarily associates with the stroma of normal breast tissue[68] and stromal fibroblasts produce more HA, which can be visualized as cell coats in a particle exclusion assay, than do epithelial cells even in vitro (Fig. 3A). In normal breast tissue, stromal production of HA is regulated by stromal branching morphogens including TGF beta, EGF and bFGF,[69] and estradiol.[70] Several in vitro studies have shown that HA can regulate the size and number of tubular outgrowths from ureteric bud and prostate epithelium in collagen gels[71,72] and is required for branching/invasion of a murine mammary carcinoma cell line, TA3.[73] We have shown that HA promotes branching of murine EPH4 breast epithelial cell line, and this effect is dependent upon the size of the HA polymer (Fig. 3B). HA fragments (average MW = 10^4 Daltons) enhance the rate tubular outgrowths from EPH4 aggregates while higher molecular HA (average MW = 10^6 Daltons) has either no effect or has an inhibitory effect (Fig. 3B). These results are consistent with an emerging paradigm for HA-mediated signaling whereby fragmentation of HA enhances its ability to activate signaling cascades (Fig. 3C).[43,74] HA has been shown to promote invasion, enhance motility,[43] promote an epithelial-to-mesenchymal transition or EMT and is required for HGF- and beta-catenin-mediated EMT.[67,75] In addition to these effects, HA/CD44 interactions mediate the anchorage indepen-dent resistance of tumor cells to apoptosis, as demonstrated by the ability of exogenous HA fragments, which compete with endogenous HA for CD44, to reduce tumor cell survival.[63] As with branching of EPH4 cells, apoptosis and EMT of breast epithelial cell lines is regulated by HA fragments.[63,67] Although the molecular basis for the activity of HA fragments vs. higher molecular HA forms has not been established definitively, a number of studies suggest that CD44 plays a role as a signal transducer[37] involved in the activation of transcriptional pro-grams favoring survival/proliferation and migration/invasion. For example, genes that are strongly up-regulated in the ureteric bud epithelium model of branching morphogenesis (see Chapter 8) include the transcription factor C/EBP, PCNA, and Myc-related transcription factor, anti-apoptotic factors such as BAG-1, BID, GAV-1 and HSP84, and invasion related genes such as CD44 itself and MMPs, in particular the stromelysins.[71] In other studies, HA has been reported to regulate expression of MMP2 and MMP9 through binding to CD44[76-78] which directly associates with MMP-14.[79] Complexes of MMP-14/CD44 and MMP-9/CD44 regu-late migration and invasion of breast cells in vitro[80,81] and can result in the activation of MMP-2,[82] which also contributes to breast cancer cell invasion.[83] How HA might contribute to the CD44/MMP-mediated invasion has not yet been clarified. The ability of HA to protect growth factors relevant to branching morphogenesis from proteolysis and to present them optimally to their receptors,[84] in a manner that is similar to heparin sulfate[85] probably contrib-utes to the effect of HA on branching morphogenesis. In addition, and of particular relevance to the focus of this review, HA participates in the development of a reactive or fibrotic stroma that is characteristic of involuting mammary gland tissue and of stroma surrounding breast tumors,[68,86,87] and which plays a role in both regulating mammary gland involution following weaning and in promoting breast tumor progression.[49,88] The functions of HA in this remod-eling tissue include regulation of collagen fibril formation and neo-angiogenesis,[66,89-91] and regulation of MMP expression including MMP-2, MMP-3 and MMP-9.[92-94]

CD44

CD44 is an HA receptor that also binds to MMP-14,[79] MMP-7[50] and MMP-9.[77] It is expressed as multiple isoforms through alternative splicing of mRNA populations.[37,39,41] The generation of splicing patterns relevant to breast branching morphogenesis (CD44s, CD44v1-10, CD44v3,8-10) is shown in Figure 4A.[95] CD44 is encoded in 20 exons; the first 4 exons and exons 16-18 are constant while exons 5-15 (also called v1-v10) and exons 19-20 are expressed variably. Exon 18 encodes the membrane spanning sequence so that variant exons 19-20 ap-pear to regulate the extent to which the cytoplasmic tail of CD44 can interact with intracellular proteins, for example cytoskeletal proteins such as the cortical actin binding proteins annexin V, cortactin and ERM proteins.[37,95] The standard form of CD44 (CD44s) is most common and is constitutively expressed. This form includes a link-type module that is responsible for binding to HA, a capability that requires activation by post-translational modification.[41,44] CD44s also binds directly to MMP-14 via sequence within the hemopexin domain of this MMP[79] and associates with MMP-7 and MMP-9.[50,77] An important function of a CD44s/HA interactions is to link HA-mediated activation of signaling pathways to the cortical actin cytoskeleton events required for cell motility and invasion.[37,41,95] CD44/MMP-14 and MMP-9 interactions have been linked to promotion of breast tumor cell invasion while CD44/MMP-7 interactions have been linked to breast epithelial cell survival during lactation.

Several CD44 variants are expressed as either heparin sulfate (HS) or chondroitin sulfate (CS) proteoglycans and these are generated by at least three separate mechanisms. CD44 vari-ant forms expressing exon3 (e.g., CD44v3, 8-10) are modified by HS chains in the variable exon 3 and these bind to HB-EGF, HGF, bFGF, MMP-7 and MMP-9. A CS GAG chain can be covalently linked to the variable exon 5 but in addition, CD44 can bind to both HS and CS GAG chains in a noncovalent association via a basic motif encoded in the variable exon 10.[96,97]

The functions of these noncovalent associations with GAG chains have not yet been dissected. The binding of CD44 proteoglycan variant forms to growth factors and MMPs is required for the resistance of breast ductal epithelial cells to apoptosis[50] and for the localization of MMPs to polarized cell lamellae.[79,81]

Like the cell type-specific compartmentalization of growth factors and their receptors in mammary tissue, CD44s and variant forms are only expressed on ductal epithelium and myo-epithelial cells.[98] Conversely, stromal cells produce HA, which is the major ligand for CD44.[68] Uniquely, myoepithelial cells can shed CD44, which has been shown to inhibit breast ductal epithelial cell proliferation in vitro, and may act as a tumor suppressor.[99,100] The CD44v6 epitope, which has been linked to breast cancer,[41] is exposed in both ductal epithelium and myoepithelial cells but is increasingly restricted during pregnancy to the myoepithelial cells and reappears in ductal epithelium during involution.[50,98] The v3 epitope is exposed on the luminal surface of lobulo-alveolar epithelium in the lactating mammary gland, and is also expressed by myoepithelial cells.[50,98]

Several in vitro studies have shown that anti-CD44 blocking antibodies inhibit the effects of HA on ureteric and prostate branching morphogenesis[71,72] (see Chapter 10) and that CD44:HA interactions are required for resistance of breast epithelial cell lines to anchorage-dependent apoptosis.[50,71] CD44 also mediates the EMT of breast epithelial cells promoted by HAS-2 overexpression, mutant active beta-catenin expression or exposure to HGF.[67] Genetic deletion of CD44 due to homologous recombination in DBA/1 mice promotes premature involution of the lactating mammary gland immediately following parturition.[50] In this study, the loss of CD44v3,8-10 resulted in premature apoptosis of the differentiated lobulo-aveolar epithelium, and this phenotype was related to reduced maturation of HB-EGF from its inactive pro-form and as a consequence, a reduced activation of the tyrosine receptor kinase ErbB4. CD44v3, 8-10 was shown to bind directly to both MMP-7 and pro-HB-EGF through its HS chain, which functions as a docking site that promotes the close association of pro-HB-EGF with its maturation factor, MMP-7, and also with its cognate receptor, erbB4. CD44/HB-EGF/MMP-7 complexes were concentrated at the apical surface of lobular alveolar cells. In the absence of CD44, MMP-7 underwent a basal redistribution as a result of an atypical association with perlecan, an HS proteoglycan restricted to the basement membrane.[50] Very little mature HB-EGF was detected in CD44-/- mammary glands whereas pro-HB-EGF was abundant throughout the breast epithelium. A similar phenotype was also observed in the uterus of CD44-/- mice.[50] These results do not exclude the possibility that other ligands for CD44 (e.g., HA and MMP-9) play a role in this mammary gland phenotype. For example, transgenic mice expressing a dominant negative erbB4 do not display the same degree of premature mammary tissue involution[101] as observed in these CD44-/- mice, and, further, genetic deletion of MMP-7 does not result in an involution phenotype of the mammary gland[102] although the lack of this MMP is probably compensated for by MMP-3 which has been reported to be up-regulated in MMP-7 -/- mice.[102] Other factors, in addition to the potential role for additional CD44 ligands such as HA, must also affect the mammary gland phenotype observed in this study since premature involution has not been observed in other strains of CD44-/- mice (e.g., BL6).[103]

Matrix Metalloproteinases

MMPs have been extensively reviewed[2,104-107] and only those that associate with, or are regulated by hyaluronan/CD44 interactions, and are involved in mammary gland morphogenesis/breast cancer will be considered in this review. These include the stromal MMPs, MMP-7, MMP-3, MMP-2 (gelatinase A), MMP-9 and a transmembrane MMP, MMP-14 (Fig. 5). All of these MMPs exhibit broad substrate specificity but this is particularly true for MMP-7 and MMP-3, which exhibit overlapping substrate specificities. For example, target proteins for both of these MMPs include ECM proteins such as collagens III, IV, V, IX, X and XI, fibronectin, laminins, tenascin and proteoglycans such as CD44; cytokines and growth factors such as

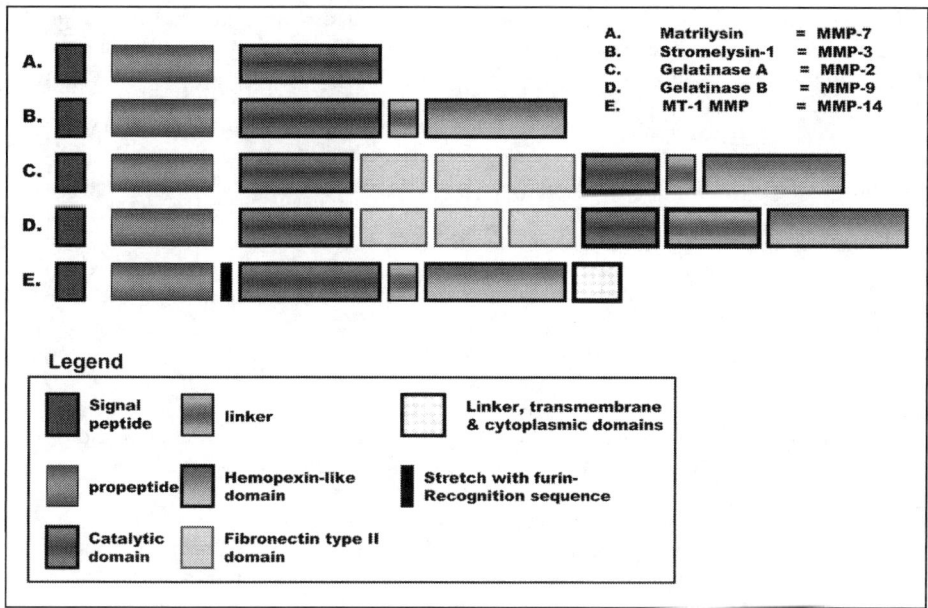

Figure 5. Diagram of domain structures and nomenclature of matrix metalloproteinases (MMPs) associated with hyaluronan and CD44. The MMPs that either associate with or are regulated by CD44 or hyaluronan include MMP-7, MMP-2, MMP-9, MMP-14 and MMP3. The domain structure of these MMPs is shown in this diagram.

TGF-beta and chemokine 4; MMPs such as MMP-2 and 9, and cell adhesion proteins such as E-cadherin. MMP-3, 7 and MMP-14-mediated proteolysis of target proteins often results in their activation. For example, MMP-promoted degradation of ECM proteins such as laminin-5 or growth factors such as TGFβs can expose cryptic, active sites that permit their interaction with specific cell surface receptors.[20,46,108-110] These activated proteins can regulate expression of other MMPs, for example, activated TGFβ promotes expression of MMP-2 or MMP-9 during branching morphogenesis.[111,112] The action of MMPs on their substrates may also release protein fragments that block function. For example MMP-3 released E-cadherin fragments block the cell-cell adhesion functions of intact E-cadherin in mammary epithelium, permitting invasion of these cells into collagen type I gels.[113] These promiscuous effects of MMPs predict that their involvement in branching morphogenesis will be multifactorial. Indeed, the morphogenetic effects of such "downstream" targets of MMP-3 or MMP-7 such as MMP-9,[21] have been related to activation of signaling pathways/gene sets involved in promoting EMT[114] and the invasion/migration of breast epithelial cells.[115-119] The development of 3-dimensional(3D) culture methods for culturing mammary epithelium, which were developed in this laboratory[120-122] and are increasingly used by other laboratories, as well as analysis of animals in which specific MMPs have either been genetically deleted or aberrantly expressed as a transgene,[21,49,88,102,123,124] has greatly aided progress in this area. In particular, specific actions and overlapping functions of MMP-3, MMP-2 and MMP-7 during breast branching morphogenesis and tumor progression have been identified using these experimental approaches.[21,49,88,102] For example, aberrant and constitutive expression of MMP-7,[49] MMP-3[88] and MMP-14[123] in mammary ductal epithelium driven by either the whey acidic protein (WAP) or MMTV promoters[49,88,125] results in tumor formation. MMP-3-mediated neoplastic transformation was noted to be associated with a chronically altered or "reactive" stroma having typical characteristics in common with stroma during involuting mammary glands.[33] MMP-2 and MMP-14, which are localized primarily to the periductal stroma of pubertal branching mammary glands but are reduced near side branches or where buds are initiating, regulate the early events in primary duct invasion of the mammary fat pad following puberty. Thus, virgin MMP-2 -/- mice exhibited an initial transient lag in duct invasion that is compensated for over time. The lag is accompanied by increased apoptosis of ductal epithelium suggesting that MMP-2 affects ductal invasion by regulating survival of end-bud cells. These effects of MMP-2 can be compensated for by other MMPs only during a brief window of morphogenesis. Unexpectedly, MMP-2 appears to repress side branching from primary ducts since a transient enhancement of side branching was observed in pubescent virgin MMP-2-/- mice. In contrast, analysis of MMP-3-/- mice suggest that it is transiently required for secondary duct formation and expansion but not primary duct expansion.[21] Consistent with these observations, an activated MMP-3 transgene expressed in ductal epithelium promotes precocious secondary branching, proliferation and differentiation during puberty but results in premature apoptosis during pregnancy. The premature involution is linked to the appearance of a reactive stroma characterized by elevated expression of tenascin, other MMPs such as MMP-9, as well as enhanced collagen deposition, and increased angiogenesis.[33] These particular studies of MMP-3 function illuminate an important principle that is relevant to our thesis of a double edged sword: the consequence of MMP activity to the ductal epithelial cells is stage- and therefore context- dependent and is associated with measurable changes in the cell's microenvironment. As is the case for MMP-3, an MMP-7 transgene expressed in mammary ductal epithelial cells also affects apoptosis during mammary tissue involution. However, the consequences of transgene expression differ in single vs. multiple pregnancies. Thus, during the first pregnancy, transgenic mice exhibited increased apoptosis of the mammary ductal epithelium during involution, a reduction in apoptosis of ductal epithelial apoptosis was observed following the third pregnancy. A resistance to apoptosis was associated with loss of FasL expression, an apoptosis-promoting gene.[126] Interestingly, CD44v3/MMP7/HB-EGF/erbB4 complexes appear on lobular alveolar epithelium during lactation[50] when expression of MMP-3 and MMP-7 are repressed.[33,102] Whether or not these CD44 complexes, which permit alveolar cell survival, affect the expression of MMPs hasn't been reported but is an intriguing possibility. In any case, an intricately timed regulation of MMP expression in the stroma (and also possibly HA accumulation), together with timed CD44 complex formation in mammary ductal epithelium, is required for the normal expansion and differentiation of the mammary gland to a lactating phenotype, and at least the MMPs are also required for the reversal to a quiescent tissue during involution. Is there any evidence that these specific molecular processes, which are clearly required for branching morphogenesis, are also integral to the malignant transformation of breast cells? If so, how do transformed cells utilize these signals to proliferate and how do they escape apoptosis that is a normal consequence of a remodeling microenvironment controlled by these MMPs?

HA, CD44 and MMP Interactions in Breast Branching Morphogenesis and Cancer

Our understanding of how malignancies originate and progress has recently undergone several important paradigm shifts that must be incorporated into any meaningful molecular analysis of events that control both breast branching morphogenesis and initiation/progression of breast tumors. These are the demonstration that the malignant phenotype is plastic even in the presence of multiple activating mutations in regulatory genes or oncogenes,[34,45] that the microenvironment can be dominant over these mutations[34,45,127] and that most of the tumorigenic capacity of tumors may reside in a minor tumor cell subset, which exhibits stem cell characteristics.[19,128,129] The studies focused upon in this review suggest that, at least in experimental models, specific stromal MMPs, HA and CD44 coordinately regulate branching morphogenesis at a minimum by controlling the growth and survival of ductal epithelium. What is the evidence for a role of these stromal genes in breast cancer, particularly in humans, and specifically in breast stem cells that may give rise to aggressive tumors?

Although a role for stroma in regulating tumor progression is less well appreciated than its role in regulating branching morphogenesis,[2] clinical links between the stroma and tumorigenesis have often been reported.[20,23,34,130-133] For example, heritable gene defects in stroma with a predisposition to cancer have been known for some time (reviewed in ref. 45). Furthermore, an association with changes in stromal characteristics, such as a predisposition to wound-like fibrosis or tissue inflammation, with increased susceptibility to a variety of cancers including breast cancer have often been cited e.g.[134,135] Experimentally, links between modification of stromal characteristics and cancer initiation has been strongly made by studies showing that aberrant expression of MMP-3, MMP-7 or MMP14 in ductal epithelium promote tumorigenesis.[49,88,123,136,137] Hyper-expression or de-regulated expression of these MMPs is also common in human breast tumors.[20,34,138,139] The accumulation of HA, which is produced in breast tumors by both the stroma and tumorigenic epithelial cells, is associated with poor differentiation of tumors, axillary lymph node positivity and short overall survival of breast cancer patients.[68] Aberrant regulation of CD44 in breast cancer biopsy samples has also been linked to patient outcome although a consistent relationship with disease progression has yet to emerge. For example, hyper-expression of CD44 has been linked to both poor and good outcomes and this has been interpreted to suggest that CD44 can act both as a tumor-progressing factor and a tumor-suppressing factor depending on context.[95,140,141] Thus, the consequence of aberrant CD44 expression to malignant progression may depend upon a variety of additional properties of the tumor cells that make assessment of expression per se an inadequate measure to assess for its role in tumorigenesis. For example, interplay amongst combinations of variant forms may determine overall CD44 function,[41,95,141] retention of myofibroblasts in tumors that express and shed a growth suppressing form of CD44 may contribute to a positive outcome,[99,142] CD44 expression in small tumor subsets (e.g., transformed progenitor cells) vs. the entire tumor[15] may be more important to tumor aggressiveness, and expression of other proteins by key malignant cells may affect CD44 function. Examples of the latter include E-cadherin, which suppresses CD44-mediated breast tumor cell invasion,[73] and RHAMM, which can counter anti-invasive properties of CD44.[143]

Evidence is increasing to suggest that progenitor cells known to be present in normal breast tissue[14,19,128,129] may be targets for malignancy in breast cancer, as is now accepted for hematopoeitic malignancies such as AML or CML.[14,144] For example, one recent study has identified a small population (2%) of tumor cells obtained from primary breast cancer biopsies that contained virtually all of the breast cancer initiating activity, as defined by an ability to form tumor cells to give rise to breast tumors when serially transplanted into severely immuno-suppressed SCID/NOD mice.[15] These highly tumorigenic cell subsets gave rise to tumors that were similar in heterogeneity, as defined by surface phenotypes, to the original primary tumor, suggesting that they arose from progenitor cells that retain the ability to differentiate into heterogeneous lineages. These results also provide clinical evidence that confirms the plasticity of the tumorigenic phenotype. Of particular relevance to the focus of this review, a high expression of CD44 was a defining surface phenotype of the tumorigenic stem cell subset surface phenotype.

The above reports are intriguing since progenitor cells from other tissues typically express CD44, utilize CD44 to adhere to the HA that is produced by stromal cells, and require an HA rich environment for their survival. This characteristic is retained with neoplastic transformation.[89,145] These results are also consistent with our thesis and considerable data[23,146] that aberrant expression of ECM receptors, such as CD44, provides both morphogenetically active normal as well as proliferating transformed breast cells with a selective advantage for growth and survival. Although the roles of MMPs during breast branching morphogenesis and breast tumor progression have been more thoroughly characterized than that of HA or CD44, evidence to date supports the existence of a functional relationship between these two classes of molecules in both branching morphogenesis and cancer. We summarize one model for how these functional interactions might contribute to branching morphogenesis and how they also

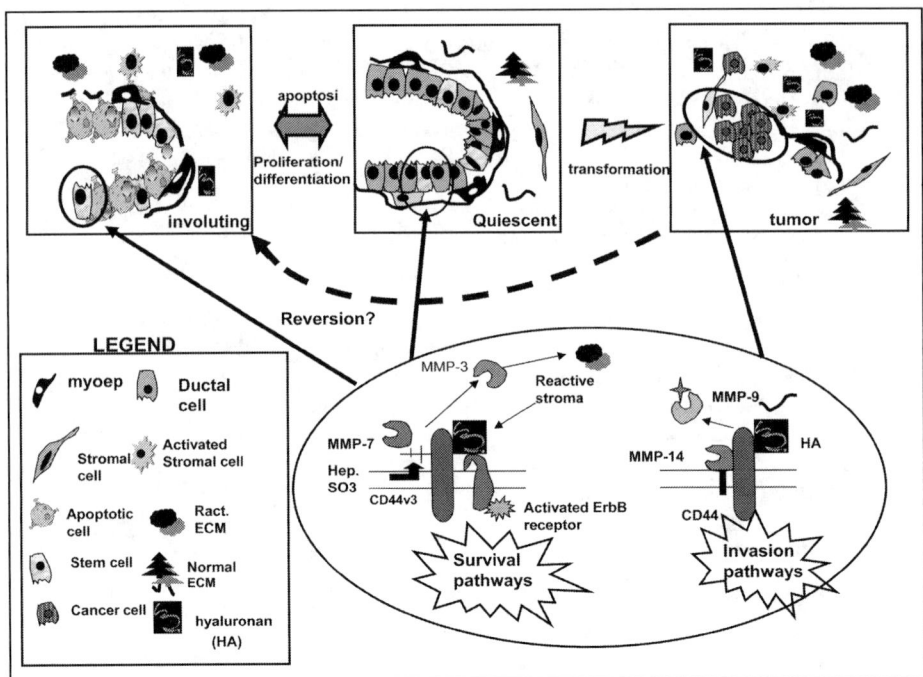

Figure 6. Model of molecular interactions proposed to regulate both morphogenesis and neoplasia. A model of some of the molecular mechanisms regulated by CD44, HA and MMPs that contribute to normal mammary gland morphogenesis yet appear to be utilized by transformed cells to form aggressive tumors. Quiescent or lactating ductal lobular aveolar cells are able to survive due to signaling through CD44v3, which also requires the functioning of MMP-7 and Erb B4. During involution that follows weaning, these normal ductal cells lose the ability to signal through this pathway. We propose that stem cells, which will replenish the ductal epithelium in subsequent pregnancies retain signaling activity through this pathway and therefore survive. As mammals age, somatic mutations accumulate in the cells of mammary tissue, and if or when mutations occur in stem cells, aggressive tumors will arise partly due to the selective advantage provided by the CD44v3/HA/MMP-7 survival pathway and partly to an ability to utilize CD44/HA/MMP-9/MMP-14 for activating invasive signaling pathways.

may promote neoplastic progression of this same tissue (Fig. 6). In this model, expression of CD44 variants that bind to HA, MMP-14 and MMP-9 are predicted to be involved in promoting side and/or primary bud branching that contribute to mammary tree expansion following puberty and during pregnancy. MMP-3 and MMP-2 activity is also transiently required for side and primary duct branching respectively.[21,33,147] The presence of HA and activation of signaling cascades through CD44 are predicted to contribute to the regulation of expression of these MMPs. Expression and release of CD44 variant forms by myofibroblasts provides one mechanism for controlling the extent of ductal epithelial cell proliferation. Expression of CD44v3, 8-10 (Fig. 4) and its association with MMP-7/erbB4/EGF complexes is upregulated as breast tissue differentiates into lobular-alveolar cells during pregnancy and following parturition. Formation of this complex is required to prevent apoptosis and to sustain a differentiation mammary tree.[50] Expression of these complexes decreases as MMP-3 expression (and possibly HA) increases following weaning, contributing to the remodeling of ECM into reactive stroma characteristic of involuting breast tissue. This modified ECM favors the apoptosis of terminally differentiated lobular-alveolar breast epithelial cells while selectively permitting the survival of progenitor cells. We speculate that these progenitor cells continue to express CD44 variant forms and this allows their survival in the presence of a remodeling or reactive stroma. The surviving progenitor cells subsequently regenerate the mammary tree of future pregnancies. These same genes could contribute to neoplastic transformation and progression of breast tumors as modeled in Figure 6. Aberrant MMP-3 expression can initiate neoplastic transformation of murine breast epithelia via an action on the stroma. The MMP-3-altered stroma has been proposed to both contribute to neoplastic transformation as a tumor promoter and permit the survival and growth of cells that become neoplastic. In a remodeling ECM, growth of even a minor population of aberrant cells could lead to de-regulation of interactions amongst breast tissue cell types, with consequent hyperplastic and dysplastic events that precede neoplastic transformation. Since mutations occur continuously in key oncogenes in most tissues, the step from a dysplastic state to a frankly neoplastic state might not be great, particularly in a chronically remodeling MMP-and HA-rich tissue.[34,45,148-150] Although it is unlikely that transformation of progenitor cells is responsible for all breast cancers, several properties of these cells could contribute to a particularly aggressive and persistent tumors. For example, if progenitor cells express CD44 isoforms that associate with key MMPs that promote a remodeling environment, they will have a selective survival advantage over other cells and also have the machinery for metastasis such as invasion and translatability. These cells would not have to undergo a lengthy evolution to become malignant but would generate highly aggressive tumors. The model shown in Figure 6 is one example of how specific genes, whose products functionally interact to regulate branching morphogenesis and are therefore essential for mammalian survival, can, with a small shift in regulation, act as seeds for individual destruction particularly in the presence of activating mutations in other genes which inevitably occurs with age.

Acknowledgements

This work was supported by funds from the US Department of Energy, Office of Biological and Environmental Research (DE-AC0376 SF00098 to MJB); the National Cancer Institute (CA64786-02 to MJB); by an Innovator Award from the US Department of Defense Breast Cancer Research Program (DAMD17-02-1-0438 to MJB); by a Concept Award from the US Department of Defense Breast Cancer Research Program (DAM17-01-01-0541 to MJB and ET), by CIHR (#MOP-57694 to ET) and the Pamela Greenaway-Kohlmeir Translational Breast Cancer Unit Salary Award (to ET).

References

1. Pierce GB. Relationship between differentiation and carcinogenesis. J Toxicol Environ Health 1977; 2(6):1335-42.
2. Fata JWZ, Bissell MJ. Branching morphogenesis. 2003; in press.
3. Hu MC, Rosenblum ND. Genetic regulation of branching morphogenesis: Lessons learned from loss-of-function phenotypes. Pediatr Res 2003; 54(4):433-8.
4. Radisky DC, H.Y, Bissell MJ. Delivering the message: Epimorphin and mammary epithelial morphogenesis. Trends Cell Biol 2003; 13(8):426-434.
5. Affolter M et al. Tube or not tube: Remodeling epithelial tissues by branching morphogenesis. Dev Cell 2003; 4(1):11-8.
6. Davies JA. Do different branching epithelia use a conserved developmental mechanism? Bioessays 2002; 24(10):937-48.
7. Hovey RC, Trott JF, Vonderhaar BK. Establishing a framework for the functional mammary gland: From endocrinology to morphology. J Mammary Gland Biol Neoplasia 2002; 7(1):17-38.
8. Silberstein GB. Postnatal mammary gland morphogenesis. Microsc Res Tech 2001; 52(2):155-62.
9. Petersen OW et al. The plasticity of human breast carcinoma cells is more than epithelial to mesenchymal conversion. Breast Cancer Res 2001; 3(4):213-7.
10. Lochter A. Plasticity of mammary epithelia during normal development and neoplastic progression. Biochem Cell Biol 1998; 76(6):997-1008.
11. Fleury V, Watanabe T. Morphogenesis of fingers and branched organs: How collagen and fibroblasts break the symmetry of growing biological tissue. C R Biol 2002; 325(5):571-83.
12. Smith GH, Chepko G. Mammary epithelial stem cells. Microsc Res Tech 2001; 52(2):190-203.

13. Metzger RJ, Krasnow MA. Genetic control of branching morphogenesis. Science 1999; 284(5420):1635-9.
14. Dick JE. Breast cancer stem cells revealed. Proc Natl Acad Sci USA 2003; 100(7):3547-9.
15. Al-Hajj M et al. Prospective identification of tumorigenic breast cancer cells. Proc Natl Acad Sci USA 2003; 100(7):3983-8.
16. Chang CC et al. A human breast epithelial cell type with stem cell characteristics as target cells for carcinogenesis. Radiat Res 2001; 155(1 Pt 2):201-207.
17. Medina D. Biological and molecular characteristics of the premalignant mouse mammary gland. Biochim Biophys Acta 2002; 1603(1):1-9.
18. Li P et al. Stem cells in breast epithelia. Int J Exp Pathol 1998; 79(4):193-206.
19. Dontu G et al. Stem cells in normal breast development and breast cancer. Cell Prolif 2003; 36(Suppl 1):59-72.
20. Wiseman BS, Werb Z. Stromal effects on mammary gland development and breast cancer. Science 2002; 296(5570):1046-9.
21. Wiseman BS et al. Site-specific inductive and inhibitory activities of MMP-2 and MMP-3 orchestrate mammary gland branching morphogenesis. J Cell Biol 2003; 162(6):1123-33.
22. Werb Z et al. Extracellular matrix remodeling and the regulation of epithelial-stromal interactions during differentiation and involution. Kidney Int Suppl 1996; 54:S68-74.
23. Roskelley CD, Bissell MJ. The dominance of the microenvironment in breast and ovarian cancer. Semin Cancer Biol 2002; 12(2):97-104.
24. Zhang YW, Vande Woude GF. HGF/SF-met signaling in the control of branching morphogenesis and invasion. J Cell Biochem 2003; 88(2):408-17.
25. Rosario M, Birchmeier W. How to make tubes: Signaling by the Met receptor tyrosine kinase. Trends Cell Biol 2003; 13(6):328-35.
26. Soriano JV et al. Roles of hepatocyte growth factor/scatter factor and transforming growth factor-beta1 in mammary gland ductal morphogenesis. J Mammary Gland Biol Neoplasia 1998; 3(2):133-50.
27. van Tuyl M, Post M. From fruitflies to mammals: Mechanisms of signalling via the Sonic hedgehog pathway in lung development. Respir Res 2000; 1(1):30-5.
28. Deugnier MA, T.J, Faraldo MM et al. The important of being a myoepithelial cell. Breast Cancer Res 2002; 4(6):224-230.
29. Bartley JC, Emerman JT, Bissell MJ. Metabolic cooperativity between mammary epithelial cells and adipocytes of mice. Am J Physio 1981; 241:C240-248.
30. Gouon-Evans V, Lin EY,Pollard JW. Requirement of macrophages and eosinophils and their cytokines/chemokines for mammary gland development. Breast Cancer Res 2002; 4(4):155-64.
31. Chilliard Y et al. Leptin in ruminants. Gene expression in adipose tissue and mammary gland, and regulation of plasma concentration. Domest Anim Endocrinol 2001; 21(4):271-95.
32. Weaver VM, L.S, Lakins JN et al. beta4 integrin-dependent formation of polarized three-dimensional architecture confers resistance to apoptosis in normal and malignant mammary epithelium. Cancer Cell 2002; 2(3):205-216.
33. Thomasset N et al. Expression of autoactivated stromelysin-1 in mammary glands of transgenic mice leads to a reactive stroma during early development. Am J Pathol 1998; 153(2):457-67.
34. Bissell MJ et al. The organizing principle: Microenvironmental influences in the normal and malignant breast. Differentiation 2002; 70(9-10):537-46.
35. Earp 3rd HS, Calvo BF,Sartor CI. The EGF receptor family—multiple roles in proliferation, differentiation, and neoplasia with an emphasis on HER4. Trans Am Clin Climatol Assoc 2003; 114:315-33, discussion 333-4.
36. Roberts AB, Wakefield LM. The two faces of transforming growth factor beta in carcinogenesis. Proc Natl Acad Sci USA 2003; 100(15):8621-3.
37. Turley EA, Noble PW, Bourguignon LY. Signaling properties of hyaluronan receptors. J Biol Chem 2002; 277(7):4589-92.
38. Ponta H, S.L, Herrlich PA. CD44: From adhesion molecules to signalling regulators. Nat Rev Mol Cell Biol 2003; 4(1):33-45.
39. Naor D et al. CD44 in cancer. Crit Rev Clin Lab Sci 2002; 39(6):527-79.
40. Yasuda M et al. CD44: Functional relevance to inflammation and malignancy. Histol Histopathol 2002; 17(3):945-50.
41. Ponta H, Sherman L, Herrlich PA. CD44: From adhesion molecules to signalling regulators. Nat Rev Mol Cell Biol 2003; 4(1):33-45.
42. Rooney P et al. The role of hyaluronan in tumor neovascularization (review). Int J Cancer 1995; 60(5):632-6.

43. Tammi MI, Day AJ, Turley EA. Hyaluronan and homeostasis: A balancing act. J Biol Chem 2002; 277(7):4581-4.
44. Day AJ, Prestwich GD. Hyaluronan-binding proteins: Tying up the giant. J Biol Chem 2002; 277(7):4585-8.
45. Bissell MJ, Radisky D. Putting tumors in context. Nat Rev Cancer 2001; 1(1):46-54.
46. Sternlicht MD, Werb Z. How matrix metalloproteinases regulate cell behavior. Annu Rev Cell Dev Biol 2001; 17:463-516.
47. de Launoit Y et al. The PEA3 group of ETS-related transcription factors. Role in breast cancer metastasis. Adv Exp Med Biol 2000; 480:107-16.
48. Benaud C, Dickson RB, Thompson EW. Roles of the matrix metalloproteinases in mammary gland development and cancer. Breast Cancer Res Treat 1998; 50(2):97-116.
49. Rudolph-Owen LA et al. The matrix metalloproteinase matrilysin influences early-stage mammary tumorigenesis. Cancer Res 1998; 58(23):5500-6.
50. Yu WH et al. CD44 anchors the assembly of matrilysin/MMP-7 with heparin-binding epidermal growth factor precursor and ErbB4 and regulates female reproductive organ remodeling. Genes Dev 2002; 16(3):307-23.
51. Delehedde M et al. Proteoglycans: Pericellular and cell surface multireceptors that integrate external stimuli in the mammary gland. J Mammary Gland Biol Neoplasia 2001; 6(3):253-73.
52. Bateman KL et al. Heparan sulphate. Regulation of growth factors in the mammary gland. Adv Exp Med Biol 2000; 480:65-9.
53. Bourhis XL et al. Autocrine and paracrine growth inhibitors of breast cancer cells. Breast Cancer Res Treat 2000; 60(3):251-8.
54. McDonald JA, Camenisch TD. Hyaluronan: Genetic insights into the complex biology of a simple polysaccharide. Glycoconj J 2002; 19(4-5):331-9.
55. Weigel PH. Functional characteristics and catalytic mechanisms of the bacterial hyaluronan synthases. IUBMB Life 2002; 54(4):201-11.
56. Itano N, Kimata K. Mammalian hyaluronan synthases. IUBMB Life 2002; 54(4):195-9.
57. DeAngelis PL. Hyaluronan synthases: Fascinating glycosyltransferases from vertebrates, bacterial pathogens, and algal viruses. Cell Mol Life Sci 1999; 56(7-8):670-82.
58. Weigel PH, Hascall VC, Tammi M. Hyaluronan synthases. J Biol Chem 1997; 272(22):13997-4000.
59. Silbert JE, Sugumaran G. Biosynthesis of chondroitin/dermatan sulfate. IUBMB Life 2002; 54(4):177-86.
60. Prydz K, Dalen KT. Synthesis and sorting of proteoglycans. J Cell Sci 2000; 113(Pt 2):193-205.
61. Hall CL, Wang FS, Turley E. Src-/- fibroblasts are defective in their ability to disassemble focal adhesions in response to phorbol ester/hyaluronan treatment. Cell Commun Adhes 2002; 9(5-6):273-83.
62. Slevin M, Kumar S, Gaffney J. Angiogenic oligosaccharides of hyaluronan induce multiple signaling pathways affecting vascular endothelial cell mitogenic and wound healing responses. J Biol Chem 2002; 277(43):41046-59.
63. Ghatak S, Misra S, Toole BP. Hyaluronan oligosaccharides inhibit anchorage-independent growth of tumor cells by suppressing the phosphoinositide 3-kinase/Akt cell survival pathway. J Biol Chem 2002; 277(41):38013-20.
64. Fujita Y et al. CD44 signaling through focal adhesion kinase and its anti-apoptotic effect. FEBS Lett 2002; 528(1-3):101-8.
65. Bourguignon LY et al. Hyaluronan-mediated CD44 interaction with RhoGEF and Rho kinase promotes Grb2-associated binder-1 phosphorylation and phosphatidylinositol 3-kinase signaling leading to cytokine (macrophage-colony stimulating factor) production and breast tumor progression. J Biol Chem 2003; 278(32):29420-34.
66. Toole BP, Wight TN, Tammi MI. Hyaluronan-cell interactions in cancer and vascular disease. J Biol Chem 2002; 277(7):4593-6.
67. Zoltan-Jones A et al. Elevated hyaluronan production induces mesenchymal and transformed properties in epithelial cells. J Biol Chem 2003.
68. Auvinen P et al. Hyaluronan in peritumoral stroma and malignant cells associates with breast cancer spreading and predicts survival. Am J Pathol 2000; 156(2):529-36.
69. Hirano S et al. Effect of growth factors on hyaluronan production by canine vocal fold fibroblasts. Ann Otol Rhinol Laryngol 2003; 112(7):617-24.
70. Russell DL et al. Hormone-regulated expression and localization of versican in the rodent ovary. Endocrinology 2003; 144(3):1020-31.
71. Pohl M et al. Role of hyaluronan and CD44 in vitro branching morphogenesis of ureteric bud cells. Dev Biol 2000; 224(2):312-25.

72. Gakunga P et al. Hyaluronan is a prerequisite for ductal branching morphogenesis. Development 1997; 124(20):3987-97.
73. Xu YYQ. E-cadherin negatively regulates CD44-hyalruonan interaction and CD44-mediated tumor invasion and branching moprhogenesis. J Biol Chem 2003; 278(mar 7):8661-8.
74. Lee JY, Spicer AP. Hyaluronan: A multifunctional, megaDalton, stealth molecule. Curr Opin Cell Biol 2000; 12(5):581-6.
75. Turley EA. The control of adrenocortical cytodifferentiation by extracellular matrix. Differentiation 1980; 17(2):93-103.
76. Saad S et al. Induction of matrix metalloproteinases MMP-1 and MMP-2 by coculture of breast cancer cells and bone marrow fibroblasts. Breast Cancer Res Treat 2000; 63(2):105-15.
77. Yu Q, Stamenkovic I. Cell surface-localized matrix metalloproteinase-9 proteolytically activates TGF-beta and promotes tumor invasion and angiogenesis. Genes Dev 2000; 14(2):163-76.
78. Spessotto P et al. Hyaluronan-CD44 interaction hampers migration of osteoclast-like cells by down-regulating MMP-9. J Cell Biol 2002; 158(6):1133-44.
79. Mori H et al. CD44 directs membrane-type 1 matrix metalloproteinase to lamellipodia by associating with its hemopexin-like domain. Embo J 2002; 21(15):3949-59.
80. Kajita M,I.Y, Chiba T et al. Membrane type 1 matrix metalloproteinase cleaves CD44 and promotes cell migration. J Cell Biol 2001; 153(5):893-904.
81. Yu QSI. Localization of matrix metalloproteinase 9 to the cell surface provides a mechanism for CD44-mediated tumor invasion. Genes Dev 1999; 13(1):35-38.
82. Imai K et al. Membrane-type matrix metalloproteinase 1 is a gelatinolytic enzyme and is secreted in a complex with tissue inhibitor of metalloproteinases 2. Cancer Res 1996; 56(12):2707-10.
83. Deryugina EI et al. Tumor cell invasion through matrigel is regulated by activated matrix metalloproteinase-2. Anticancer Res 1997; 17(5A):3201-10.
84. Vincent T et al. Hyaluronic acid induces survival and proliferation of human myeloma cells through an interleukin-6-mediated pathway involving the phosphorylation of retinoblastoma protein. J Biol Chem 2001; 276(18):14728-36.
85. Pellegrini L. Role of heparan sulfate in fibroblast growth factor signalling: A structural view. Curr Opin Struct Biol 2001; 11(5):629-34.
86. Boyd NF et al. Mammographic densities and breast cancer risk. Cancer Epidemiol Biomarkers Prev 1998; 7(12):1133-44.
87. Rowley DR. What might a stromal response mean to prostate cancer progression? Cancer Metastasis Rev 1998; 17(4):411-9.
88. Sternlicht MD et al. The stromal proteinase MMP3/stromelysin-1 promotes mammary carcinogenesis. Cell 1999; 98(2):137-46.
89. Bullard KM, Longaker MT, Lorenz HP. Fetal wound healing: Current biology. World J Surg 2003; 27(1):54-61.
90. Turino GM, Cantor JO. Hyaluronan in respiratory injury and repair. Am J Respir Crit Care Med 2003; 167(9):1169-75.
91. Neal MS. Angiogenesis: Is it the key to controlling the healing process? J Wound Care 2001; 10(7):281-7.
92. Park MJ et al. PTEN suppresses hyaluronic acid-induced matrix metalloproteinase-9 expression in U87MG glioblastoma cells through focal adhesion kinase dephosphorylation. Cancer Res 2002; 62(21):6318-22.
93. Han F et al. Effects of sodium hyaluronate on experimental osteoarthritis in rabbit knee joints. Nagoya J Med Sci 1999; 62(3-4):115-26.
94. Takahashi K et al. The effects of hyaluronan on matrix metalloproteinase-3 (MMP-3), interleukin-1beta(IL-1beta), and tissue inhibitor of metalloproteinase-1 (TIMP-1) gene expression during the development of osteoarthritis. Osteoarthritis Cartilage 1999; 7(2):182-90.
95. Bourguignon LY. CD44-mediated oncogenic signaling and cytoskeleton activation during mammary tumor progression. J Mammary Gland Biol Neoplasia 2001; 6(3):287-97.
96. Henke CA, T.U, Mickelson DJ et al. CD44-related chondroitin sulfate proteoglycan, a cell surface receptor implicated with tumor cell invasion, mediates endothelial cell migration on fibrinogen and invasion into a fibrin matrix. J Clin Invest 1996; 97(11):2541-2552.
97. Hayes GM et al. Identification of sequence motifs responsible for the adhesive interaction between exon v10-containing CD44 isoforms. J Biol Chem 2002; 277(52):50529-34.
98. Hebbard L et al. CD44 expression and regulation during mammary gland development and function. J Cell Sci 2000; 113(Pt 14):2619-30.
99. Alpaugh ML et al. Myoepithelial-specific CD44 shedding contributes to the anti-invasive and antiangiogenic phenotype of myoepithelial cells. Exp Cell Res 2000; 261(1):150-8.

100. Lee MC et al. Myoepithelial-specific CD44 shedding is mediated by a putative chymotrypsin-like sheddase. Biochem Biophys Res Commun 2000; 279(1):116-23.
101. Jones FE et al. ErbB4 signaling in the mammary gland is required for lobuloalveolar development and Stat5 activation during lactation. J Cell Biol 1999; 147(1):77-88.
102. Rudolph-Owen LA et al. Coordinate expression of matrix metalloproteinase family members in the uterus of normal, matrilysin-deficient, and stromelysin-1-deficient mice. Endocrinology 1997; 138(11):4902-11.
103. Toelg C, B.M, Turley EA. Unpublished data 2003.
104. Nagase II, Woessner Jr JF. Matrix metalloproteinases. J Biol Chem 1999; 274(31):21491-4.
105. Egeblad M, Werb Z New functions for the matrix metalloproteinases in cancer progression. Nat Rev Cancer 2002; 2(3):161-74.
106. Chang C, Werb Z. The many faces of metalloproteases: Cell growth, invasion, angiogenesis and metastasis. Trends Cell Biol 2001; 11(11):S37-43.
107. Lochter A et al. The significance of matrix metalloproteinases during early stages of tumor progression. Ann N Y Acad Sci 1998; 857:180-93.
108. Van den Steen PE et al. Biochemistry and molecular biology of gelatinase B or matrix metalloproteinase-9 (MMP-9). Crit Rev Biochem Mol Biol 2002; 37(6):375-536.
109. Mueller MM, Fusenig NE. Tumor-stroma interactions directing phenotype and progression of epithelial skin tumor cells. Differentiation 2002; 70(9-10):486-97.
110. Fillmore HL, VanMeter TE,Broaddus WC. Membrane-type matrix metalloproteinases (MT-MMPs): Expression and function during glioma invasion. J Neurooncol 2001; 53(2):187-202.
111. Schenk S et al. Binding to EGF receptor of a laminin-5 EGF-like fragment liberated during MMP-dependent mammary gland involution. J Cell Biol 2003; 161(1):197-209.
112. Seiki M. The cell surface: The stage for matrix metalloproteinase regulation of migration. Curr Opin Cell Biol 2002; 14(5):624-32.
113. Noe V et al. Release of an invasion promoter E-cadherin fragment by matrilysin and stromelysin-1. J Cell Sci 2001; 114(Pt 1):111-118.
114. Tester AM et al. MMP-9 secretion and MMP-2 activation distinguish invasive and metastatic sublines of a mouse mammary carcinoma system showing epithelial-mesenchymal transition traits. Clin Exp Metastasis 2000; 18(7):553-60.
115. Hazan RB et al. Exogenous expression of N-cadherin in breast cancer cells induces cell migration, invasion, and metastasis. J Cell Biol 2000; 148(4):779-90.
116. Rolli M et al. Activated integrin alphavbeta3 cooperates with metalloproteinase MMP-9 in regulating migration of metastatic breast cancer cells. Proc Natl Acad Sci USA 2003; 100(16):9482-7.
117. Mira E et al. Insulin-like growth factor I-triggered cell migration and invasion are mediated by matrix metalloproteinase-9. Endocrinology 1999; 140(4):1657-64.
118. Koshikawa N et al. Role of cell surface metalloprotease MT1-MMP in epithelial cell migration over laminin-5. J Cell Biol 2000; 148(3):615-24.
119. Rozanov DV et al. Mutation analysis of membrane type-1 matrix metalloproteinase (MT1-MMP). The role of the cytoplasmic tail Cys(574), the active site Glu(240), and furin cleavage motifs in oligomerization, processing, and self-proteolysis of MT1-MMP expressed in breast carcinoma cells. J Biol Chem 2001; 276(28):25705-14.
120. Barcellos-Hoff MH et al. Functional differentiation and alveolar morphogenesis of primary mammary cultures on reconstituted basement membrane. Development 1989; 105(2):223-35.
121. Petersen OW et al. Interaction with basement membrane serves to rapidly distinguish growth and differentiation pattern of normal and malignant human breast epithelial cells. Proc Natl Acad Sci USA 1992; 89(19):9064-8.
122. Schmeichel KL, Bissell MJ. Modeling tissue-specific signaling and organ function in three dimensions. J Cell Sci 2003; 116(Pt 12):2377-88.
123. Ha HY et al. Overexpression of membrane-type matrix metalloproteinase-1 gene induces mammary gland abnormalities and adenocarcinoma in transgenic mice. Cancer Res 2001; 61(3):984-90.
124. Itoh T et al. Reduced angiogenesis and tumor progression in gelatinase A-deficient mice. Cancer Res 1998; 58(5):1048-51.
125. Sympson CJ et al. Targeted expression of stromelysin-1 in mammary gland provides evidence for a role of proteinases in branching morphogenesis and the requirement for an intact basement membrane for tissue-specific gene expression. J Cell Biol 1994; 125(3):681-93.
126. Vargo-Gogola T et al. Matrilysin (matrix metalloproteinase-7) selects for apoptosis-resistant mammary cells in vivo. Cancer Res 2002; 62(19):5559-63.
127. Kenny PA, Bissell MJ. Tumor reversion: Correction of malignant behavior by microenvironmental cues. Int J Cancer 2003; 107(5):688-95.

128. Welm B et al. Isolation and characterization of functional mammary gland stem cells. Cell Prolif 2003; 36(Suppl 1):17-32.
129. Petersen OW et al. Epithelial progenitor cell lines as models of normal breast morphogenesis and neoplasia. Cell Prolif 2003; 36(Suppl 1):33-44.
130. Boudreau N, Myers C. Breast cancer-induced angiogenesis: Multiple mechanisms and the role of the microenvironment. Breast Cancer Res 2003; 5(3):140-6.
131. Shekhar MP, Pauley R, Heppner G. Host microenvironment in breast cancer development: Extracellular matrix-stromal cell contribution to neoplastic phenotype of epithelial cells in the breast. Breast Cancer Res 2003; 5(3):130-5.
132. Radisky D, Muschler J, Bissell MJ. Order and disorder: The role of extracellular matrix in epithelial cancer. Cancer Invest 2002; 20(1):139-53.
133. Werb Z et al. Extracellular matrix remodeling as a regulator of stromal-epithelial interactions during mammary gland development, involution and carcinogenesis. Braz J Med Biol Res 1996; 29(9):1087-97.
134. van Golen KL. Inflammatory breast cancer: Relationship between growth factor signaling and motility in aggressive cancers. Breast Cancer Res 2003; 5(3):174-9.
135. Hasebe T et al. Highly proliferative intratumoral fibroblasts and a high proliferative microvessel index are significant predictors of tumor metastasis in T3 ulcerative-type colorectal cancer. Hum Pathol 2001; 32(4):401-9.
136. Cunha GR, Hom YK. Role of mesenchymal-epithelial interactions in mammary gland development. J Mammary Gland Biol Neoplasia 1996; 1(1):21-35.
137. Dong LJ, Chung AE. The expression of the genes for entactin, laminin A, laminin B1 and laminin B2 in murine lens morphogenesis and eye development. Differentiation 1991; 48(3):157-72.
138. Sloan EK, Anderson RL. Genes involved in breast cancer metastasis to bone. Cell Mol Life Sci 2002; 59(9):1491-502.
139. Rudolph-Owen LA, Matrisian LM. Matrix metalloproteinases in remodeling of the normal and neoplastic mammary gland. J Mammary Gland Biol Neoplasia 1998; 3(2):177-89.
140. Herrera-Gayol A, Jothy S. Adhesion proteins in the biology of breast cancer: Contribution of CD44. Exp Mol Pathol 1999; 66(2):149-56.
141. Naot D, Sionov RV, Ish-Shalom D. CD44: Structure, function, and association with the malignant process. Adv Cancer Res 1997; 71:241-319.
142. Peterson RM et al. Perturbation of hyaluronan interactions by soluble CD44 inhibits growth of murine mammary carcinoma cells in ascites. Am J Pathol 2000; 156(6):2159-67.
143. Wang CTZ, Moore IInd DH, Zhao Y et al. The overexpression of RHAMM, a hyaluronan-binding protein that regulates ras signaling, correlates with overexpression of mitogen-activated protein kinase and is a significant parameter in breast cancer progression. Clin Cancer Res 1998; 4(3):567-576.
144. Passegue E et al. Normal and leukemic hematopoiesis: Are leukemias a stem cell disorder or a reacquisition of stem cell characteristics? Proc Natl Acad Sci USA 2003; 100(Suppl 1):11842-9.
145. Smadja-Joffe F et al. CD44 and hyaluronan binding by human myeloid cells. Leuk Lymphoma 1996; 21(5-6):407-20, color plates following 528.
146. Erickson AC, Barcellos-Hoff MH. The not-so innocent bystander: The microenvironment as a therapeutic target in cancer. Expert Opin Ther Targets 2003; 7(1):71-88.
147. Sternlicht MD, W.Z. How matrix metalloproteinases regulate cell behavior. Annu Rev Cell Dev Biol 2001; 17:463-516.
148. Macaluso M, Paggi MG, Giordano A. Genetic and epigenetic alterations as hallmarks of the intricate road to cancer. Oncogene 2003; 22(42):6472-8.
149. Berx G, Van Roy F. The E-cadherin/catenin complex: An important gatekeeper in breast cancer tumorigenesis and malignant progression. Breast Cancer Res 2001; 3(5):289-93.
150. Elenbaas B et al. Human breast cancer cells generated by oncogenic transformation of primary mammary epithelial cells. Genes Dev 2001; 15(1):50-65.

CHAPTER 8

Branching Morphogenesis in Mammalian Kidneys

Jamie A. Davies

Introduction

B
ranching morphogenesis is an important mechanism for the development of the permanent kidneys of reptiles, mammals and birds. Branching of renal epithelia is similar to that seen in the other organs described in this book[1] but organogenesis of kidneys has unique features that, at the expense of some complication, offer an opportunity to address deep questions of both developmental and evolutionary biology. Understanding the development of branched epithelia in the kidney is also important medically because abnormalities of these epithelia are responsible for a number of serious diseases which, at an incidence of more than 1:800, are amongst the most common human congenital abnormalities.[2,3]

The branched epithelium of the kidney, the collecting duct system, exists mainly to channel urine to a common drainage duct, the ureter. This contrasts with the other organs described in this book, such as the mammary, salivary and prostate glands, in which cells derived from the branching epithelium are responsible for producing the secretions of the organ, as well as for channelling them to the outside world. In the mammary gland, for example, cells derived directly from the branching milk ducts produce mammary alveoli that secrete milk into the ducts (see Chapter 7). In kidneys, the main 'secretion' (urine) is made by nephrons, which are epithelial structures derived not directly from the collecting ducts but rather by a mesenchymal-epithelial transition in the tissue that surrounds them. These nephrons then connect to nearby ducts and drain in to them. The functions of 'secretion' and drainage are therefore almost separate in the kidney ('almost' because the collecting ducts do play a role in modifying the contents of urine, particularly acid-base balance).

The use of a branched drainage structure arose rather late in the evolutionary history of vertebrate excretory systems; the permanent kidneys of agnatha, fish and amphibians are unbranched (or show only rudimentary branching of the fusion type—see Chapter 1) and highly-branched kidneys arose only with animals whose entire life cycle can be spent out of water. This late acquisition of branching is not unique to the excretory system—the airways of lungs, for example, are highly-branched in mammals and birds but hardly branch at all in most reptiles.[4] What is unusual is that reptiles, mammals and birds still make the primitive forms of kidney (pronephroi and mesonephroi) during their embryonic lives and construct their branched kidneys (metanephroi) as completely new organs.

The first morphological sign of mammalian kidney development is the emergence of an epithelial tube called the ureteric bud. This forms as an outgrowth from an existing epithelial tube, the nephric duct, that runs down the cranio-caudal axis of the body and drains the temporary kidneys of the embryo (pro- and mesonephroi) (Fig. 1). The ureteric bud grows towards and invades an adjacent area of intermediate mesoderm, the metanephrogenic mesenchyme. Once in that mesenchyme, the bud begins to branch and continues to grow to create a

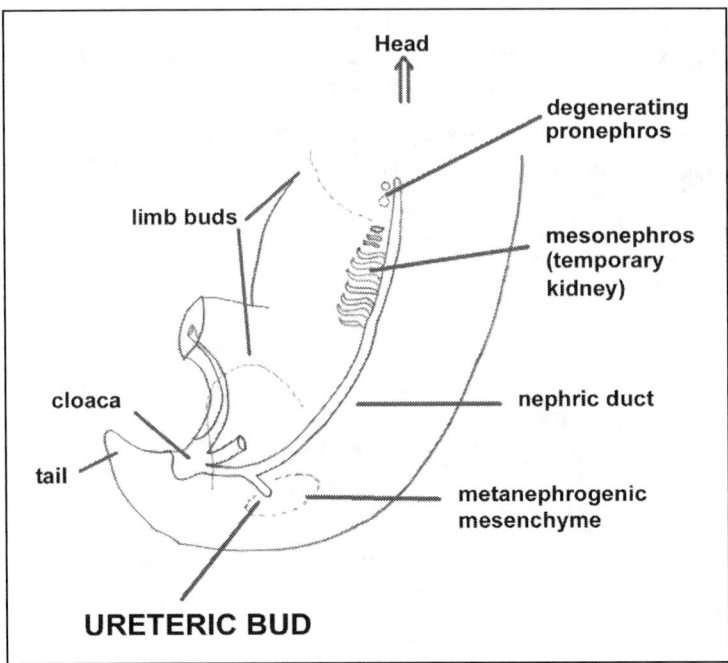

Figure 1. The general arrangement of the developing excretory system of an E10.5 mouse embryo.

tree-like arrangement of tubules (Fig. 2). As it grows, the epithelial tree induces nearby groups of mesenchymal cells to differentiate into nephrons which will later connect to it. The nephrons command a blood supply (to form the glomerular capillaries and the counter-current multiplication system) and also a nerve supply. Since the shape of the collecting duct controls the positions at which nephrons form, and the nephrons control the blood and nerve supplies, it is fair to say that the branching of the ureteric bud/ collecting duct determines the anatomy of the entire organ. The rest of this chapter will be dominated by discussion of the mechanisms and regulation of ureteric bud branching, in view of its importance, but I shall discuss the blood system at the end.

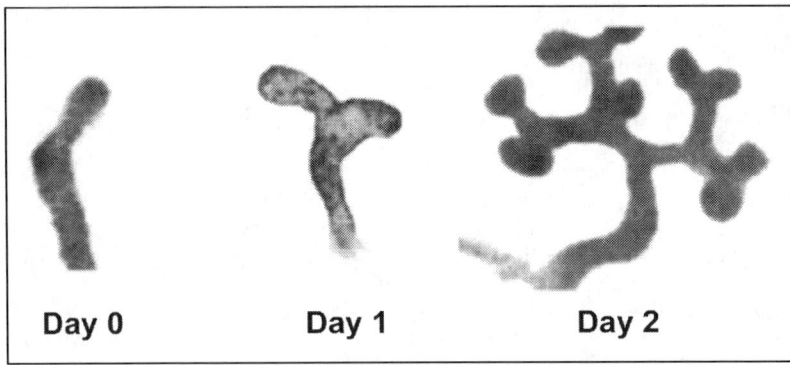

Figure 2. The branching of ureteric bud in organ culture (stained with anti-calbindin-D-28K, a marker for ureteric bud in this system[113]).

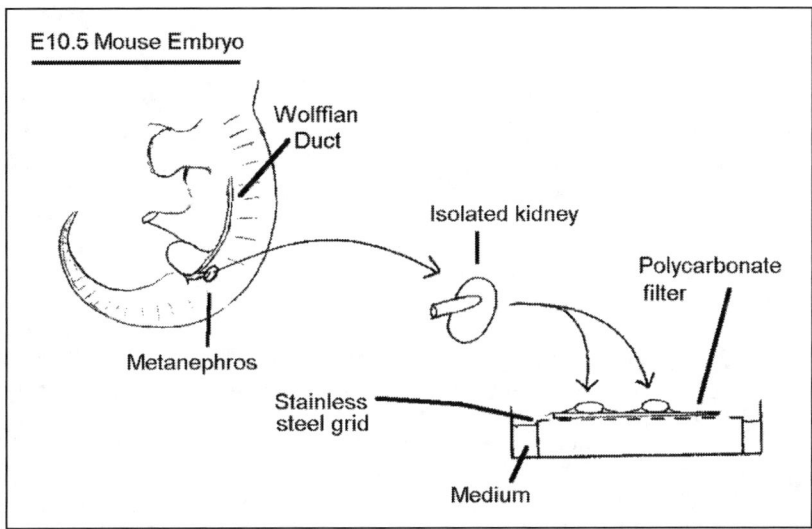

Figure 3. The standard method for mouse kidney organ culture.

Models for Studying Branching in the Kidney

Embryonic kidneys are unusually accessible for study, partly because they will grow well in organ culture and partly because mammalian embryos can rely completely on their mother's circulation for excretion so that renal abnormalities do not cause an early termination of foetal development.

One obvious 'model' for study of renal development is growth in vivo, of either completely normal animals or of mutants. This method has the advantage that development is seen in a realistic situation, but it has the disadvantages of poor access and of separating direct effects of a mutation on kidney development from indirect effects caused by abnormal development of other embryonic systems. It is, however, the only method suitable for the analysis of later events, because no culture system has yet managed to support kidney development up to mature stages.

Organ culture of isolated kidney rudiments has been a powerful model for studying mammalian organogenesis since the 1950s, when it was found that rudiments, isolated by microdissection at embryonic day 10.5 and cultured on a supporting filter at a gas-medium interface, would grow organotypically and would reproduce the first few days of renal development[5] (Fig. 3). Analysis of kidneys growing in culture can help to separate systemic effects of a mutation from local effects autonomous to the kidney.[6] Cultured kidneys are also accessible to antibodies, pharmacological reagents and exogenous growth factors, a fact that has been central to a large number of studies on the functions of particular molecules in renal development. Recently, we have developed a technique, based on small interfering RNAs (siRNAs), for inhibiting specific genes in cultured kidneys at a time of the experimenter's choosing;[7] this should greatly facilitate the molecular analysis of renal development. Isolated kidneys can be dissected into their component tissues (eg ureteric bud and mesenchyme), and these tissues can then be recombined and will develop normally—this allows experimenters to study the development of a ureteric bud from a mutant animal in the context of a wild-type mesenchyme and vice-versa. Such tissue recombination experiments have been valuable in determining precisely which tissues are affected by a mutation.[6]

Recently, a culture system has been developed in which isolated ureteric buds will grow and branch apparently normally in a 3-dimensional gel (in the presence of appropriate growth and survival factors).[8] This allows experimenters to study the regulation of ureteric bud morphogenesis in the absence of feedback loops operating in the mesenchyme, which used to be thought to be essential for branching to occur normally.[9,10] Finally, there is a culture system in which cell lines generated from ureteric bud or from mature collecting ducts are suspended in a 3-dimensional collagen matrix, in which they multiply and form cysts with the same polarity (apical domain facing the lumen) as a normal ureteric bud. When these cysts are treated with appropriate growth factors (such as HGF), they produce processes which then branch in a manner alleged, by those who use this system, to be similar to that of a real ureteric bud.[11] A great advantage of the cell-line model is that the cells can be transfected before use so that advanced genetic manipulations can be performed on them, something that is difficult for intact kidneys.

There is, as in other systems, some tension between proponents of simple culture models and those who insist that only experiments carried out in vivo are truly informative. Progress is usually fastest, however, when a combination of all techniques can be used so that simple models can generate hypotheses quickly and these can then be tested in, for example, transgenic animals. Regulation of ureteric bud branching by the GDNF signalling system (described later) is an excellent example of a story that draws on cell lines, isolated ureteric buds, cultured kidneys and transgenic animals and is much stronger for the combination.[12-16]

There is also an occasional tendency for commentators to reject claims that a molecule expressed in kidneys and shown to have an effect in organ culture is a regulator of renal development, if knockout of that molecule in vivo has no detectable phenotype. This rejection is based on a misunderstanding of the likely properties of a regulatory network. The few biological networks to have been studied mathematically have been found to possess the same general 'scale-free' architecture as man-made networks such as the Internet.[17] They would therefore be expected to show the similar responses to damage. Deleting random elements of such a network leads to its 'graceful failure;' a gradual reduction in efficiency with increasing numbers of deleted elements rather than a cataclysmic collapse.[18] Even knocking out up to 5% of the components randomly makes little difference to such a network as a whole (this is why the Internet is tolerant of the random failures of hardware that happen all the time). Only by targeting the few very critical elements can single deletions bring about serious damage. The lack of an obvious effect when any one particular renal gene is knocked out does not therefore mean that that gene has no role in the regulation of kidney development, but only that the gene is question is not one of those few critical network elements. It is to be hoped that an increasing understanding of biological networks will lay this confusion to rest, particularly as genetic experiments identify more and more partially-penetrant phenotypes which may well just be the expression of the declining efficiency seen in 'graceful failure'.

The Ureteric Bud Tip As an 'Organizer' of the Kidney

In many developing systems, one specific component seems to play such an important role in regulating the behaviour of all of the others that it is considered to be an 'organizer'. That is not to say that the other components do nothing or that the organizer is completely autonomous, but rather that most of the important regulatory pathways, even those originating elsewhere, pass via and are integrated by that organizer. The first organizers to be described were those of gross body structure, such as the dorsal lip of the frog blastopore,[19] but subsequent studies have identified organizers of more local development, such as the enamel knots that control the development of teeth.[20,21]

The tips of the branching ureteric bud/ collecting duct system control renal development to such an extent that they too seem to be organizers. Their many activities will be described in more detail below, but in brief summary; they are the main site of cell division in the ureteric bud, they are responsible for branching of the bud, they receive and integrate mesenchyme-derived signals that control bud morphogenesis, they originate signals that control mesenchyme development, proliferation and apoptosis, they induce differentiation of nephrons and they probably originate signals that ensure the correct spacing of collecting duct branches. In short, if the concept of an organizer has any validity in the kidney, then the bud tip has by far the strongest credentials for the role. It is of course possible that the whole language of 'organizing centres' is inappropriate and that further analysis of renal development will reveal it to be under the control of a much more distributed network, but studies of various types of real-world networks suggests that the presence of key integrating nodes is common. In networks as diverse as bacterial metabolism, interacting *Drosophila* proteins, littoral food webs, the World Wide Web hyperlinks and flight paths between airports, there are a few nodes through which very high traffic (/information/ energy) flows;[22-25] in networks controlling development, these would be called organizers.

In this chapter, I shall begin by describing the cell- and molecular-biological features of the ureteric bud tips, concentrating on those most closely connected to morphogenesis (reviews about other aspects of renal development may be found elsewhere[26,9,27]). I will then go on to explain how some of these features of the bud, combined with those of surrounding cells, might produce feedback systems that regulate branching morphogenesis.

Cell Biology of the Branch Tips

The tips of the ureteric bud/ collecting duct system are slightly bulb-shaped, when not actually branching. They are composed of a rather disorganised epithelium, which seems to have more than one layer (at least in rats) and which is not surrounded by the obvious, continuous basement membrane that can be seen around the stalks when examined by electron microscopy.[28] The tip cells also show rather few intercellular junctions compared with those of stalks and (at least in rabbits) do not express some of the cadherins that are expressed in the stalks.[29] This rather disorganised arrangement of the terminal epithelium may be an adaptation to allow rapid cell rearrangement during branching morphogenesis. Alternatively, or perhaps additionally, it may facilitate recruitment of extra cells from the surrounding mesenchyme.[30]

The process of branching alters the morphology of the tips cyclically, as emergence of new branches alternates with elongation. In mouse kidneys developing in organ culture, branching seems to take place mainly by simple bifurcation of the tips of the growing collecting duct tree. In microdissected human kidneys, though, the pattern seems to be more complicated, consisting of the emergence of a new tip just behind the old one followed by the bifurcation of this new tip only, resulting in a three-pointed structure consisting of the original tip and two new ones[31,32] (Fig. 4). The new tips elongate and, later, a new tip emerges just proximally to each and immediately bifurcates, the process repeating about 15 times in humans.[33] It is not clear whether the apparent difference between mouse culture data and human microdissection data results from a difference between species, a difference between behaviour in vivo and in vitro, or a difference between the early generations of branching seen in culture and the later generations that would have been represented in the human samples.

The morphogenetic mechanisms by which branching takes place are probably the least understood aspects of the ureteric bud tip. The kidney does not appear to use the clefting mechanism seen, for example, in the salivary gland (see Chapters 9, 12) in which bands of collagen divide an expanding ampulla into lobes that then extend, allowing the cycle to repeat.[34] Apart from the fact that there is no morphological evidence for clefting in the kidney, tissue inhibitors of metalloproteinases (TIMPs), which protect collagen from degradation, inhibit branching in the kidney[35,36] but do exactly the opposite in salivary glands.[37] This contrast suggests a fundamentally different mechanism may be at work. Rather than being generated by cleavage of an existing ampulla, new branch points seem to grow outwards from an existing ureteric bud tip, usually at a direction perpendicular to that of the original and 180 degrees away, at least initially (this may be seen from the excellent time-lapse sequences of the Costantini lab[38]). There are various possible mechanisms that might drive this emergence, but none have been investigated in any detail.

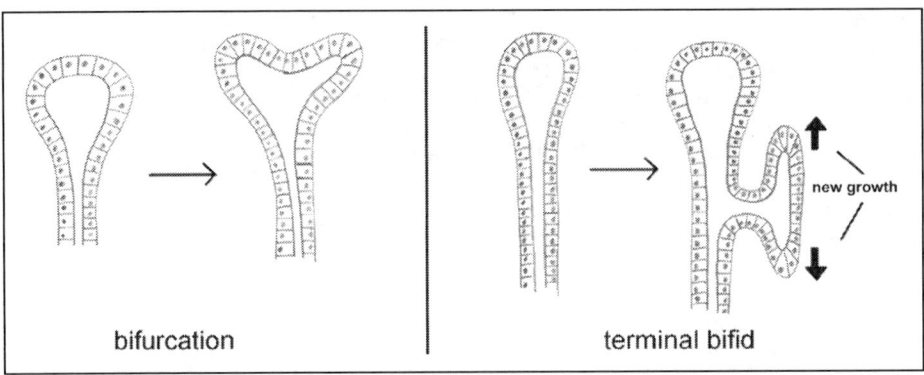

Figure 4. The two types of branching in the kidney: bifurcation, as seen in organ culture, and terminal bifid branching observed in later collecting duct development in humans. It is not yet clear whether the difference reflects developmental stage, species, or whether development is in vivo or in vitro.

One mechanism for branching that may be used in the kidney is epithelial folding driven by local contraction of apical actin, which would make cells wedge-shaped and therefore cause that part of the epithelium to curve outwards (Fig. 5). There are numerous bands of apical actin at the tip[39] and inhibitors of actin polymerisation and of myosin-mediated actin contraction inhibit branching in cultured mouse kidneys; these data support the mechanism, but the many roles of myosin makes such an experiment difficult to interpret. The relative lack of intercellular junctions in this area also makes the actin-contraction model less attractive, since the forces would have to be transmitted cell-to-cell by these junctions. This mechanism is used in other examples of epithelial development, though, such as invagination of the neural tube[40,41] or folding of the colon[42] so it should not be dismissed without further work.

Another possible way in which new tips are created, perhaps made more likely by the rather disorganized nature of the tip epithelium, is locomotion of epithelial cells. Fibroblast-like

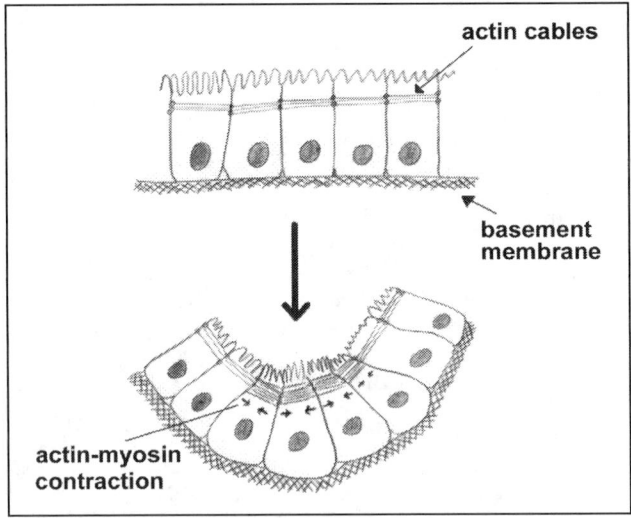

Figure 5. Bending of an epithelium by actin-myosin driven apical contraction.

locomotion of epithelia, using lamellipodia and filopodia, has been observed in wound healing in vertebrates,[43] in dorsal closure in *Drosophila*[44] and also in a variety of culture models. There has been no direct observation of lamellipodia and filopodia structures in ureteric bud cells, but when small amounts of cytochalasin are applied to kidneys growing in culture to compromise the integrity of the actin-adhaerens junction system, migratory cells do stream out specifically from the ureteric bud tips.[45] This suggests that these cells may be primed for migration, although it certainly does not prove the point. In most cells, the balance between actin being organised for contractile stress fibres or for lamellipodia and filopodia is controlled by competition between the small GTPases, Rac, Rho and cdc42. It would be very interesting to examine the activation states of these pathways during ureteric bud morphogenesis.

New tips might also perhaps be produced simply by localised cell proliferation and directional cytokinesis. Culture of mouse kidney rudiments in medium containing the thymidine analogue, bromodeoxyuridine (BrdU), reveals that there is much more cell proliferation in the tips than in the stalks as long as morphogenesis is taking place normally, but not when branching is inhibited by a variety of treatments.[39,46] It is not clear whether cell proliferation is a direct mechanism of morphogenesis or just a necessary accompaniment, although it is striking that a local increase in proliferation is seen in the nephric duct before any obvious morphological sign of ureteric bud emergence, suggesting that elevated proliferation is not simply a consequence of morphogenesis.[46] Proliferation in the tips is controlled, either directly or indirectly, by the MAP-kinase signalling pathway.[47]

Molecular Markers of 'Tip' Character

Until recently, it has not been obvious whether the special behaviour of tip cells reflects an unique state of gene expression in these cells or just different behaviours of otherwise identical cells being driven simply by the different shapes and stresses in the tissue. In recent years, however, evidence has been obtained for differential gene expression between tip and stalk, suggesting that tip cells are in a specific state of differentiation. The most striking marker for the tip cells is expression of *wnt11* gene, which encodes a signalling protein: *wnt11* is expressed only by the cells at the very tip and vanishes as soon as they leave for the stalk.[48-50] The as-yet uncharacterised glycoprotein that bears a ligand for the *Dolichos biflorus agglutinin* lectin has a reciprocal expression pattern, staining the stalks of the ureteric bud but not extending into the tip (Sweeney, Michael, Davies unpublished). Other genes, such as *ret* and *ros*, have been reported to be expressed in 'tips', but the regions described as 'tips' in these reports extend much further into the stalk than the region described as a 'tip' in this Chapter, and staining for the proteins confirms a less-restricted pattern.[39]

During normal growth, it is natural to assume that new stalk cells differentiate from tip cells that are 'left behind' by the advancing tip. In support of this view are the facts that most bud proliferation takes place in the tip, and that the ureteric bud begins as nothing but a 'tip' so it is difficult to see from where else stalk cells could come. The 'natural' direction of differentiation, from tip to stalk, may be reversible, though, and this might have important implications for the shape of the collecting duct tree. In a recent series of experiments, we have used microdissection to separate the tip and stalk regions of a once-branched ureteric bud (discarding the intermediate portion whose status may be ambiguous) and have cultured each in the presence of embryonic kidney mesenchyme. Isolated tips, which begin Wnt11-positive and DBA-negative, slowly generate a branched structure which includes DBA-positive stalks; this is not surprising given the 'normal' direction of differentiation. Isolated stalks, which begin DBA-positive and Wnt11-negative, seem to generate new tip regions which are Wnt11-positive (Sweeney, Michael and Davies, unpublished). This surprising result suggests that stalk cells can reverse their differentiation to become tip cells again, and may explain why cell-lines derived from mature collecting ducts are able to form branching structures when cultured in collagen gels.[51,52] Furthermore, the apparent ability of stalk to generate new tips when an existing tip has been removed suggests that the presence of a nearby tip normally inhibits stalk-to-tip differentiation; this may be an important mechanism for spacing the branches of the tree, and will be discussed in more detail below.

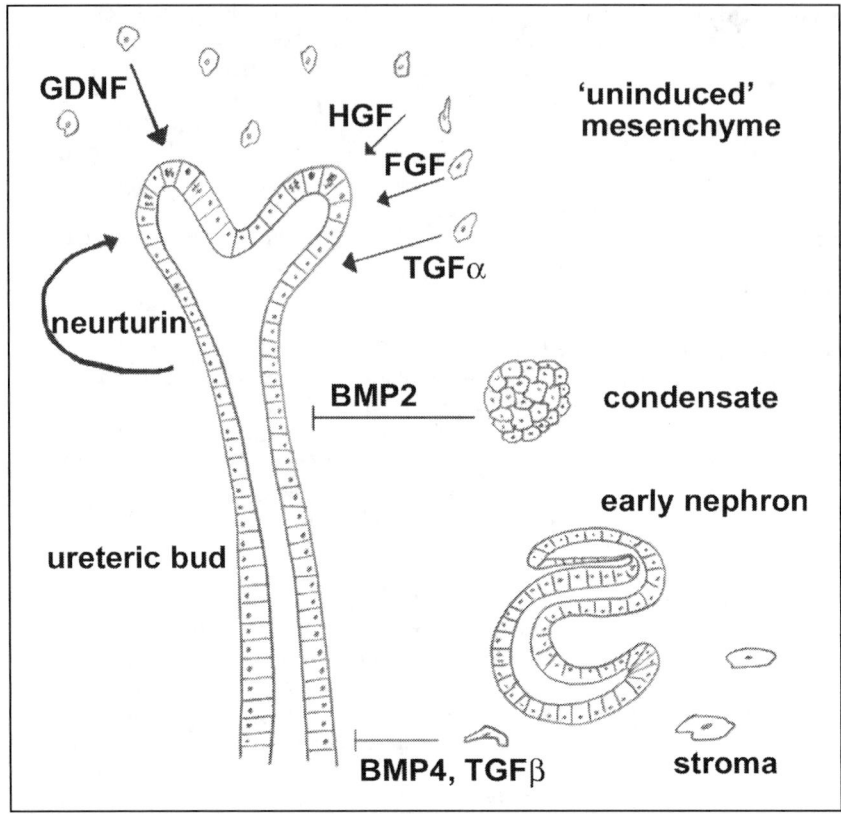

Figure 6. Regulatory inputs to the ureteric bud. Inputs that encourage branching morphogenesis are shown by ↓ and inputs that inhibit branching by ⊥.

Regulatory Inputs to the Bud Tips

Cells of the ureteric bud tip bear a large number of receptors for signalling molecules that are made by surrounding tissues (Fig. 6). One of the most important is the receptor system for glial cell-line derived neurotrophic factor (GDNF), which is made by the 'uninduced' mesenchymal cells into which the bud tips invade. The receptor complex for GDNF is composed of the Ret receptor tyrosine kinase, a GFRα1 co-receptor and 2-O-sulphated heparan sulphate glycosaminoglycan, which also seems to serve as a co-receptor.[53,54] Transgenic mice that are defective in any of these components, or cultured organs that have been subjected to treatments that remove or inhibit any of these components, cause failure of ureteric bud branching.[55-60] Conversely, provision of exogenous GDNF in culture increases the amount of branching, up to a point, and local sources of GDNF can cause the production of ectopic ureteric buds from the Wolffian duct.[61,62] GDNF therefore seems to serve an analogous role to that played by FGF7/10 signalling through FGFRIIIb in the branching epithelia of lungs, salivary glands, mammary glands, prostate and pancreas in that it seems to be the principal driver of branching. GDNF cannot be the only signal capable of inducing the emergence of the ureteric bud itself, though, because some *gdnf-/-* mice do manage to produce a ureteric bud although it does not then go on to branch.[63,64]

The Ret receptor tyrosine kinase also acts as a receptor for neurturin, produced by the bud itself, and persephin, both of which are members of the GDNF family and both of which stimulate ureteric bud branching in organ culture. The ureteric bud also bears the Met receptor tyrosine kinase which is stimulated by hepatocyte growth factor (HGF) from the surrounding mesenchyme, the EGF receptor which is stimulated by TGFα from the bud itself, and FGF receptors which respond to mesenchyme-derived FGF: all of these stimulate bud growth and branching.[65,66,67,68] Different FGFs evoke different responses from ureteric buds in simple culture systems; FGFs1 and 10 induce long branches with ampullary tips, and FGFs 2 and 7 induce a more general and less organised proliferation.[69] Each of these receptor systems can signal intracellularly via the MAP-kinase and PI-3-kinase pathways, and there is evidence that implicates each of these pathways in the stimulation of ureteric bud branching.[39,70]

As well as having receptor tyrosine kinases, the bud expresses receptor serine/threonine kinases for BMPs, TGFβ and activin.[71-73] These receptors, which generally signal via Smad proteins, all tend to inhibit branching (although BMP7 is a little complicated, encouraging branching when applied at low concentrations but inhibiting it when applied at higher concentration, at least in complete cultured organs). The ability of TGFβ and BMPs to inhibit branching is a feature not just of kidneys, but also of lungs, pancreas and salivary, mammary and prostate glands.[74]

The ureteric bud bears also receptors for matrix components, particularly integrins. At least some of these, for example integrins containing the α8 chain, are necessary for normal morphogenesis.[75] Integrin α8 can associate with a variety of ligands (e.g., fibronectin, virtonectin, tensacin), but a kidney-specific matrix component called nephronectin seems to be a particularly important α8 ligand for development of the ureteric bud.[76] Connections within the matrix itself are important too, and inhibiting interactions between matrix components such as laminin and nidogen inhibits branching.[77] It is not clear whether the matrix provides an instructive or merely a permissive role in regulating bud morphogenesis.

Many of the receptors described above are not expressed exclusively in the tips, but also extend some way along the stalks. Some, such as Ret, extend back a relatively short distance while others, such as the serine/threonine kinases, are expressed along much of the length of the system.

Morphoregulatory Outputs from the Ureteric Bud Tips

As well as bearing receptors for signalling molecules coming from elsewhere, the ureteric bud is the origin of a number of important signals (Fig. 7). Some of these, such as neurturin, appear to act on the bud itself[78] but most act on the cells that surround it. At least three bud-derived signals, TGFα, FGF2 and TIMP2, act as survival factors and mitogens for cells of the mesenchymal blastema into which the ureteric bud grows.[79-84] Without these factors, the mesenchyme dies by apoptosis[85] but with them it proliferates enough to maintain growth of the organ rudiment as well as to contribute cells to various pathways of differentiation. It is interesting that some function of TIMP2 other than its well-known ability to inhibit metalloproteinase activity seems to be required in this context, because pharmacological inhibitors of the same metalloproteinases cannot subsititute for TIMP2.[86] The bud tips (and only the tips) produce Wnt11, a signalling protein that stimulates production of GDNF by the surrounding mesenchyme.[87-90]

The ureteric bud also makes an inductive signal, which may include the above growth factors and which probably includes as-yet unidentified components as well, that cause groups of mesenchyme cells to condense together and differentiate into nephrons. This inductive process is still not understood well—it is not even certain whether the mesenchyme is a homogenous population, the cells of which have the choice between remaining blastemal, differentiating into nephrons or into stroma, or whether the nephron and stroma lineages are distinct from the start. Once they have been induced to form nephrons, cells draw together into a tight cell condensate which then epithelializes to form a cyst. The cyst, the sequence of developmental events followed by these cells is reasonably well understood. They aggregate undergoes a stereotyped series of morphogenetic movements to produce a comma-shaped body, and S-shaped

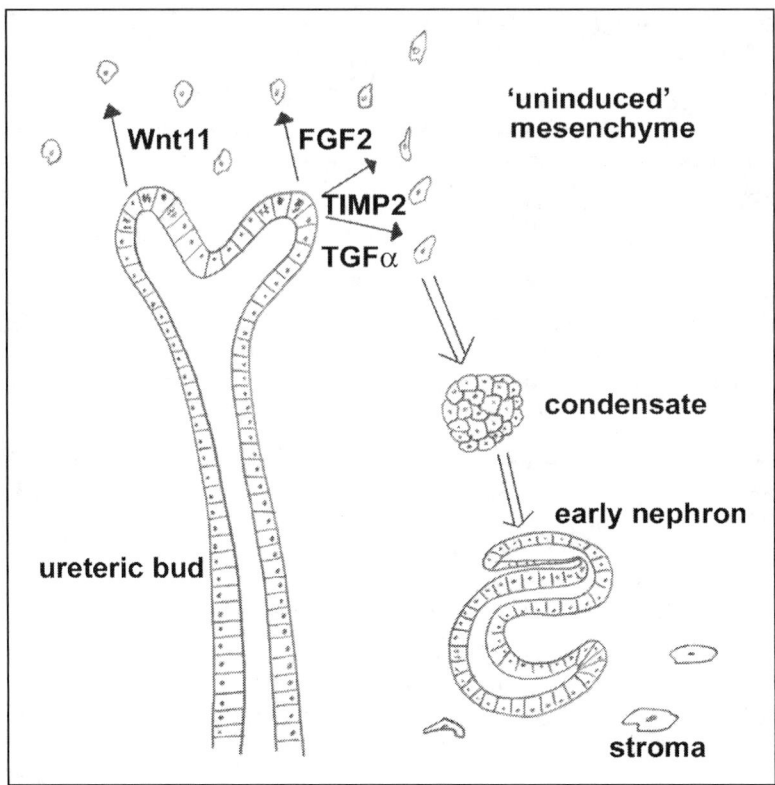

Figure 7. Regulatory outputs from the ureteric bud. The main outputs of the bud prevent mesenchymal apoptosis and drive the differentiation of mesenchyme to nephrons.

body and then the mature nephron. The changes in gene expression that take place during this process have been reviewed extensively elsewhere[9,91,92] so, because they are not really to do with branching, they will not be described in detail here. One aspect of them that is relevant to the branching of the ureteric bud is that cells at different stages of nephron and stroma differentiation produce different sets of growth factors, and these signal back to the ureteric bud (see the section on feedback, below).

Some 'outputs' of the ureteric bud may play a direct role in morphogenesis (rather than an indirect one by signalling to other tissues). One example is MT1-MMP, which is a membrane-bound metalloproteinase that cleaves and activates other metalloproteinases, that can in turn digest components of the extracellular matrix.[93] Ths digestion process may modulate signalling, by releasing matrix-bound growth factors, but it may also be important in simply clearing a path along which the ureteric bud can advance. In organ culture, metalloproteinases such as MMP9 are essential for ureteric bud branching.[94] This is also true for isolated ureteric buds in culture and for cell-line models.[95] The activation of metalloproteases is itself controlled by factors coming from the mesenchyme (/mesenchyme substitute), suggesting another level of feedback.[96]

Feedback Loops That Control Bud Branching

The simple descriptions of signalling to and from the ureteric bud tip presented above highlight two key facts; tip branching normally takes place only in the presence of the correct mesenchyme-derived signals, and signals coming from the tip cause the mesenchyme to alter its expression of signalling proteins. These facts combine to create a feedback system that could potentially control the shape and extent of ureteric bud branching in the complete organ.

The main positive regulators of ureteric bud development, GDNF, HGF etc, are made by the mesenchymal blastema before it is invaded by the bud, and therefore before it is induced to differentiate into nephrons. Once the bud has begun to branch, the only parts of the kidney that are in this branch-promoting state would be those at its periphery, beyond the current reach of the bud. Branch-promoting factors would therefore be expected to form a gradient, increasing centrifugally and potentially guiding the ureteric bud tips outwards (although the presence of such a gradient has not yet been proven by direct measurement of protein concentrations). As soon as they begin to condense and to differentiate into nephrons, mesenchymal cells lose their expression of GDNF and HGF.[97,98] The processes of induction and mesenchymal differentiation take time (in culture, obvious condensation of mesenchyme takes place about 18 hours after contact with the inducing tissue[99]), so the mesenchymal cells that have just been reached by and induced by contact with the bud will still produce GDNF for a short while. By the time their expression of GDNF has been lost, the bud will have moved on a little way to invade fresh mesenchyme. The result of this is that, while the virgin mesenchyme that lies just ahead of the bud tip will be a rich source of GDNF, and that immediately around it will still have some, the mesenchyme that lies behind it will produce none.

The main negative regulators of ureteric bud branching, BMP2, BMP4 and TGFβ, are expressed only by (ex-)mesenchymal cells that are reaching more advanced states of differentiation. BMP2 is made by cells as they condense and become epithelial,[100] while TGFβ is expressed by the bud itself and by mesenchymal cells that have differentiated into mature stroma instead of nephrons, and BMP4 by both stroma and developing nephrons.[101,102] This pattern of expression probably reinforces the message given by the decline of GDNF expression in differentiating cells: the ureteric bud is provided with encouragement to grow and branch from ahead, but receives strong "keep out" signals from the zones that it has already induced into nephrogenesis (Fig. 8). This would be important, because invasion of groups of cells that are already forming nephrons by new bud branches would probably result in a tangled mess rather than orderly morphology.

Invasion and new branching are therefore confined mainly to the zone of virgin mesenchyme at the cortex of the developing organ and will take place proximally to that only if there happens to be a region of mesenchyme that has escaped earlier invasion by the bud and that therefore still expresses GDNF and does not yet express branch inhibitors. Even in the proximal zone, the subtle gradients of concentrations of branch activators and inhibitors may be used to direct branching so that a series of essentially two-dimensional branching events builds a tree that fills up three-dimensional space (the two-dimensional nature of organ culture makes this three-dimensional space-filling process difficult to observe by time-lapse photography).

It is perhaps surprising, in view of the potential feedback loops described above, that ureteric bud branching will still occur in the absence of normal mesenchymal differentiation. This can be seen to some extent in kidneys in which mesenchymal differentiation has been inhibited by specific treatments,[103] but is shown most dramatically by the ability of isolated ureteric buds to develop in three-dimensional matrices when provided with appropriate growth factors (the combination of matrix and growth factors acting as a mesenchyme-substitute.[104] Although the branching seen in such circumstances is not exactly of the form seen in real organs, the bud is clearly capable of organising itself into tips and stalks, and of spacing out branch points appropriately. This suggests that the bud has an intrinsic mechanism for suppressing the formation of tips (branches) that are too close together, and for avoiding collisions. The most obvious way in which this might occur would be for tips to suppress tip formation in nearby cells, and for both tips and stalks to secrete a repellant that prevents new branches from colliding with them. There is not yet any direct evidence for such inhibitory or repulsive mechanisms. The most promising hint is the presence in the ureteric bud of murine homologues of the *Droshopila* gene *sprouty*.[105] In *Drosophila*, *sprouty* suppresses the formation of tips in the branching

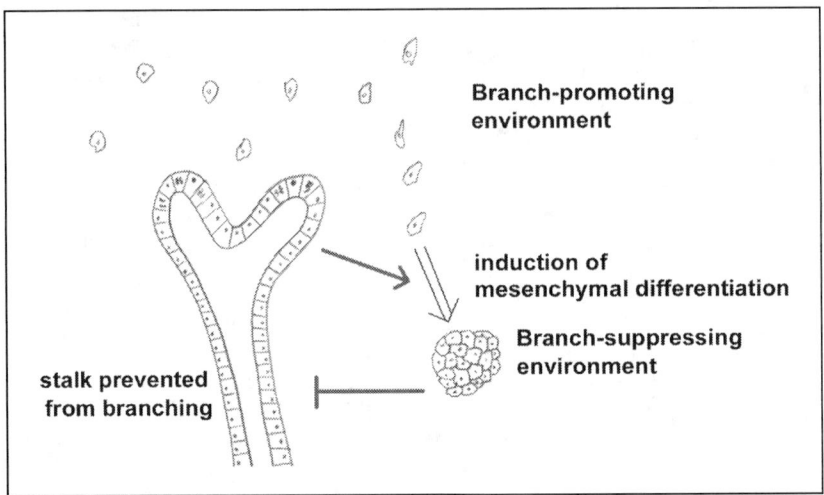

Figure 8. Hypothetical feedback loop that restricts branching activity to the tips.

tracheal system.[106] If *mSproutys* work in a similar manner in mice, this may be the basis of lateral inhibition by tip cells, and release from this inhibition may be the mechanism that allows regeneration of tips from stalks, described earlier.

As well as being driven by genetic mechanisms, the branching behaviour of the ureteric bud, whether in isolation or in the mesenchyme, may be controlled by the mechanical/ fluid dynamic forces of viscous fingering. These are discussed at length in Chapter 12 so will not be described here.

Branching of the Renal Blood System

The blood system of the mature kidney follows a branched arrangement. Arterial blood flows towards the kidney via the renal artery, which branches directly from the abdominal aorta and receives 20% of the cardiac output.[107] In humans, the renal artery divides into segmental arteries just before these arteries enter the kidney; in mice there is only one segment. Once in the kidney, the segmental arteries divide again to form interlobar arteries, which give rise to arcuate arteries that serve the nephrons themselves. Arterioles take blood from the arcuate arteries to a tight complex of specialized capillaries in the glomerulus, and a second arteriole conducts blood from the glomerulus to a second capillary network that surrounds other parts of the nephron; finally it drains, via venules and a vein network that follows the general arrangement of the arterial one, to the vena cava.

In general, two different processes can give rise to embryonic blood systems; angiogenesis produces blood vessels by branching from pre-existing vessels, in a manner morphologically similar to the branching of epithelia, and vasculogenesis produces them from mesenchyme-like precursor cells already present in the tissues. In kidneys, there is evidence for both processes. Grafting embryonic kidney rudiments to sites already rich in a blood supply, such as the chorioallantoic membrane of a bird egg, results in the formation of a blood system derived from the host tissue by angiogenesis.[108] Culture of isolated kidney rudiments in low oxygen, or in the presence of exogenous VEGF, causes the formation of primitive blood systems by vasculogenesis.[109-111]

Developing nephrons, especially the glomeruli, are strong sources of VEGF and attract endothelial precursors, this attraction being susceptible to being blocked by antibodies against VEGF.[112] The final arrangement of fine blood vessels within the glomerulus is complex, and probably arises by intussusceptive division of vessels as described in more detail in Chapter 6. The fine capillaries of kidneys probably arise by vasculogenesis, while the largest vessels are more likely to arise by angiogenesis from the great vessels of the rest of the body. Unfortunately, very little is known about how these processes cooperate in renal development, or how the blood system aligns with that of the epithelia, although migration of vessels along the ureteric bud and its major branches is one obvious possibility.

Conclusions and Perspectives

Understanding the branching of the ureteric bud is, then, key to understanding the morphogenesis of the kidney as a whole. The last two decades have identified many signals are produced by or that converge on to the tips of the ureteric buds. While more of these signals no doubt remain to be discovered, it is perhaps more important to focus now on the largely uncharacterised mechanisms that couple these signals to morphological change. Some potential mechanisms have been implied by recent data, including roles for cell division, modulation of cell adhesion and cell locomotion: it would be useful to test these carefully, and to assess the extent to which known modulators of these processes, such as small GTPases, play a role in kidney development. Another very promising area of research is that of the feedback systems that seem to restrict 'tip' character to limited areas of the epithelium, and may be responsible for ensuring that the entire kidney is supplied with collecting ducts without any zones being over-supplied: it will be interesting to learn whether the same systems operate in all branching epithelia.

Understanding the ureteric bud is important for medical reasons as well: dysmorphologies of the bud produce clinically-devastating cystic diseases, and it is possible that the invasive behaviour of renal carcinomas results in part from inappropriate reactivation of the invasive mechanisms used in development. A deeper knowledge of these processes might allow them to be modulated medically, and offer the possibility or treatment, or at least amelioration, of debilitating and life-threatening medical conditions.

Acknowledgements

I would like to thank my colleague, Dr Jane Armstrong, for her helpful comments and suggestions, and The Leverhulme Trust, The Anatomical Society and BBSRC for funding work presented here as 'unpublished data'.

References

1. Davies JA. Do different branching epithelia use a conserved developmental mechanism? Bioessays. 2002; 24:937-948.
2. Woolf AS, Winyard JDP, Hermanns MM et al. Maldevelopment of the human kidney and lower urinary tract:an overview. In: Vize PD, Woolf AS, Bard JBL, eds. The Kidhey—From Normal Development to Congenital Disease. Academic Press, 2003:377-393.
3. Mulroy S, Boucher C, Winyard P et al. Cystic renal diseases. In: Vize PD, Woolf AS, Bard JBL, eds. The Kidhey: From Normal Development to Congenital Disease. Academic Press, 2003:433-450.
4. George JC, Shah RV. Evolution of air sacs in sauropsida. J Anim Morphol Physiol 1965; 12:255-263.
5. Grobstein C. Inductive epitheliomesenchymal interaction in cultured organ rudiments of the mouse. Science 1953; 118:52-55.
6. Kreidberg JA, Sariola H, Loring JM et al. WT-1 is required for early kidney development. Cell 1993; 74:679-691.
7. Davies JA, Ladomery M, Hohenstein P et al. Development of an siRNA-based method for repressing specific genes in renal organ culture and its use to show that the Wt1 tumour suppressor is required for nephron differentiation. Hum Mol Genet 2004; 13:235-246.
8. Qiao J, Sakurai H, Nigam SK. Branching morphogenesis independent of mesenchymal-epithelial contact in the developing kidney. Proc Natl Acad Sci USA 1999; 96:7330-7335.
9. Davies, JA, Fisher CE. Genes and proteins in renal development. Exp Nephrol. 2002; 10:102-13.
10. Davies JA. Do different branching epithelia use a conserved developmental mechanism? Bioessays. 2002; 24:937-948.

11. Montesano R, Schaller G, Orci L. Induction of epithelial tubular morphogenesis in vitro by fibroblast-derived soluble factors. Cell 1991; 66:697-711.
12. Davies JA, Yates EA, Turnbull JE. Structural determinants of heparan sulphate modulation of GDNF signalling. Growth Factors 2003; 21:109-119.
13. Moore MW, Klein RD, Farinas I et al. Renal and neuronal abnormalities in mice lacking GDNF. Nature 1996; 382:76-79.
14. Pichel J, Shen L, Sheng HZ et al. Defects in enteric innervation and kidney development in mice lacking GDNF. Nature 1996; 382:73-76.
15. Sainio K, Suvanto P, Davies J. et al. Glial-cell-line-derived neurotrophic factor is required for bud initiation from ureteric epithelium. Development 1997; 124:4077-4087.
16. Majumdar A. Vainio S. Kispert A. et al. Wnt11 and Ret/Gdnf pathways cooperate in regulating ureteric branching during metanephric kidney development. Development 2003; 130:3175-3185.
17. Jeong H, Tombor B, Albert R et al. The large-scale organization of metabolic networks. Nature. 2004; 407:651-654.
18. Albert R. Jeong H. Barabasi AL. Error and attack tolerance of complex networks. Nature 2000; 406:378-382.
19. Spemann H, Mangold H. Induction of emrbyonic primordia by implantation of organizers from a different species. In: Foundations of Experimental Embryology. New York: Hafner, 1924:144-184.
20. Jernvall J, Kettunen P, Karavanova I et al.Evidence for the role of the enamel knot as a control center in mammalian tooth cusp formation:non-dividing cells express growth stimulating Fgf-4 gene. Int J Dev Biol 1994; 38:463-469.
21. Thesleff I, Keranen S, Jernvall J. Enamel knots as signaling centers linking tooth morphogenesis and odontoblast differentiation. Adv Dent Res 2001; 15:14-18.
22. Giot L, Bader JS, Brouwer C et al. A protein interaction map of Drosophila melanogaster. Science. 2003; 302:1727-1736.
23. Wang XF, Chen G. Complex Networks: small-world, scale-free and beyond. IEEE Ciruites & Systems Magazine 2003; 1:6-20.
24. Jeong H, Tombor B, Albert R. The large-scale organization of metabolic networks. Nature. 2000; 407:651-654.
25. Albert R, Hawoong J, Barabasi A-L. Diameter of the World Wide Web. Nature. 401:130.
26. Sariola, H. Nephron induction revisited:from caps to condensates. Curr Opin Nephrol Hypertens 2002; 11:17-21.
27. Vize PD, Woolf AS, Bard JBL. The Kidney: From Normal Development to Congenital Disease. Academic Press, 2003.
28. Qiao J, Cohen D, Herzlinger D. The metanephric blastema differentiates into collecting system and nephron epithelia in vitro. Development 1995; 121:3207-3214.
29. Thomson BB, Biemesderfer D, Aronson PS. Developmental regulation of Ksp-cadherin expression in rabbit kidneys. J Am Soc Nephrol 1995; 6:711.
30. Qiao J, Cohen D, Herzlinger D. The metanephric blastema differentiates into collecting system and nephron epithelia in vitro. Development 1995; 121:3207-3214.
31. al Awqati Q, Goldberg MR. Architectural patterns in branching morphogenesis in the kidney. Kidney Int 1998; 54:1832-1842.
32. al Awqati Q, Goldberg MR. Architectural patterns in branching morphogenesis in the kidney. Kidney Int 1998; 54:1832-1842.
33. al Awqati Q, Goldberg MR. Architectural patterns in branching morphogenesis in the kidney. Kidney Int 1998; 54:1832-1842.
34. Nakanishi Y, Ishii T. Epithelial shape change in mouse embryonic submandibular gland:modulation by extracellular matrix components. Bioessays 1989; 11:163-167.
35. Pohl M, Sakurai H, Bush KT et al. Matrix metalloproteinases and their inhibitors regulate in vitro ureteric bud branching morphogenesis. Am J Physiol Renal Physiol 2000; 279:F891-F900.
36. Barasch J, Yang J, Qiao J et al. Tissue inhibitor of metalloproteinase-2 stimulates mesenchymal growth and regulates epithelial branching during morphogenesis of the rat metanephros. J Clin Invest 1999; 103:1299-1307.
37. Hayakawa T, Kishi J, Nakanishi Y. Salivary gland morphogenesis:possible involvement of collagenase. Matrix Suppl 1992; 1:344-351.
38. Srinims S, Goldberg MR, Watanabe T et al. Expression of green fluorescent protein in the uneteric bud of transgenic mice: A new bid for the analysis of unuteric bud morphogenesis. Dev Genet 1999; 24:241-251.
39. Fisher CE, Michael L, Barnett MW et al. Erk MAP kinase regulates branching morphogenesis in the developing mouse kidney. Development 2001; 128:4329-38.

40. Lee HY, Kosciuk MC, Nagele RG et al. Studies on the mechanisms of neurulation in the chick:possible involvement of myosin in elevation of neural folds. J Exp Zool 1983; 225:449-457.
41. Jacobson AG. Normal neurulation in amphibians. Ciba Found Symp 1994; 181:6-21.
42. Colony PC, Conforti JC. Morphogenesis in the fetal rat proximal colon:effects of cytochalasin D. Anat. Rec 1993; 235:241-252.
43. Jacinto A, Martinez-Arias A, Martin P.Mechanisms of epithelial fusion and repair. Nat.Cell Biol. 2001; 3:E117-E123.
44. Wood W, Jacinto A, Grose R et al. Wound healing recapitulates morphogenesis in Drosophila embryos. Nat Cell Biol 2002; 4:907-912.
45. Davies J. Intracellular and extracellular regulation of ureteric bud morphogenesis. J Anat. 2001; 198:257-264.
46. Michhael L, Davies JA. Pattern and regulation of cell proliferation during morphogenesis of the mouse ureteric bud. J Anatomy 2004; 204:241-255..
47. al Awqati Q, Goldberg MR. Architectural patterns in branching morphogenesis in the kidney. Kidney Int 54:1998; 1832-1842.
48. Kispert A, Vainio S, Shen L et al. Proteoglycans are required for maintenance of Wnt-11 expression in the ureter tips. Development 1966; 122:3627-3637.
49. Gavin BJ, McMahon JA, McMahon AP. Expression of multiple novel Wnt-1/int-1-related genes during fetal and adult mouse development. Genes Dev 1990; 4:2319-2332.
50. Lako M, Strachan T, Bullen P et al. Isolation, characterisation and embryonic expression of WNT11, a gene which maps to 11q13.5 and has possible roles in the development of skeleton, kidney and lung. Gene 1998; 219:101-110.
51. Montesano R, Schaller G, Orci L. Induction of epithelial tubular morphogenesis in vitro by fibroblast-derived soluble factors. Cell 1991; 66:697-711.
52. Sakurai H. Nigam SK. Transforming growth factor-beta selectively inhibits branching morphogenesis but not tubulogenesis. Am J Physiol 1997; 272:F139-F146.
53. Davies JA, Yates EA, Turnbull JE. Structural determinants of heparan sulphate modulation of GDNF signalling. Growth Factors 2003; 21:109-119.
54. Saarma M. Sariola H. Other neurotrophic factors:glial cell line-derived neurotrophic factor (GDNF). Microsc Res Tech 1999; 45:292-302.
55. Pichel JG, Shen L, Sheng HZ et al. Defects in enteric innervation and kidney development in mice lacking GDNF. Nature 1996; 382:73-76.
56. Moore MW, Klein RD, Farinas I et al. Renal and neuronal abnormalities in mice lacking GDNF. Nature 1996; 382:76-79.
57. Schuchardt A, D'Agati V, Larsson-Blomberg L et al. Defects in the kidney and enteric nervous system of mice lacking the tyrosine kinase receptor Ret. Nature 1994; 367:380-383.
58. Schuchardt A, D'Agati V, Pachnis V et al. Renal agenesis and hypodysplasia in ret-k- mutant mice result from defects in ureteric bud development. Development. 1996; 122:1919-1929.
59. Enomoto H, Araki T, Jackman A et al.GFR alpha1-deficient mice have deficits in the enteric nervous system and kidneys. Neuron 1998; 21:317-324.
60. Bullock SL, Fletcher JM, Beddington RS et al. Renal agenesis in mice homozygous for a gene trap mutation in the gene encoding heparan sulfate 2-sulfotransferase. Genes Dev 1998; 12:1894-1906.
61. Sainio K, Suvanto P, Davies J et al. Glial-cell-line-derived neurotrophic factor is required for bud initiation from ureteric epithelium. Development 1997; 124:4077-4087.
62. Vega QC, Worby CA, Lechner MS et al.Glial cell line-derived neurotrophic factor activates the receptor tyrosine kinase RET and promotes kidney morphogenesis. Proc Natl Acad Sci USA 1996; 93:10657-10661.
63. Pichel JG, Shen L, Sheng HZ et al.Defects in enteric innervation and kidney development in mice lacking GDNF. Nature 1996; 382:73-76.
64. Moore MW, Klein RD, Farinas I et al. Renal and neuronal abnormalities in mice lacking GDNF. Nature 1996; 382:76-79.
65. Woolf AS, Kolatsi-Joannou M, Hardman P et al.Roles of hepatocyte growth factor/scatter factor and the met receptor in the early development of the metanephros. J Cell Biol 1995; 128:171-184.
66. Hrabe de Angelis MH, Flaswinkel H, Fuchs H et al.Genome-wide, large-scale production of mutant mice by ENU mutagenesis. Nat Genet 2000; 25:444-7.
67. Bernardini N, Mattii L, Bianchi F et al. TGF-alpha mRNA expression in renal organogenesis:a study in rat and human embryos. Exp Nephrol 2001; 9:90-98.
68. Coles HS, Burne JF, Raff MC. Large-scale normal cell death in the developing rat kidney and its reduction by epidermal growth factor. Development 1993; 118:777-784.
69. Qiao J, Bush KT, Steer DL et al. Multiple fibroblast growth factors support growth of the ureteric bud but have different effects on branching morphogenesis. Mech Dev 2001; 109:123-135.

70. Tang MJ, Cai Y, Tsai SJ. et al. Ureteric bud outgrowth in response to RET activation is mediated by phosphatidylinositol 3-kinase. Dev Biol 2002; 243:128-136.

71. Ritvos O, Tuuri T, Eramaa M et al. Activin disrupts epithelial branching morphogenesis in developing glandular organs of the mouse. Mech Dev. 1995; 50:229-245.

72. Martinez G, Loveland KL, Clark AT et al. Expression of bone morphogenetic protein receptors in the developing mouse metanephros. Exp Nephrol 2001; 9:372-379.

73. Lehnert SA, Akhurst RJ. Embryonic expression pattern of TGF beta type-1 RNA suggests both paracrine and autocrine mechanisms of action. Development 1988; 104:263-273.

74. Davies JA. Do different branching epithelia use a conserved developmental mechanism? Bioessays 2002; 24:937-948.

75. Muller U, Wang D, Denda S et al. Integrin alpha8beta1 is critically important for epithelial-mesenchymal interactions during kidney morphogenesis. Cell 1997; 88:603-613.

76. Brandenberger R, Schmidt A, Linton J, et al. Identification and characterization of a novel extra-cellular matrix protein nephronectin that is associated with integrin alpha8beta1 in the embryonic kidney. J Cell Biol 2001; 154:447-458.

77. Ekblom P, Ekblom M, Fecker L et al. Role of mesenchymal nidogen for epithelial morphogenesis in vitro. Development 1994; 120:2003-2014.

78. Davies JA, Millar CB et al. Neurturin:an autocrine regulator of renal collecting duct development. Dev Genet 1999; 24:284-292.

79. Bernardini N, Mattii L, Bianchi F et al. TGF-alpha mRNA expression in renal organogenesis:a study in rat and human embryos. Exp Nephrol 2001; 9:90-98.

80. Coles HS, Burne JF, Raff MC. Large-scale normal cell death in the developing rat kidney and its reduction by epidermal growth factor. Development 1993; 118:777-784.

81. Dono R, Zeller R. Cell-type-specific nuclear translocation of fibroblast growth factor-2 isoforms during chicken kidney and limb morphogenesis. Dev Biol 1994; 163:316-330.

82. Perantoni AO, Dove LF, Karavanova I. Basic fibroblast growth factor can mediate the early inductive events in renal development. Proc Natl Acad Sci USA 1995; 92:4696-4700.

83. Barasch J, Yang J, Ware CB et al. Mesenchymal to epithelial conversion in rat metanephros is induced by LIF. Cell 1999; 99:377-386.

84. Barasch J, Yang J, Qiao J et al. Tissue inhibitor of metalloproteinase-2 stimulates mesenchymal growth and regulates epithelial branching during morphogenesis of the rat metanephros. J Clin Invest 1999; 103:1299-1307.

85. Coles HS, Burne JF, Raff MC. Large-scale normal cell death in the developing rat kidney and its reduction by epidermal growth factor. Development 1993; 118:777-784.

86. Barasch J, Yang J, Qiao J et al. Tissue inhibitor of metalloproteinase-2 stimulates mesenchymal growth and regulates epithelial branching during morphogenesis of the rat metanephros. J Clin.Invest. 1999; 103:1299-1307.

87. Gavin BJ, McMahon JA, McMahon AP. Expression of multiple novel Wnt-1/int-1-related genes during fetal and adult mouse development. Genes Dev 1990; 4:2319-2332.

88. Majumdar A, Vainio S, Kispert A et al. Wnt11 and Ret/Gdnf pathways cooperate in regulating ureteric branching during metanephric kidney development. Development 2003; 130:3175-3185.

89. Lako M, Strachan T, Bullen P et al.Isolation, characterisation and embryonic expression of WNT11, a gene which maps to 11q13.5 and has possible roles in the development of skeleton, kidney and lung. Gene 1998; 219:101-110.

90. Kispert A, Vainio S, Shen L et al.Proteoglycans are required for maintenance of Wnt-11 expression in the ureter tips. Development 1996; 122:3627-3637.

91. Dekel B. Profiling gene expression in kidney development. Nephron Exp Nephrol 2003; 95:e1-e6

92. Sariola H. Nephron induction revisited:from caps to condensates. Curr Opin Nephrol Hypertens 2002; 11:17-21.

93. Ota K, Stetler-Stevenson WG, Yang Q et al. Cloning of murine membrane-type-1-matrix metalloproteinase (MT-1-MMP) and its metanephric developmental regulation with respect to MMP-2 and its inhibitor. Kidney Int 1998; 54:131-142.

94. Lelongt B, Trugnan G, Murphy G et al. Matrix metalloproteinases MMP2 and MMP9 are produced in early stages of kidney morphogenesis but only MMP9 is required for renal organogenesis in vitro. J Cell Biol. 1997; 136:1363-1373.

95. Pohl M, Sakurai H, Bush KT et al. Matrix metalloproteinases and their inhibitors regulate in vitro ureteric bud branching morphogenesis. Am J Physiol Renal Physiol 2000; 279:F891-F900.

96. Pohl M, Sakurai H, Bush KT et al. Matrix metalloproteinases and their inhibitors regulate in vitro ureteric bud branching morphogenesis. Am J Physiol Renal Physiol 2000; 279:F891-F900.

97. Sainio K, Suvanto P, Davies J et al. Glial-cell-line-derived neurotrophic factor is required for bud initiation from ureteric epithelium. Development 1997; 124:4077-4087.

98. Woolf AS, Kolatsi-Joannou M, Hardman P et al. Roles of hepatocyte growth factor/scatter factor and the met receptor in the early development of the metanephros. J Cell Biol 1995; 128:171-184

99. Davies JA, Garrod DR. Induction of early stages of kidney tubule differentiation by lithium ions. Dev Biol 1995; 167:50-60.

100. Lyons KM, Hogan BL, Robertson EJ. Colocalization of BMP 7 and BMP 2 RNAs suggests that these factors cooperatively mediate tissue interactions during murine development. Mech Dev 1995; 50:71-83.

101. Raatikainen-Ahokas A, Hytonen M, Tenhunen A et al. BMP-4 affects the differentiation of metanephric mesenchyme and reveals an early anterior-posterior axis of the embryonic kidney. Dev Dyn 2000; 217:146-158.

102. Clark AT, Young RJ, Bertram JF. In vitro studies on the roles of transforming growth factor-beta 1 in rat metanephric development. Kidney Int 2001; 59:1641-1653.

103. Davies JA, Ladomery M, Hohenstein P et al. Development of an siRNA-based method for repressing specific genes in renal organ culture and its use to show that the Wt1 tumour suppressor is required for nephron differentiation. Hum Mol Genet 2004; 13:235-246.

104. Qiao J, Sakurai H, Nigam SK. Branching morphogenesis independent of mesenchymal-epithelial contact in the developing kidney. Proc Natl Acad Sci USA 1999; 96:7330-7335.

105. Zhang S, Lin Y, Itaranta P et al. Expression of Sprouty genes 1, 2 and 4 during mouse organogenesis. Mech Dev 2001; 109:367-370.

106. Hacohen N, Kramer S, Sutherland D et al. sprouty encodes a novel antagonist of FGF signaling that patterns apical branching of the Drosophila airways. Cell 1998; 92:253-263.

107. Woolf AS, Yuan HT. Development of kidney blood vessels. In: Vize PD, Woolf AS, Bard JBL, eds. The Kidney: From Normal Development to Congenital Disease. Academic Press, 2003.

108. Donovan MJ, Natoli TA, Sainio K et al. Initial differentiation of the metanephric mesenchyme is independent of WT1 and the ureteric bud. Dev Genet 1999; 24:252-262.

109. Tufro-McReddie A, Norwood VF, Aylor KW et al. Oxygen regulates vascular endothelial growth factor-mediated vasculogenesis and tubulogenesis. Dev Biol 1997; 183:139-149.

110. Tufro A, Norwood VF, Carey RM et al. Vascular endothelial growth factor induces nephrogenesis and vasculogenesis. J Am Soc Nephrol 1999; 10:2125-2134.

111. Kolatsi-Joannou M, Li XZ, Suda T et al. Expression and potential role of angiopoietins and Tie-2 in early development of the mouse metanephros. Dev Dyn 2001; 222:120-126.

112. Tufro A. VEGF spatially directs angiogenesis during metanephric development in vitro. Dev Biol. 2000; 227:558-566.

113. Davies J. Control of calbindin-D28K expression in developing mouse kidney. Dev Dyn 1994; 199:45-51.

CHAPTER 9

Embryonic Salivary Gland Branching Morphogenesis

Tina Jaskoll and Michael Melnick

Abstract

Salivary submandibular gland (SMG) morphogenesis is regulated by the functional integration of stage-specific growth factor-, cytokine- and transcription factor-mediated signaling which mediates specific patterns of cell proliferation, cell quiescence, apoptosis, and histodifferentiation. We describe the stage-specific distribution of protein components of key signaling pathways during embryonic SMG development and correlate their protein expression patterns with cell proliferation and apoptosis. We then review the critical role of the Fibroblast Growth Factor (FGF), Hedgehog (Hh) and Ectodysplasin (Eda) signaling pathways and discuss how they may interact within the context of a nonlinear genetic network.

Introduction

Morphogenesis of complex organs such as the submandibular salivary gland (SMG) requires cooperation and coordination of multiple signaling pathways to regulate cell proliferation, quiescence, apoptosis, and histodifferentiation.[1-4] Its development is regulated by the functional integration of stage-specific growth factor-, cytokine- and transcription factor-mediated signaling which mediates these cellular events.[1,5-11] In this review, we will focus on those signaling pathways which have been shown to play important and essential morphoregulatory roles during SMG organogenesis and discuss their functional relationships in the context of a dynamic, nonlinear genetic network.

Mouse submandibular salivary gland organogenesis is initiated with a thickening of the oral epithelium of the mandibular arch around E11 and is best conceptualized in stages: *Prebud, Initial Bud, Pseudoglandular, Canalicular* and *Terminal Bud* Stages (Fig. 1). In the *Prebud* Stage, SMG development begins as a thickening of the primitive oral cavity epithelium adjacent to the tongue. During the *Initial Bud* Stage, this thickening epithelium grows down into the first branchial (mandibular) arch mesenchyme to form the initial SMG bud. With continued epithelial proliferation and downgrowth, the SMG primordium becomes a solid, elongated epithelial stalk terminating in a bulb. The primordium branches by repeated furcation in the distal ends of successive buds to produce a bush-like structure comprised of a network of epithelial branches and terminal epithelial buds (the *Pseudoglandular* Stage). These branches and buds hollow out by epithelial cell apoptosis during the *Canalicular* and *Terminal Bud* Stages to form the ductal system and presumptive acini (for details, see refs. 7, 12). Epithelial cell proliferation is found in all stages, even after well-defined lumen formation in the *Terminal Bud* Stage. By contrast, epithelial apoptosis begins with the onset of lumen formation in the *Canalicular* Stage. Moreover, our studies suggest that ductal canalization is primarily due to caspase 8-mediated apoptosis whereas p53 primarily mediates terminal bud lumina formation.[1,7,13-14] In addition, although we found three pro-survival/anti-apoptotic proteins (NF-κB, bcl2, and

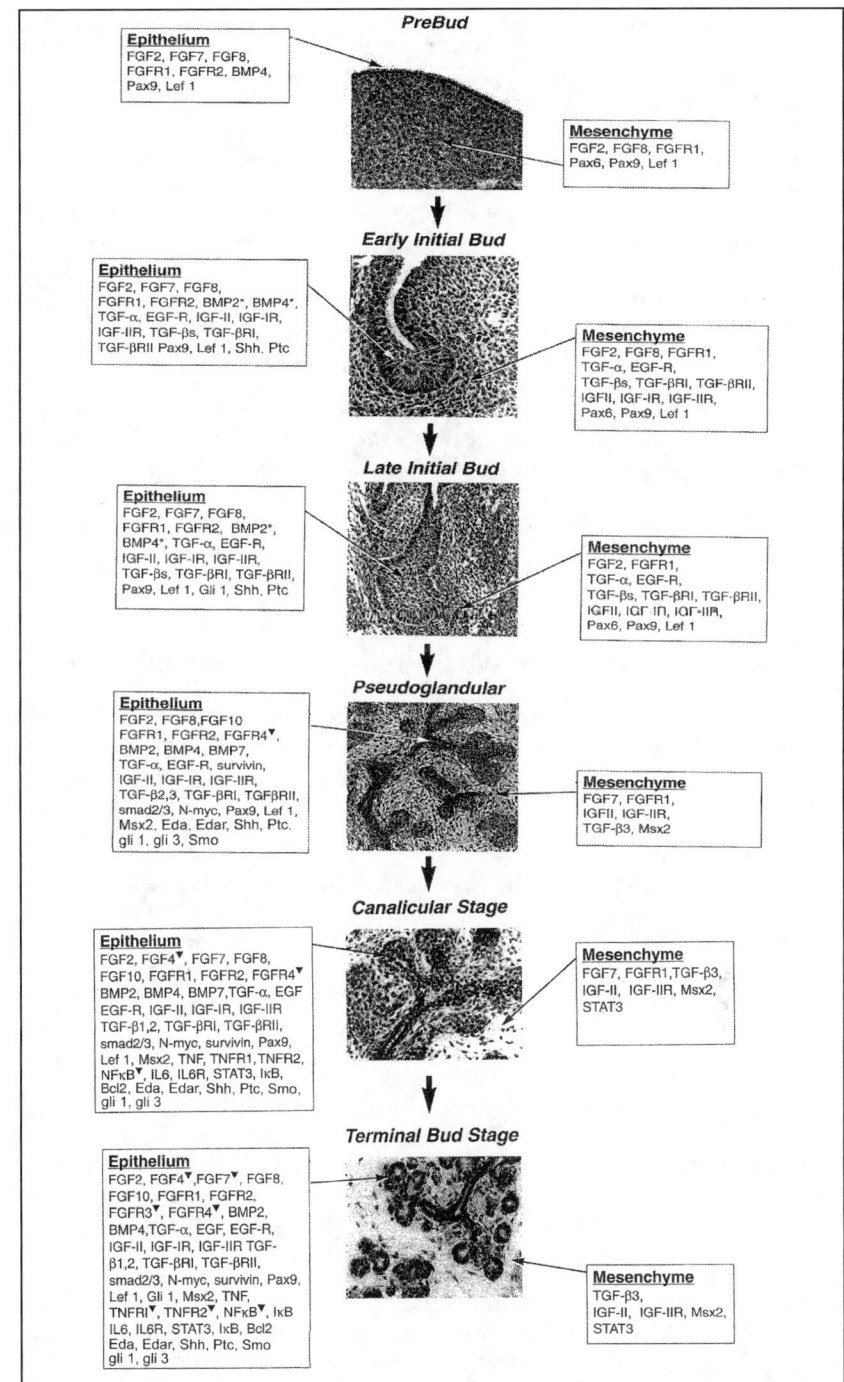

Figure 1. Schematic representation of staged SMG development. The important cell-specific distribution patterns of protein components of multiple signal transduction pathways are shown in the boxes. * epithelial stalk localization; ▼ factors or receptors found in the center of epithelial terminal buds.

survivin) in lumen-bounding epithelial cells during canalization, the presence of only nuclear-localized (activated) survivin in the layer of surviving cells after the completion of lumen formation suggests that survivin mediates terminal bud cell survival. [15] Based on our data, we conclude that apoptosis-mediated lumen formation and cell survival are achieved through several different pathways.

To delineate the molecules and signaling pathways involved in the induction and regulation of embryonic SMG morphogenesis, we first determined the stage-specific distribution of relevant growth factors and cytokines (and their receptors) and transcription factors and correlated their protein expression patterns with cell proliferation and apoptosis (Fig. 1; see refs. 7, 12, 16-17). Our identification of the protein components of the Fibroblast Growth Factor (FGF), Epidermal Growth Factor (EGF), Hedgehog (Hh), Ectodysplasin (Eda), Insulin-like Growth Factor (IGF), Transforming Growth Factor-β (TGF-β) and Tumor Necrosis Factor (TNF) mediated signaling cascades suggests their importance during embryonic SMG development. The marked overlaps in multiple ligand and receptor localization patterns suggest redundancy in their functions. In addition, the nearly exclusive epithelial localization of protein components of key signaling cascades in *Pseudoglandular* Stage and older SMGs suggests that, in these stages, branching morphogenesis and histodifferentiation are primarily mediated by epithelial-epithelial interactions and not epithelial-mesenchymal interactions.

SMG organogenesis is due to the functional integration of parallel and broadly related signaling pathways (for review see refs. 1, 7, 17). To begin to understand the complex interactions within this dynamic signaling network, one must first determine the contribution of individual pathways and identify those which are important and necessary for embryonic SMG development. Analyses of knock-out, transgenic and mutant mice provide insights into which of the signaling pathways present in the developing SMG play essential morphogenetic roles[8-10] (for review see refs. 7, 17). Of particular note are the following genotype/phenotype observations: (1) SMG aplasia in *Fgfr2-IIIb-/-, Fgf10-/-*, and *Fgf8* conditional mutant mice; [10,16-19] (2) SMG dysplasia in *Edar1* (downless) mutant mice; [8] and (3) SMG dysplasia in *Shh-/-* mice.[9] These results suggest that the FGF/FGFR, ShhPtc, and Eda/Edar signaling cascades are critical for SMG organogenesis.

FGF/FGFR Signaling

The FGF family includes at least 23 members which have been shown to have diverse biological functions, including cell proliferation, epithelial branching and histodifferentiation (for review see refs. 20-23). FGF function is through a family of five transmembrane tyrosine-kinase receptors (FGFRs). Alternate splicing of *Fgfr1, Fgfr2*, and *Fgfr3* generates receptor isoforms (e.g., IIIb and IIIc) with different ligand-binding specificities. Ligand binding to the appropriate receptor results in receptor dimerization, activation of the intrinsic tyrosine kinase activity and autophosphorylation which activates several intracellular cascades, including the Ras pathway, Src family tyrosine kinases, phosphoinositide 3-kinase/AKT (PI3K/AKT), the PLC-γ/PKC (phospholopase-Cγ/protein kinase C) pathway, and the STAT3 pathway (Fig. 2; for review see refs. 21-23). Although FGF receptors have the ability to activate multiple pathways simultaneously, the variable pleitrophic effects of different FGF/FGFR signaling pathways are associated with clearly distinguishable signaling which induce different downstream targets according to a cell's need. For example, the RAS pathway has been shown to be critical for cell proliferation and differentiation whereas the PI3K/AKT pathway plays a major prosurvival/anti-apoptotic role.[6,24-25]

Regarding FGFR2 signaling, gene targeting has clearly demonstrated that both FGFR2 isoforms, FGFR2-IIIb and FGFR2-IIIc, play critical roles during SMG organogenesis. Although an initial SMG bud is present in *Fgfr2-IIIb-/-* mice (De Moerlooze, Jaskoll and Melnick, unpublished), no gland is found in E14.5 and older mutant mice.[18] The observation of a similar SMG phenotype in *Fgf10* null mice,[19] as well as the absence of abnormalities in *Fgf7* null mice,[26] indicate that FGF10 is the ligand responsible for the *Fgfr2-IIIb*

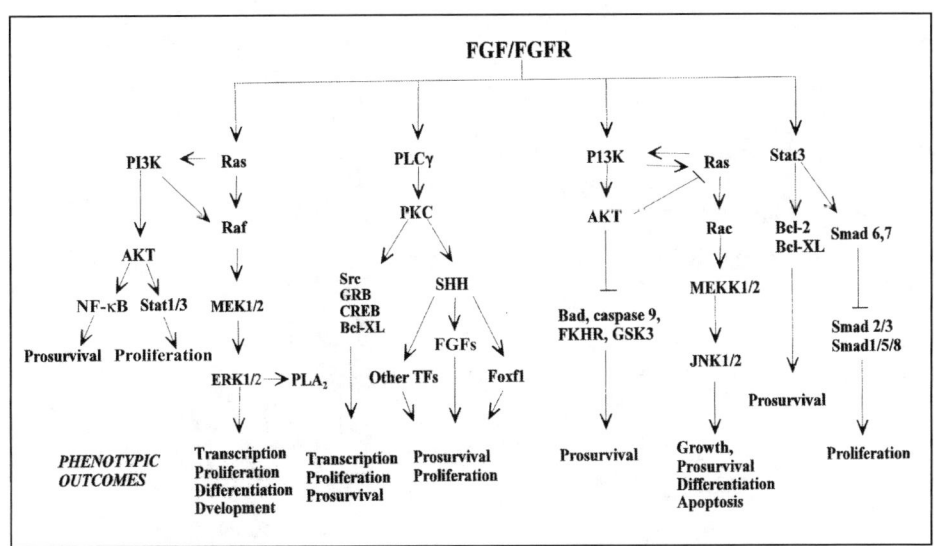

Figure 2. FGF/FGFR signaling pathway. Known and putative connections are based on published results (e.g., refs. 23-25, 59-62).

mutant phenotype. Since the absence of a SMG primordium in E14.5 and older *Fgfr2-IIIb* mutant mice made it impossible to determine if FGF10/FGFR2-IIIb signaling regulates branching morphogenesis, we then investigated the effect of enhanced or reduced FGF10/FGFR2-IIIb signaling on SMG development in vitro. Exogenous FGF10 peptide supplementation induced a significant 55% (P<0.01) increase in epithelial branching in embryonic SMGs compared to control. Abrogation of FGFR2-IIIb signaling with *anti-Fgfr2-IIIb* oligonucleotides (ODNs) resulted in a significant dose-dependent decrease in branching morphogenesis compared to sense controls (Fig. 3). Together these results indicate that FGF10/FGFR2-IIIb signaling plays a critical role for SMG branching morphogenesis and cell survival, but not initial bud formation.

Since *Fgfr2-IIIc* null mutations are embryolethal, we analyzed the role of FGFR2-IIIc signaling in *Fgfr2-IIIc* deficient SMGs.[16] *Fgfr2-IIIc+/Δ* mutant SMGs are smaller, with fewer terminal buds compared to wildtype glands. We then confirmed the functional importance of the FGFR2-IIIc pathway for branching morphogenesis using our organ culture system. *Anti-Fgfr2-IIIc* ODN-mediated reduction in FGFR2-IIIc signaling resulted in a significant dose-dependent decrease in branching compared to sense control (Fig. 4). Based on these results, we conclude that FGFR2-IIIc signaling plays an important role during SMG branching morphogenesis. What remained unclear was whether, like the FGFR2-IIIb pathway discussed above, FGFR2-IIIc signaling is critical and essential for embryonic SMG cell survival.

To address this question, we turned our attention to which FGF ligand likely induces the FGFR2-IIIc signal during embryonic SMG development. Although FGF8 isoforms have been shown to bind with high affinity to FGFR2-IIIc, FGFR3-IIIc and FGFR4,[27,28] the absence of FGFR3 and FGFR4 from *Initial Bud* Stage and older SMGs [16] indicates that FGF8 mediates its signal through FGFR2-IIIc. Moreover, although FGF2, FGF4, FGF6 and FGF8 and FGF9 bind with high affinity to FGFR2-IIIc (see reviews, refs. 20-22), the normal SMG phenotype in *Fgf4* null mice (A. Moon, E.J. Park, L. Francis, unpublished), the relatively normal appearance of *Fgf2* and *Fgf6* null mice,[29-30] and the absence of *Fgf9* transcripts from embryonic SMGs [31] indicate that FGF8 is the important FGFR2-IIIc ligand for SMG morphogenesis. Thus, the abnormal SMG phenotype seen with interrupted FGFR2-IIIc signaling in vivo and in vitro is primarily due to diminished FGF8/FGFR2-IIIc signaling.

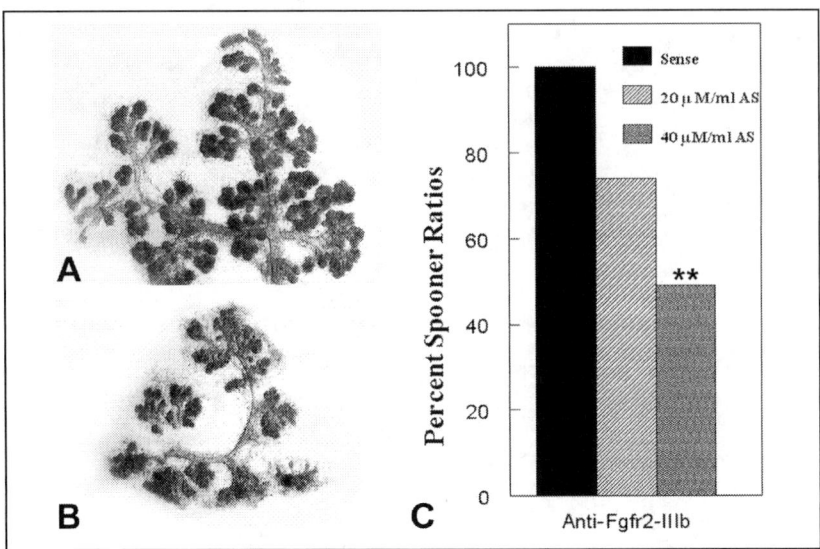

Figure 3. Reduced FGFR2-IIIb signaling in vitro results in decreased SMG branching morphogenesis. Paired E14 SMG primordia were cultured in sense (A) or antisense (B) ODNs for 48 hours and Spooner ratios (bud number at 48 hrs/bud number at 0 hr) were determined for each explant. C) A comparative representation of the percent change in Spooner branching ratios with antisense treatment relative to sense control. Mean Spooner ratios were determined, the data were arcsin transformed to insure normality and homoscedasticity, and compared by paired t-test for all embryos studied as previously described.[1] A minimum of 4 explants/treatment were evaluated. ** P<0.01.

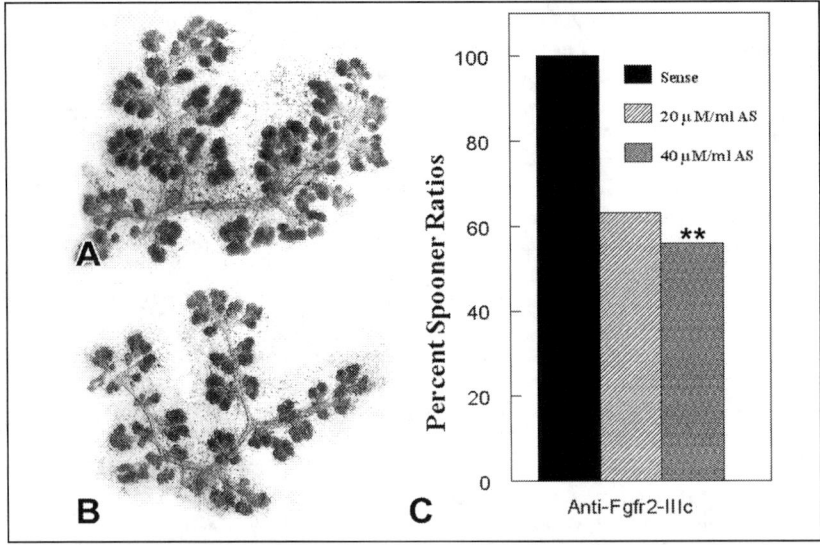

Figure 4. Reduced FGFR2-IIIc signaling in vitro results in decreased SMG branching morphogenesis. Paired E14 SMG primordia were cultured in sense (A) or antisense (B) ODNs for 48 hours and Spooner ratios were determined for each explant. C) A comparative representation of the percent change in Spooner branching ratios associated with antisense treatment relative to sense control. A minimum of 4 explants/treatment were analyzed. **P<0.02.

We then investigated the SMG phenotypes in *Fgf8* hypomorphic and tissue-specific conditional mutant mice.[10] Like the phenotype seen in *Fgfr2-IIIc* deficient mice, we see hypoplastic glands in *Fgf8* hypomorphic mice. By contrast, SMG aplasia is seen in *Fgf8* conditional mutant mice in which *Fgf8* expression was completely ablated from the first pharyngeal arch ectoderm. Our observation of SMG hypoplasia and aplasia in *Fgf8* hypomorphs and conditional mutant mice, respectively, indicates that FGF8/FGFR2-IIIc signaling plays an essential role for branching and survival of *Pseudoglandular* Stage and older SMGs, acting in a dose-dependent manner. Importantly, the functional presence of other endogenous FGF/FGFR pathways (e.g., FGF10/FGFR2-IIIb) or other signaling pathways (e.g., TGF-α/EGF/EGFR, IGRII/IGFR1) could not prevent the complete death of embryonic SMG cells in the *Fgf8* conditional mutant mice. These results indicate that the FGF10/FGFR2-IIIb and FGF8/FGFR2-IIIc signaling pathways induce unique and specific downstream cascades during embryonic SMG development that cannot be compensated by other pathways.

In addition, since FGFR1 is the only other FGF receptor found at all stages of embryonic SMG development (Fig. 1) and FGFR1 and FGFR2 have the ability to induce similar downstream cascades, we postulated that FGF/FGFR1 signaling also plays an important role during embryonic SMG development. However, the early embryonic death of *Fgfr1* null mice [32] precluded any investigation into the role of FGFR1 signaling in later stages of development. Recently, our collaborator Mohammed Hajihosseini generated transgenic mice which express a mutant *Fgfr1* gene and showed a direct relationship between mutant *Fgfr1* copy number and severity of defects.[33] We investigated the SMG phenotype in these *Fgfr1* transgenic mice and found smaller glands composed of fewer epithelial branches in mutants compared to wildtype mice (Fig. 5). The observation of distinct lumina in terminal buds in both wildtype and mutant glands indicates that altered levels of FGFR1-mediated signaling inhibits epithelial branching but not histodifferentiation. In vitro studies are consistent with these findings. Abrogation of FGFR1 signaling in vitro with antisense ODNs[34] or SU5402 peptide treatment,[34] Jaskoll and Melnick, unpublished) results in a substantial decrease in branching. Taken together, these in vivo and in vitro studies indicate that the FGF/FGFR1 signaling cascade is important for SMG branching morphogenesis.

Since the IIIb and IIIc isoforms of FGFR1 bind different ligands (FGF1, 2, 4, 5, and 6 bind to FGFR1-IIIb,[35] FGF1, 2 and 10 bind to FGFR1-IIIb[28]) and may elicit different downstream responses during embryonic SMG development, we then investigated the effect of reduced FGFR1-IIIb and FGFR1-IIIc signaling in vitro. Exogenous soluble FGFR1-IIIb-Fc or FGFR1-IIIc-Fc chimera was added to the culture medium to competitively bind endogenous

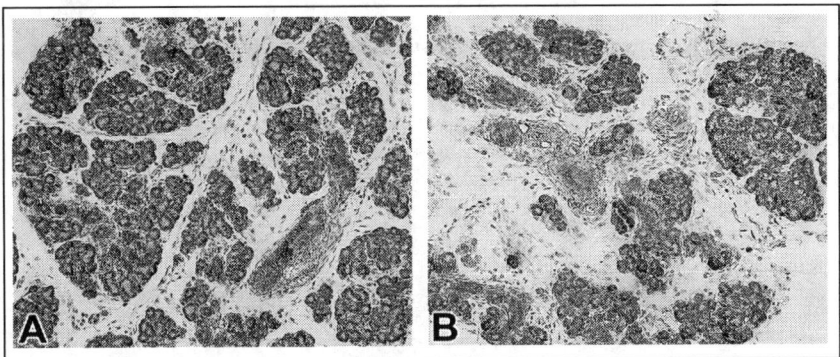

Figure 5. *Fgfr1* mutant mice exhibit SMG hypoplasia. A) *Terminal Bud* Stage SMGs characterized by distinct lumina within ductal and terminal buds are seen in wildtype mice. B) *Fgfr1* mutant SMGs are smaller and composed of fewer branches than the wildtype gland. The presence of lumina within their terminal buds indicate that mutant glands have also achieved the *Terminal Bud* Stage.

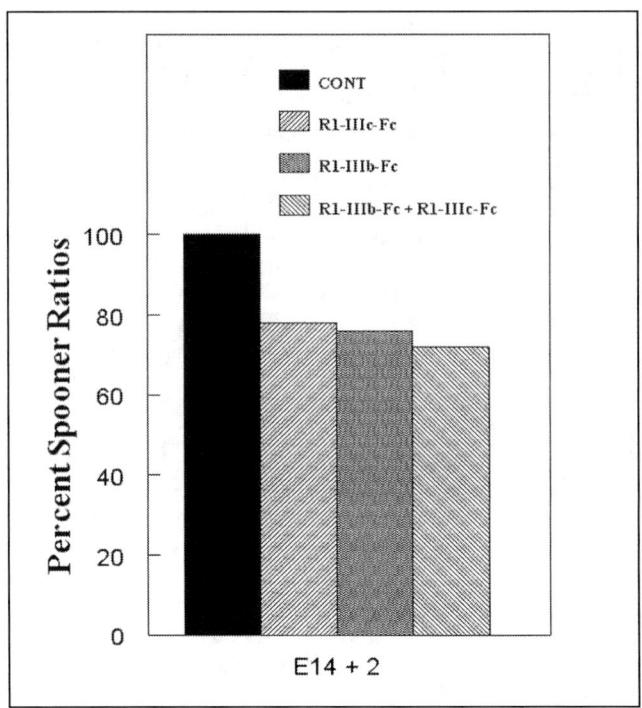

Figure 6. Interrupted FGFR1-IIIb or FGFR1-IIIc signaling significantly decreases SMG branching morphogenesis in vitro. Paired E14 SMG primordia were cultured for 2 days in 3 ng/ml IgG-Fc (CONT), 3 ng/ml FGFR1-IIIb-Fc (R1-IIIb) or 3 ng/ml FGFR1-IIIc-Fc (R1-IIIc). To determine a possible synergistic effect, paired E14 SMG primordia were cultured in 3 ng/ml FGFR1-IIIb + 3 ng/ml FGFR1-IIIc-Fc (R1-IIIb + IIIc) or 6 ng/ml IgG-Fc (CONT). Spooner ratios were determined for each explant and treatments compared by paired t-test as previously described. A minimum of 4 explants/treatment were analyzed. A significant 24% (P<0.005) decrease in branching was seen with FGFR1-IIIb-Fc chimera treatment and a significant 22% (P<0.002) decrease in branching was seen with FGFR1-IIIc-Fc chimera treatment. The combination of FGFR1-IIIb-Fc and FGFR1-IIIc-Fc chimera treatment resulted in a significant 28% (P<0.05) decrease in branching compared to IgG controls.

FGFR1-IIIb or FGFR1-IIIc ligands. Abrogated FGFR1-IIIb or FGFR1-IIIc signaling resulted in a highly significant ~23% decrease in branching morphogenesis compared to controls (Fig. 6). Embryonic SMGs primordia were also cultured in a combination of FGFR1-IIIb-Fc + FGFR1-IIIc-Fc chimeras. This combined treatment reduction of 28% reduction compared to control (Fig. 6) is not different from that seen with either isoform chimera alone; thus, there is no synergism. One possible explanation is that both receptor isoforms share identical ultimate downstream pathways during embryonic SMG organogenesis; thus interruption of either FGFR1-IIIb or FGFR1-IIIc signaling (or both) interrupts the same downstream cascade. What remains to be determined are the specific downstream targets of FGFR1-IIIb and FGFR1-IIIc signaling and whether they are unique and required for SMG development.

Hedgehog Signaling

Sonic hedgehog (Shh) is a member of the hedgehog (Hh) family of signaling molecules that induces cell survival, proliferation, differentiation and pattern formation in various embryonic tissues (for review see ref. 36). The cellular response to Hh is controlled by two transmembrane

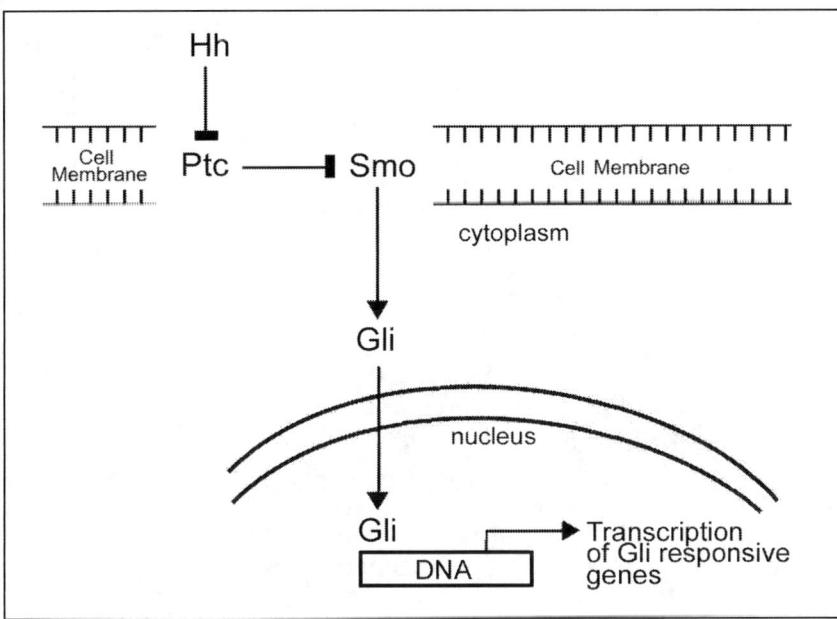

Figure 7. The hedgehog signaling pathway. In the absence of Hh ligand, Ptc inhibits the activity of Smo to block the downstream signaling cascade. Hh binding to Ptc relieves Smo from its inhibition to initiate a signaling cascade that results in the activation of the Gli family of transcription factors and subsequent transcription of Gli responsive genes.

proteins, Patched 1 (Ptc) and Smoothened (Smo) (Fig. 7). Ptc acts as a negative regulator of the Hh signal whereas Smo is a positive activator. Hh binding to Ptc relieves the inhibition of Smo signaling to initiate a signaling cascade that results in the activation of target genes via the Gli family of transcription factors, including genes associated with cell cycle progression.[36-38]

Data from knockout mice clearly indicate that the Shh signaling cascade is essential for many aspects of mammalian embryogenesis, including neural tube, craniofacial, limb, and kidney development (for review see ref. 10). Given that (1) *Shh* null mice are characterized by cyclopia, holoprosenchepaly and the virtual absence of mandibular derivatives,[39] (2) Shh is essential for neural crest cell survival,[40] and (3) the SMG initial bud develops as an invagination of the oral epithelium into the underlying neural crest-derived mesenchyme of the mandibular arch,[12] we postulated that the *Shh* null SMG would be absent. Much to our surprise, we found a severely dysplastic gland in *Shh-/-* mice.[9] The presence of a small, severely abnormal SMG primordium, as well as our observation that enhanced Shh signaling induced, and abrogated signaling decreased, SMG branching morphogenesis in vitro,[9] indicate that Shh signaling is critical for embryonic SMG branching morphogenesis, most likely by regulating epithelial cell proliferation.

Eda/Edar Signaling

Ectodysplasin (Eda) and its receptor (Edar) are members of the TNF superfamily shown to play critical roles during the development of ectodermal derivatives, including teeth, hair and sweat glands.[41-42] Eda-A1, the biologically active isoform, binds specifically to Edar activate NF-κB translocation into the nucleus to regulate the transcription of NF-κB responsive genes, including genes associated with DNA repair, cell cycle progression, cell

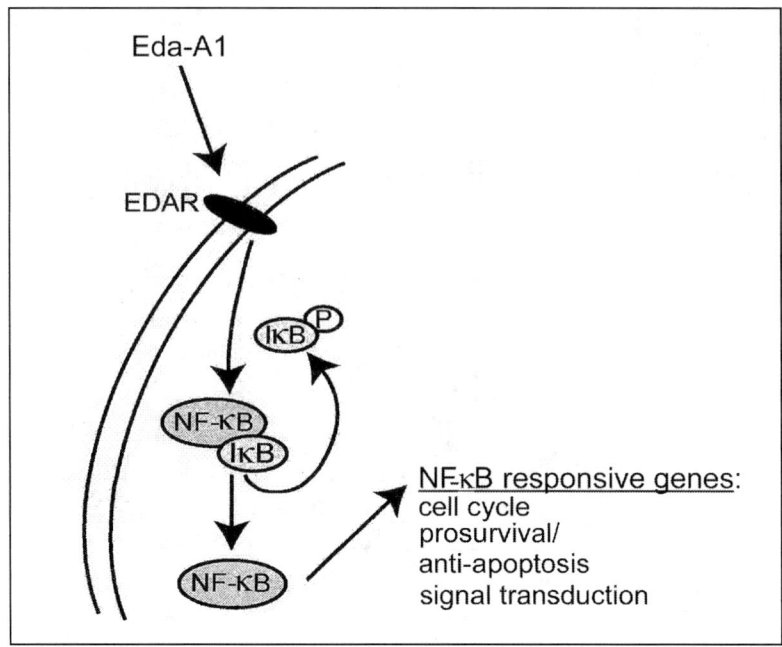

Figure 8. The Eda/Edar signaling pathway. EDA-A1 ligand binds to EDAR to activate NF-κB. NF-κB activity is regulated by cytoplasmic IκB proteins which interact with NF-κB and prevent its nuclear translocation. Eda/Edar signaling activates IκB kinase (IKK) to phosphorylate the IκBs, resulting in the release of NF-κB which then translocates to the nucleus. Activated NF-κB then induces downstream target genes, including genes associated with DNA repair, cell cycle progression, survival/anti-apoptosis, and signal transduction.

survival, and apoptosis (Fig. 8; see refs. 43-44). (In this review, Eda is used to designate the biologically active ligand, Eda-A1). Loss of Eda or Edar function in humans results in hypohidrotic ectodermal dysplasia which is characterized by absent or hypoplastic teeth, hair, and sweat glands;[41-42,45-46] similar phenotypes are seen in Tabby (*EdaTa*; Eda-less) and downless (*Edardl*; Edar-less) mutant mice.[41-42,45] Our previous demonstration that TNF/TNFR signaling is important for balancing mitogenesis and apoptosis during embryonic SMG development[13] and that the NF-κB cascade (the primary downstream pathway of Eda/Edar signaling) plays an essential role for SMG epithelial cell proliferation and survival[1] suggested that the Eda/Edar pathway is essential for SMG development. Thus, we analyzed Tabby (*EdaTa*) and downless (*Edardl*) mutant mouse SMGs and found mutation-specific phenotypic abnormalities.[8] The Tabby SMG is hypoplastic whereas the downless gland is dysplastic. Since Eda ligand is present in downless mice, their more severe abnormal SMG phenotype indicates that no other receptor can compensate for Edar's functional absence. The decrease or absence of SMG ducts, acini and mucin protein in Tabby and downless SMGs, respectively, indicate that Eda/Edar signaling is essential for branching morphogenesis, lumina formation, and histodifferentiation. Complementary in vitro studies provide additional mechanistic data in support of our conclusion. Enhanced Eda/Edar signaling in vitro induces, and abrogated signaling decreases, branching morphogenesis as well as the activation of NF-κB.[8] What remains to be determined are which signaling cascades downstream of the Eda/Edar/NF-κB signal are important for SMG organogenesis.

Downstream Targets Rescue SMG Phenotypes

To help delineate the morphogenetic role of any individual signaling pathway, developmental biologists have analyzed the effect of reduced/absent signaling on the expression of a select number of candidate growth factors or receptors (e.g., see refs. 47-49) and determined if supplementation with a downstream factor could rescue the abnormal phenotype in vitro (e.g., see refs. 50-52). Since FGF8 is a target downstream of the Shh signal,[9,49] we determined if FGF8 peptide supplementation in vitro could rescue the abnormal SMG phenotype seen with decreased Shh/Ptc signaling.[9] FGF8 supplementation restored SMG branching morphogenesis to normal. Similarly, exogenous Shh and FGF10, physiologic downstream targets of FGF8/FGFR2-IIIc signaling,[10,47,53] rescued the abnormal SMG phenotype seen with decreased FGF8/FGFR2-IIIc signaling in vitro.[10] In addition, FGF7, FGF10 and BMP7, physiologic downstream targets of FGF/FGFR1 signaling, rescued the SMG phenotype seen with decreased FGF/FGFR1 in vitro.[34] These rescue experiments indicate that enhancement of a downstream signaling pathway can compensate for decreased signaling in a single pathway and can restore SMG branching morphogenesis. Nevertheless, the observation of abnormal SMG phenotypes in the transgenic and mutant mice discussed above indicate that such compensatory signaling does not occur during normal in vivo SMG organogenesis.

A Nonlinear Genetic Network Regulates SMG Organogenesis

Over the last decade, it has become increasingly apparent that organogenesis is the programmed expression of regulatory genes coupled to downstream structural genes and epigenetic events.[1-3] This process is dependent on the functional integration of diverse signal transduction pathways that transmit information between and within cells. The cellular and extracellular components may be visualized as a Connections Map which details the functional relationships within and between pathways (Fig. 9). Specific growth factor- and cytokine-mediated signal transduction pathways are parallel and largely functionally redundant; that is, some pathways differentially and combinatorially compensate for the dysfunction of a given individual pathway. However, other pathways have unique and nonredundant functions since no other pathways(s) can compensate in full for their absence/dysfunction. These nonredundant pathways are the key signaling cascades which regulate the cell dynamics of the developing SMG.

In this genetic network, ligand/receptor binding initiates the activation of wide array of broadly related, not independent, transcription factor cascades. Many of these signaling cascades converge at a single factor, a so-called "hub," which then induces a myriad of downstream cellular responses. A prime example is NF-κB (Fig. 9). A large number of pathways activate NF-κB which, in turn, induces more than 150 downstream responses.[54] Regarding NF-κB's role during embryonic SMG development, we have previously demonstrated that interruption of NF-κB activation in vitro results in a small gland characterized by a highly significant decrease in cell proliferation and a significant increase in apoptosis, as well as the altered expression of a multiple genes and proteins.[1] These include signal transduction, translation control (checkpoint), cell cycle, and apoptosis genes and proteins that are downstream from the Eda/Edar, TNF/TNFR, FGF/FGFR, TGF-α/EGF/EGFR, and IGFII/IGFRI signals (Fig. 9).

The identification of how these genetic pathways functionally interact during SMG organogenesis in vivo provide insight into whether an individual pathway is redundant or unique (non redundant). Such information can be obtained from knock-out, transgenic and mutant mice. The observation of normal SMGs in *TGF-β1-/-*, *TGF-β2-/-*, *TGF-β3-/-*, *Msx2-/-*, *Pax9-/-*, *Hoxa5-/-*, *Gli1-/-*, *Gli2-/-*, and *Gli3-/-* mice indicate that other signaling cascades completely compensate for the absence of each single pathway to restore the phenotype to normal. Thus, these pathways are not the sine qua non of embryonic SMG development; therefore they are redundant (Table 1).

In contrast, the absence of SMGs in *Fgf8* conditional mutants, *Fgf10-/-* and *Fgfr2-IIIb-/-* mice indicates that endogenous levels of other FGFR-mediated or parallel signaling pathways

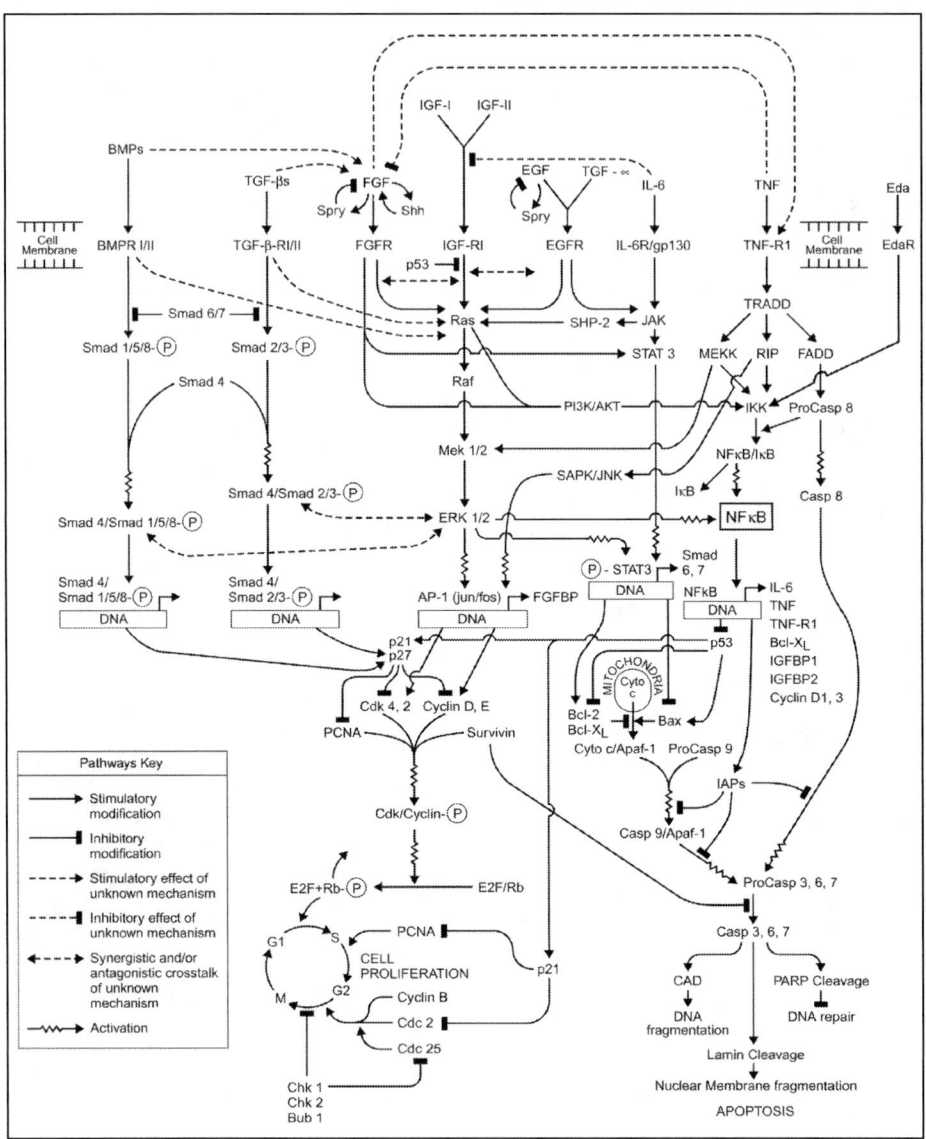

Figure 9. Connections Map. This signaling map reflects the pathways investigated in our laboratory. Known and putative connections are based on work in our laboratory and the laboratories of others.

(TGF-α/EGF/EGFR, IGF/IGFR) could not compensate for the absence of FGF10/FGFR2-IIIb or FGF8/FGFR2-IIIc signaling. Significantly, the failure of endogenous levels of FGF10/ FGFR2-IIIb signaling in *Fgf8* conditional mutants or FGF8/FGFR2-IIIc signaling in *Fgf10* null mice to prevent the complete death of SMG cells indicates that the FGF8 signal transduction pathway induces specific and unique downstream responses, different from those mediated by FGF10 signaling. Thus, we conclude that the FGF8/FGFR2-IIIc and FGF10/ FGFR2-IIIb pathways can be characterized as sine qua non pathways, i.e., they are essential

Table 1. Molecular pathways: Functional relationships

Redundant Pathways	Partially Redundant Pathways	NonRedundant Pathways
Gli 1, 2 or 3	Eda/Edar	FGF10/FGFR2-IIIb
Hoxa5	Shh/Ptc	FGF8/FGFR2-IIIc
Msx2	TGFα/EGF/EGFR	
Pax9	BMP7	
TGF-β1/ TGF-RII	Pax6	
TGF-β2/ TGF-RII	FGF/FGFR1 (?)	
TGF-β3/ TGF-RII	NF-κB (?)	

and non redundant (Table 1). It is interesting to note that, although EGFR is a receptor tyrosine kinase which can also induce responses identical to those downstream of FGF/FGFR signaling,[5-6,11,15,55] our observation of hypoplastic (and not aplastic) glands in *Egfr* and *Tace* null mice (for review see ref. 17) indicates that, in this case, other pathways compensate, although only partially, for the absence of TGF-α/EGF/EGFR signaling (Table 1). This again points to the uniqueness of FGF8 and FGF10 signaling in SMG development.

Further, our observation of a dysplastic gland in downless (*Edardl*; Edar-less) mutants indicates that other pathways cannot completely compensate for Edar's functional absence; thus, the Eda/Edar pathway is only partially redundant (Table 1). Since multiple pathways have been shown to activate NF-κB (Fig. 9; ref. 54), it is likely that this partial compensatory affect is due to NF-κB activation through pathways other than Eda/Edar. However, the absence of abnormal SMGs in TNF or TNFR1 null mice[56-58] and the presence of endogenous levels of TNF/TNFR signaling in downless mutants indicate that TNF/TNFR-mediated activation of NF-κB is not responsible for the rescued phenotype. In addition, although our in vitro functional studies indicate that NF-κB is an important network "hub" during SMG organogenesis, the unavailability of appropriate *Nfkb* mutant mice makes it impossible to definitively ascertain if the NF-κB cascade is essential and non redundant.

Regarding other signaling cascades, the presence of a dysplastic SMG in *Shh-/-* and a hypoplastic SMG in B*mp7-/-*and *Pax6-/-* mice indicate that other parallel pathways only partially compensate for the absence of Shh/Ptc, BMP7 or Pax6 signaling (Table 1). The more severely abnormal SMG phenotype in *Shh* null mice indicates that other endogenous parallel pathways compensate to a lesser degree for the absence of the Shh/Ptc cascade than the absence of BMP7 or Pax6 cascade. Finally, given the absence of *Fgfr1* knock-out and tissue-specific conditional null mice, the precise role of the FGFR1-mediated pathway remains unclear. However, functional underexpression or overexpression of FGFR1 signaling indicates that, at the very least, FGF/FGFR1 homeostasis is essential for normal SMG morphogenesis.

Concluding Remarks

With our present knowledge, we can begin to model the regulation of SMG branching morphogenesis (Fig. 10). The initiation of branching and the progression from the *Initial Bud* Stage to the *Pseudoglandular* Stage is dependent on FGF8/FGFR2-IIIc and FGF10/FGFR2-IIIb signaling. Subsequent epithelial branching and progression from the *Pseudoglandular* Stage to the *Canalicular* Stage also require additional signaling through the Shh/Ptc and Eda/Edar pathways. Finally, since TGF-α/EGF/EGFR, BMP7 and Pax6 signaling regulate the rate of branching morphogenesis and histodifferentiation, these additional signal transduction pathways are all essential for progression from the *Canalicular* Stage to the *Terminal Bud* Stage.

In this review, then, we have provided a glimpse into the dynamic, nonlinear network which regulates SMG organogenesis. In studying the ontogeny of the embryonic SMG, the key is to

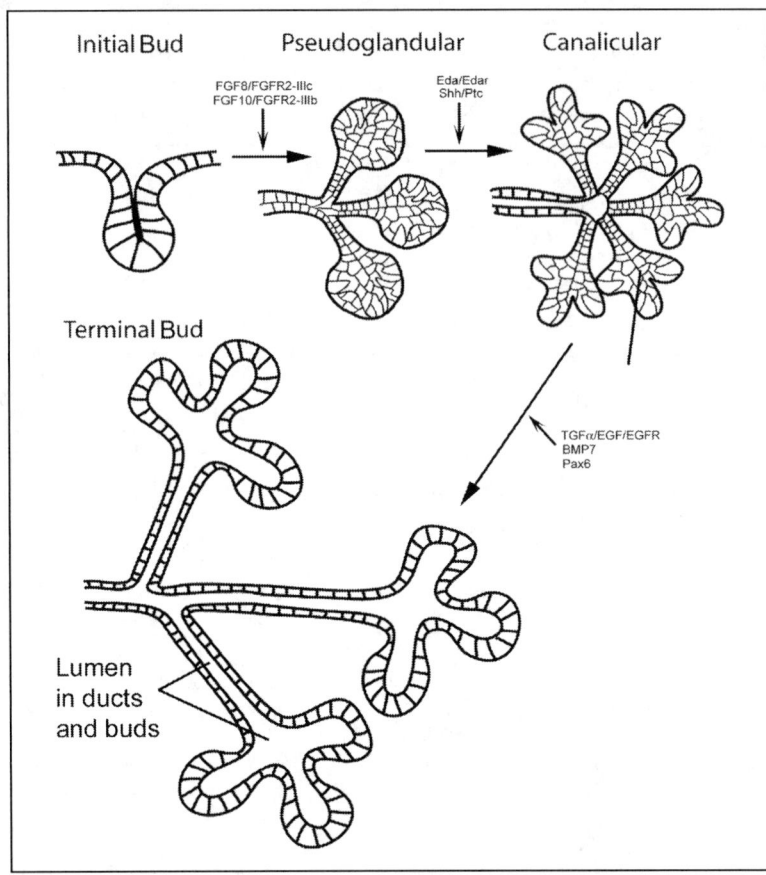

Figure 10. Control of embryonic SMG branching morphogenesis. The initiation of branching and progression from the *Initial Bud* to the *Pseudoglandular* Stage is dependent on FGF8/FGFR2-IIIc and FGF10/FGFR2-IIIb signaling. Continued epithelial branching and progression from the *Pseudoglandular* to the *Canalicular* Stage also requires Shh/Ptc and Eda/Edar signaling. Since TGF-α/EGF/EGFR, BMP7 and Pax6 continued branching morphogenesis, lumen formation, and histodifferentiation, the progression from the *Canalicular* Stage to the *Terminal Bud* Stage is dependent of TGF-α/EGF/EGFR. BMP7 and Pax6 signaling pathways. The data presented is based on SMG phenotypes in mutant and null mice. For details, see references 7 and 17.

understand the functional coordination and cooperation between parallel signaling pathways to achieve the differentiated gland. The major challenge remaining is to identify the full complement of broadly related network components and to understand how they interact to regulate the cellular dynamics of SMG development. To accomplish this daunting goal requires a systems biology approach; that is, the integration of transcriptomics, proteomics, and bioinformatics.

Acknowledgements

We would like to thank Dan Witcher and Pablo Bringas, Jr. for help preparing the figures. We would also like to acknowledge the other scientists investigating salivary gland organogenesis. Due to space limitation, we were unable to cite all relevant studies. This work was supported by NIH R01 DE91142 and RO1 DE14535.

References

1. Melnick M, Chen H, Zhou Y et al. The functional genomic response of developing embryonic submandibular glands to NF-κB inhibition. BMC Developmental Biology 2001; 1:15.
2. Davidson EH, Rast JP, Oliveri P et al. A genomic regulatory network for development. Science 2002; 295:1669-1678.
3. Davidson EH, McClay DR, Hood L. Regulatory gene networks and the properties of the development process. Proc Natl Acad Sci USA 2003; 100:1475-1480.
4. Gardner TS, Bernardo DD, Lorenz D et al. Inferring genetic networks and identifying compound mode of action via expression profiling. Science 2003; 301:102-105.
5. Kashimata MW, Sakagami HW, Gresik EW. Intracellular signaling cascades activated by the EGF receptor and/or integrins, with potential relevance for branching morphogenesis of the fetal mouse submandibular gland. Eur J Morphol 2000b; 38:269-275.
6. Koyama N, Kashimata M, Sakagami H et al. EGF-stimulated signaling by means of PI3K, PLCγ1, and PKC isozymes regulates branching morphogenesis of the fetal mouse submandibular gland. Devel Dyn 2003; 227:216-226.
7. Melnick M, Jaskoll T. Mouse submandibular gland morphogenesis: A paradigm for embryonic signal processing. Crit Rev Oral Biol 2000; 11:199-215.
8. Jaskoll T, Zhou Y-M, Trump G et al. Ectodysplasin receptor-mediated signaling is essential for embryonic submandibular salivary gland development. Anat Rec 2003; 271A:322-331.
9. Jaskoll T, Leo T, Witcher D et al. Sonic Hedgehog Signaling plays an essential role during embryonic salivary gland epithelial branching morphogenesis. Devel Dynam 2004; (in press).
10. Jaskoll T, Witcher D, Leo T et al. FGF8 dose-dependent regulation of embryonic submandibular salivary gland morphogenesis. Devel Biol 2004; (in press).
11. Larsen M, Hoffman MP, Sakai T et al. Role of PI3-kinase and PIP₃ in submandibular gland branching morphogenesis. Devel Biol 2003; 255:178-191.
12. Jaskoll T, Melnick M. Submandibular gland morphogenesis: Stage-specific expression of TGF-alpha, EGF, IGF, TGF-beta, TNF and IL-6 signal transduction in normal mice and the phenotypic effects of TGF-beta2, TGF-beta3, and EGF-R null mutations. Anat Rec 1999; 256:252-268.
13. Melnick M, Chen H, Zhou Y-M et al. Embryonic mouse submandibular salivary gland morphogenesis and the TNF/TNF-R1 signal transduction pathway. Anat Rec 2001; 262:318-320.
14. Melnick M, Chen H, Zhou Y et al. Interleukin-6 signaling and embryonic mouse submandibular salivary gland morphogenesis. Cells Tiss Org 2001; 168:233-245.
15. Jaskoll T, Chen H, Zhou Y-M et al. Developmental expression of survivin during embryonic submandibular salivary gland development. BMC Developmental Biol 2001; 1:5.
16. Jaskoll T, Zhou Y-M, Chai Y et al. Embryonic submandibular gland morphogenesis: Stage-specific protein localization of FGFs, BMPs, Pax 6 and Pax 9 and abnormal SMG phenotypes in Fgf/R2-IIIc⁺/Δ, BMP7⁻/⁻ and Pax6⁻/⁻ mice. Cells Tiss Org 2002; 270:83-98.
17. Jaskoll T, Melnick M. Embryonic SMG Development. < http://www. usc.edu/hsc/ dental/odg/ index.htm>).
18. De Moerlooze L, Spencer-Dene B, Revest J-M. An important role for the IIIb isoform of fibroblast growth factor receptor 2 (FGFR2) in mesenchymal-epithelial signaling during mouse organogenesis. Development 2000; 127:483-492.
19. Ohuchi H, Hori Y, Yamasaki M et al. FGF10 acts as a major ligand for FGF receptor 2 IIIb in mouse multi-organ development. Bioch Biophys Res Com 2000; 277:643-649.
20. McKeehan WL, Wang F, Kan M. The heparin-sulfate fibroblast growth factor family-diversity of structure and function. Prog Nucl Acid Res Mol Biol 1999; 59:135-176.
21. Ornitz DM. 2000. FGFs, heparan sulfate and FGFRs. Complex interactions essential for development. Bioessays 2000; 22:108-112.
22. Ornitz DM, Itoh N. Fibroblast growth factors. Genome Biol 2002; 5:1-12.
23. Szebenyi G, Fallon J. Fibroblast growth factors as multifunctional signaling factors. International Review of Cytology 1999; 185:45-106.
24. Brunet A, Bonni A, Zigmond MJ et al. Akt promotes cell survival by phosphorylating and inihibiting a forkhead transcription factor. Cell 1999; 96:857-868.
25. Chen Y, Li X, Eswarakumar VP et al. Fibroblast growth factor (FGF) signaling through PI 3 kinase and AKT/PKB is required for embryoid body differentiation. Oncogene 2000; 19:3750-3756.
26. Guo L, Degenstein L, Fuchs E. Keratinocyte growth factor is required for hair development but not for wound healing. Genes Dev 1996; 10:165-175.
27. MacArthur CA, Lawshe A, Xu J et al. FGF-8 isoforms activate receptor splice forms that are expressed in mesenchymal regions of mouse development. Development 1995; 121:3603-3613.
28. Ornitz DM, Xu J, Colvin JS. 1996. Receptor specificity of the fibroblast growth factor family. J Biol Chem 1996; 271:15292-15297.

29. Miller DL, Ortega S, Bashayan O et al. Compensation by fibroblast growth factor 1 (FGF1) does not account for the mild phenotypic defects observed in FGF2 null mice. Mol Cell Biol 2000; 20:2260-2268.

30. Fiore F, Planche J, Gibier P et al. Apparent normal phenotype in Fgf6-/- mice. Int J Dev Biol 1997; 41:639-642.

31. Colvin JS, Feldman B, Nadeau JH et al. Genomic organization and embryonic expression of the mouse fibroblast growth factor 9 gene. Dev Dynam 1999; 216:72-88.

32. Deng CX, Wynshaw-Boris A, Shen MM et al. Murine FGFR-1 is required for early postimplantation growth and axial organization. Genes Dev 1994; 8:3045-3057.

33. Hajihosseini MK, Lalioti MD, Arthaud S et al. Skeletal development is regulated by fibroblast growth factor receptor 1 signaling. Development 2004; 131:325-335.

34. Hoffman MP, Kidder BL, Steinberg ZL et al. Gene expression profiles of mouse submandibular gland development:FGFR1 regulates branching morphogenesis in vitro through BMP- and FGF-dependent mechanisms. Development 2002; 129:5767-5778.

35. Beer H-D, Vindevoghel L, Gait MJ et al. Fibroblast growth factor (FGF) receptor 1-IIIb is a naturally occurring functional receptor for FGFs that is preferentially expressed in the skin and the brain. J Biol Chem 2000; 275:16091-16097.

36. McMahon AP, Ingham PW, Tabin C. Developmental roles and clinical significance of hedgehog signaling. Curr Top Dev Bio 2003; 53:1-114.

37. Kenney AM, Cole MD, Rowitch DH. N-myc upregulation by sonic hedgehog signaling promotes proliferation in developing cerebellar neuron precursors. Development 2003; 130:15-26.

38. Lowry JA, Stewart GA, Lindey S et al. Sonic hedgehog promotes cell cycle progression in activated CD4(+) T lymphocytes. J Immunol 2002; 169:1869-1875.

39. Chiang C, Litingtung Y, Lee E et al. Cyclopia and defective axial patterning in mice lacking Sonic Hedgehog gene function. Nature 1996; 383:407-413.

40. Ahlgren SC, Thackur V, Bronner-Fraser M et al. Sonic Hedgehog rescues cranial neural crest cells from cell death induced by ethanol exposure. Proc Natl Acad Scienc 2002; 99:10476-10481.

41. Monreal AW, Zonana J, Ferguson B. Identification of a new splice form of the EDA1 gene permits detection of nearly all X-linked hypohidrotic ectodermal dysplasia mutations. Am J Hum Genet 1998; 63:380-389.

42. Srivastava AK, Pispa J, Hartung AJ et al. The Tabby phenotype is caused by mutation in mouse homologue of the EDA gene that reveals novel mouse and human exons and encodes a protein (ectodysplasin-A) with collagenous domains. Proc Natl Acad Sci 1997; 94:13069-13074.

43. Yan M, Wang L-C, Hymowitz SG et al. Two-amino acid molecular switch in an epithelial morphogen that regulates binding to two distinct receptors. Science 2000; 290:523-526.

44. Kumar A, Eby MT, Sinha S et al. The ectodermal dysplasia receptor activates the nuclear factor κB, JNK, and cell death pathways and binds to ectodysplasin A. J Biol Chem 2001; 276:2668-2677.

45. Pinheiro M, FreireMaia N. Ectodermal dysplasias: A clinical classification and a causal review. Am J Med Genet 1994; 53:153-162.

46. Sundberg J. 1994. The Downless (dl) and Sleek (Dl sleek) mutations, Chromosome 10. In: Maibach H, ed. Handbook of Mouse Mutations with Skin and Hair Abnormalities. Boca Raton, FL: CRC Press.

47. Moon AM, Capecchi MR. Fgf8 is required for outgrowth and patterning of the limbs. Nat Genet 2000; 26:455-459.

48. Frank DF, Fotheringham LK, Brewer JA et al. An Fgf8 mouse mutant phenocopies human 22q11 deletion syndrome. Development 2002; 129:4591-4601.

49. Aoto K, Nishimura TEK, Motoyama J. Mouse GLI3 regulates Fgf8 expression and apoptosis in the developing neural tube, face and limb bud. Dev Biol 2002; 251:320-332.

50. Taya Y, O'Kane S, Ferguson MW. Pathogenesis of cleft palate in TGF-beta3 knockout mice. 1999; 126:3869-3879.

51. Zhao J, Chen H, Wang YL et al. Abrogation of tumor necrosis factor-alpha converting enzyme inhibits embryonic lung morphogenesis in culture. Int J Dev Biol 2001; 4:623-631.

52. Sarkar L, Cobourne M, Naylor S et al. Wnt/Shh interactions regulate ectodermal boundary formation during mammalian tooth development. Proc Natl Acad Sci USA 2000; 9:4520-4524.

53. Macatee TL, Hammond BP, Arenkial BR et al. Ablation of specific expression domains reveals discrete functions of ectoderm- and endoderm-derived FGF8 during cardiovascular and pharyngeal development. Development 2003; 130.

54. Pahl HL. Activators and target genes of Rel/NF-kappaB transcription factors. Oncogene 1999; 18:6853-686652.

55. Kashimata M, Sayeed S, Ka A et al. The ERK-1/2 signaling pathway is involved in the stimulation of branching morphogenesis of fetal mouse submandibular glands by EGF. Develop Biol 2000; 220:183-196.

56. Marin MW, Dunn A, Grail D et al. Characterization of tumor necrosis factor-deficient mice. Pro Natl Acad Sci USA 1997; 94:8093-8098.

57. Peschon JJ, Torrance DS, Stocking KL et al. TNF receptor-deficient mice reveal divergent roles for p55 and p75 in several models of inflammation. J Immunol 1998; 160:943-952.

58. Rothe J, Mackay F, Bluethmann H et al. Phenotypic analysis of TNFR1-deficient mice and characterization of TNFR1-deficient fibroblasts in vitro. Cir Shock 1994; 44:51-56.

59. Cross MJ, Claesson-Welsh L. FGF and VEGF function in angiogenesis: Signaling pathways, biological responses and therapeutic inhibition. Trends Pharm Sci 2001; 22:201-201.

60. Kodaki T, Woscholski R, Hallberg B et al. The activation of phosphatidylinositol 3-kinase by Ras. Curr Biol 1994; 9:798-806.

61. Lu H-C, Swindell EC, Sierralta WD et al. Evidence for a role of protein kinase C in FGF signal transduction in the developing chick limb bud. Development 2002; 128:2451-2460.

62. Klint P, Kanda S, Kloog Y et al. Contribution of Src and Ras pathways in FGF-2 induced endothelial cell differentiation. Oncogene 1999; 18:3354-3364.

CHAPTER 10

Branching Morphogenesis of the Prostate

A.A. Thomson and P.C. Marker

Introduction

The prostate is a male sex accessory organ whose development is regulated by androgens and mesenchymal/epithelial interactions. The organ comprises branched epithelial ducts within a stroma consisting of fibroblasts and smooth muscle as well as other components such as vasculature and nerves. The function of the prostate is to produce secretions that make up part of the seminal fluid, though it is not certain if these are essential for fertility or sperm function. There has been considerable interest in the identification of molecules and pathways that regulate prostatic growth, due to their relevance in prostatic disease. Few studies have focussed directly on prostatic branching though some have identified factors or pathways that play a role in prostatic growth and branching morphogenesis. Several pathways have been identified that appear to influence growth of the prostate and the process of branching morphogenesis simultaneously. However, genetic evidence suggests that prostatic growth and prostatic branching morphogenesis are processes that are independently regulated. The prostate is not one of the organs widely used for studies of branching morphogenesis, though this seems unfortunate as there are many factors which suggest that this organ would be an excellent model for the study of branching. These are: the ease with which these organs can be grown in vitro, the fact that the prostate is not required for viability, and the late genesis and growth of this organ relative to others during organogenesis.

Anatomical and Endocrine Aspects of Prostate Development

The prostate is an organ that is found only in male mammals and surrounds the urethra at the base of the bladder. The function of the prostate is to add various components to the seminal fluid, and these are emptied directly into the urethra. The contraction of smooth muscle within and surrounding the prostate leads to the expulsion of products made in the branched secretory acini through a branched ductal system and thence the urethra. The secretory epithelia show a tall columnar morphology while epithelia of the ducts show a more flattened morphology.

The formation of the prostate occurs relatively late in gestation when compared to the genesis of most organs. In humans the prostate starts to form between 10-12 weeks of gestation while in mice prostatic development begins at E16.5 and in rats at E17.5. At present, the onset of prostatic organogenesis is defined by the induction of epithelial buds from the urethra, however, it is likely that formation of a condensed mesenchyme involved in organ induction precedes this. During prostatic bud induction and subsequent branching morphogenesis, epithelial buds grow as solid cords which later differentiate and undergo canalisation.

Androgens are required for prostatic development, and it is important to note that the androgen receptor (AR) is not expressed in epithelia during organ induction but is expressed in the mesenchyme.[1] This indicates that essential components of bud induction (in response to androgens) are found within the mesenchyme, though the identity of these signals is presently

Branching Morphogenesis, edited by Jamie A. Davies.

unknown. The importance of mesenchymal to epithelial signalling was established many years ago using tissue recombination studies, which clearly demonstrated that AR was required in the mesenchymal compartment for prostate development.[2] AR is expressed in prostatic epithelia at later stages of prostatic growth and in adulthood, though epithelial AR is neither necessary nor sufficient for normal organ growth.[3] This data also indicates that any involvement of the AR in branching is likely to be restricted to androgen signalling in the mesenchyme. Further details of genetic studies implicating androgens and the AR in prostatic development are described in the section on mutations that alter prostatic branching morphogenesis. The dependence of prostatic growth and branching upon androgens might make the prostate an excellent system in which to study branching, since the ability to inhibit or augment growth and branching allows for thorough examination of different pathways in these processes. Another important issue is that, while it is clear that androgens stimulate the growth and branching of the prostate, it is not known if androgens act upon growth or branching independently. Thus, the effects on branching may be mediated via general effects upon growth and it remains to be determined whether branching is directly affected by androgens.

In rodents, the prostate consists of anatomically distinct lobes (termed dorsal, lateral, ventral and anterior)[4] while in humans it is more compact and composed of different zones that differ subtly in their histology.[5] The different lobes of the rodent prostate exhibit functional differences in terms of the secretory proteins that they make and also differ in their branching patterns.[6] This has allowed for some characterisation of how the separate branching patterns arise, using tissue recombination studies. This work has shown that aspects of epithelial identity (such as secretory protein profile) are specified by mesenchymal signals and suggests that the distinct branching patterns are also mesenchymally defined.[7,8]

Signals from the mesenchyme play a key role in defining branching pattern, however in some circumstances there are limitations of the epithelial response to inductive mesenchyme. For example, the mesenchyme of the seminal vesicle will induce prostatic branching in urethral or bladder epithelium. The seminal vesicle shows an epithelial architecture which is both highly folded as well as branched, yet the mesenchyme of the seminal vesicle will elicit a true branching pattern with a heterologous epithelium (such as bladder). It is worth noting that the seminal vesicle epithelium is derived from the mesoderm (pronephric/wolffian duct) while the urethral and bladder epithelium are endoderm derived. There are two important conclusions to be drawn from this work. Firstly, that some elements of branching pattern can be defined by the epithelium and secondly that instructive branching interactions can occur between tissues of different germ layer origin (mesoderm and endoderm).

Epithelial Bud Induction

Prostatic bud induction from the epithelia of the urethra may be regarded as the initial branching event of prostatic growth and it is emerging that bud induction and subsequent growth and branching may have significant differences. For example, fibroblast growth factor 10 (FGF10) is required for the formation and growth of the prostate, but there is induction of a few prostatic buds in FGF10 null mice.[9] This suggests that FGF10 may be required for prostatic growth but is not essential for bud induction, even though fewer prostatic buds are apparent in the FGF10 null mouse. Similarly, the sonic hedgehog pathway has recently been shown to be important for prostatic growth and branching,[10] but is not required for prostatic bud induction.[11] While bud induction versus subsequent branching morphogenesis have been clearly separated in other systems, these processes have only recently been separated in prostatic organogenesis.

Signals from the mesenchyme are essential for the induction and growth of the prostate but there is little information regarding the identity of these molecules or how they are regulated. Recent work has shown that smooth muscle may play an important role in regulating prostatic induction.[12] During prostatic induction it appears that a layer of smooth muscle forms between the mesenchyme and epithelium, and that the formation of this layer

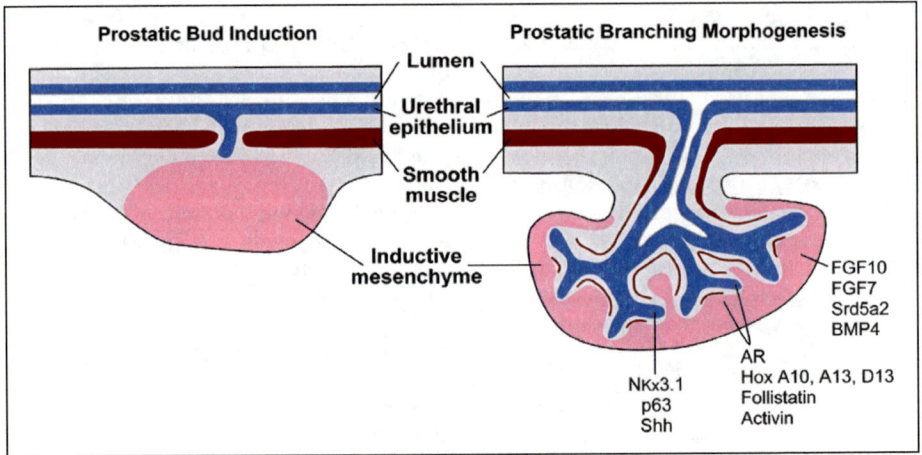

Figure 1. Schematic representation of ventral prostatic induction and subsequent growth and branching morphogenesis. On the left side bud induction in males is illustrated, and the role that smooth muscle plays in regulating interactions between nascent buds and inductive mesenchyme is shown. Androgens delay the development of the smooth muscle allowing organ induction to take place in males. In females, the smooth muscle layer forms more rapidly which precludes organ induction. On the right side, early branching events of the prostate are shown. It is proposed that smooth muscle also plays a role in regulating growth and branching as epithelial buds are directly juxtaposed to the mesenchyme whereas maturing ducts are surrounded by a layer of smooth muscle. Developing prostatic buds grow and branch as solid epithelial cords made up of undifferentiated epithelia. The cords undergo epithelial differentiation and canalisation after branching. Examples of factors involved in prostatic growth and branching are shown, in relation to their pattern or expression in mesenchmye, epithelium, or both compartments.

is sexually dimorphic. In females, the smooth muscle layer develops but in males testosterone inhibits the formation of this layer. The inhibition of the smooth muscle layer is proposed to allow continued signalling between the mesenchyme and epithelium, and instructive interactions leading to bud stabilisation and growth. A schematic diagram of prostatic bud induction and subsequent growth is shown in Figure 1. Additionally, it is possible that smooth muscle also plays a role in branching morphogenesis, as growing epithelial buds lack adjacent smooth muscle but mature ducts are surrounded by layers of well differentiated smooth muscle.[13] Such a pattern of discontinuous smooth muscle distribution is also observed in the developing lung.[14,15] At present it is not known precisely how smooth muscle might affect mesenchymal/epithelial interactions though there are several possibilities. As mesenchyme differentiates into smooth muscle it is very likely that there is a change in the expression of growth or branching regulatory molecules and it is also possible that the smooth muscle acts as a barrier to the diffusion of regulatory molecules from the mesenchyme. Additionally, it is possible that smooth muscle is a source of molecules that are inhibitory to growth or branching. It is also possible that there is a combination of these mechanisms active which are involved in the mediating the effects of smooth muscle.

Growth Factor Signalling in Prostate Branching Morphogenesis

Various studies have examined the function of growth factor signalling in prostate branching morphogenesis, and this section will discuss those that have used organ culture studies of prostatic rudiments grown in vitro. This is a highly tractable system which allows for the addition of growth factors, blocking antibodies, or small molecules that activate or inhibit receptor signalling. Figure 2 shows an example of day-of-birth rat prostatic organs grown in vitro. Testosterone clearly stimulates growth and branching morphogenesis, and in this example the

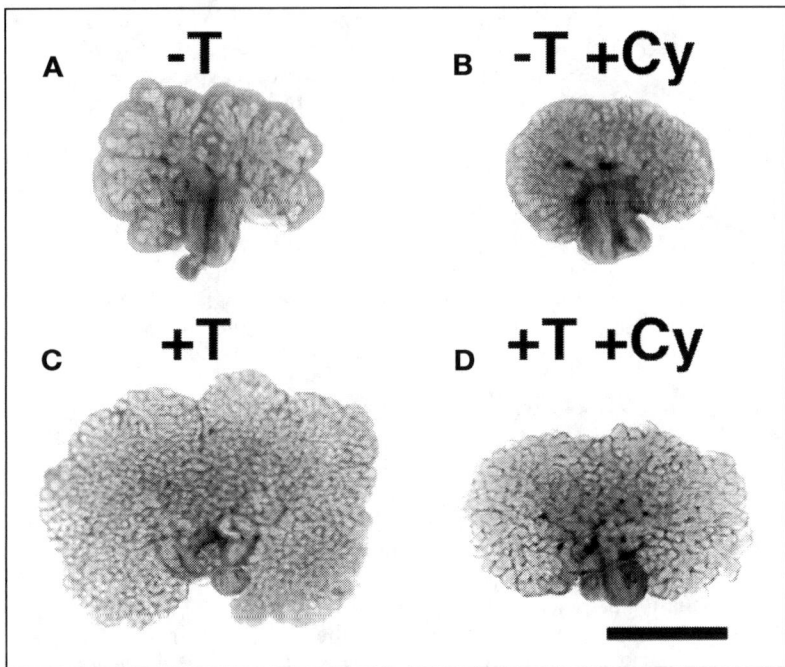

Figure 2. An organ culture system for testing pathways involved in prostatic growth and branching. The system uses day-of-birth (0 day old) rat ventral prostate, although other ages, species and lobes of the prostate can be used. Cultures are grown in defined media under serum free conditions, and recombinant proteins, antibodies, or small molecules can be added to examine their effects upon prostate morphogenesis. Comparison of panels A versus C shows the effect of testosterone (T) on ventral prostate growth and branching, while panels B and D show similar conditions but with the addition of a steroidal alkyloid (cyclopamine, Cy) which inhibits signalling of Shh and related ligands. Effects upon branching can be examined by counting tip numbers at the periphery of the organ (described in detail in Freestone et al 2003).

hedgehog (Hh) signalling pathway has been blocked using a small molecule (cyclopamine) in either the presence or absence of testosterone. The advantages of the organ culture system are the ability to perform a functional test of a given pathway, however a significant problem is that inhibition or stimulation are applied throughout the organ and are not restricted to sites of active branching morphogenesis within the organ. Most studies have not examined branching in detail but have looked at effects on the whole organ, though a few have studied branching directly. These limitations notwithstanding, growth factor pathways studied have included fibroblast growth factors (FGFs), Transforming Growth Factors betas (TGFβs), activin and inhibins, and sonic hedgehog (Shh).

The effects of FGF7 (also known as KGF) and FGF10 have been examined in ventral prostate rudiments grown in vitro.[16,17] Both of these factors are made by the mesenchyme and act via a receptor which is restricted to the epithelium (FGFR2IIIb) and act as paracrine factors. Addition of recombinant FGF7 or 10 stimulated the growth of the prostate and did not appear to result in any distortion of branching pattern. In contrast, addition of TGFβ results in the inhibition of prostatic growth as well as changes in the branch number.[18] Addition of TGFβ resulted in a decrease in ductal tip number, when the total number of branch tips were counted. However, it is quite difficult to define individual tips in the whole organ, and it is possible that TGFβ may have differential effects at the periphery of the organ versus the middle. It appears that, when TGFβ is added to organ cultures of ventral prostate, there is an increase in epithelial tip number at the periphery and an overall reduction in size (D. Tomlinson and AAT, unpublished observations). Another member of the TGFβ superfamily which inhibits prostatic growth is activin. Cancilla et al showed that addition of activin A inhibited the growth of the prostate, and led to little branching morphogenesis. Addition of follistatin (an antagonist of activin signalling) led to organ growth and branching in the absence of testosterone.[19] These data suggest that activin signalling may be a modulator of growth and branching of the prostate.

Shh signalling has been studied in relation to prostatic growth and recent studies indicate that this pathway is involved in branching and epithelial differentiation. Initial studies suggested that Shh was required for the formation of the prostate,[10] but more recent data indicate that formation of the prostate is not dependent upon Shh signalling.[11] In addition, inhibition of Shh signalling causes an increase in epithelial tips at the periphery of the organ, suggesting that Shh may be an inhibitor of the branching process (Fig. 2).

Genetic Analysis of Prostatic Branching

The phenotypes created by spontaneous and engineered mutations in mice and humans have provided direct evidence for the roles of several genes in prostatic branching morphogenesis. In humans, mutations affecting the prostate were identified because they result in androgen insensitivity syndrome (AIS) and pseudovaginal perineoscrotal hypospadias (PPSH). In mice, spontaneous mutations affecting the prostate were identified because of external phenotypes at other anatomical locations. These mutations include Testicular feminization (Tfm) and Hypodactyly (Hd). The remaining mutations affecting prostatic branching morphogenesis were engineered in mice using homologous recombination and transgenic approaches. A clear advantage of using a genetic approach to define the roles of particular genes in prostatic branching morphogenesis is that genetics can prove that a gene is required for branching morphogenesis in the context of the intact mouse or human. A potential disadvantage of using a genetic approach is that many mutations cause defects in other organ systems that can complicate the interpretation of prostatic phenotypes. The prostatic phenotypes created by spontaneous and engineered mutations in mice and humans are summarized in Table 1 and discussed in greater detail in the sections that follow.

Mutations That Affect Endocrine Signalling

Several circulating hormones regulate prostatic development. The most important of these are the androgens.[20] Cellular response to systemic androgens is mediated by nuclear androgen receptors that are activated by testosterone or dihydrotestosterone. The requirement for the *androgen receptor (AR)* gene in prostatic epithelial budding and branching morphogenesis is shown by the phenotypes caused by AIS mutations in humans and the Tfm mutation in mice.[21,22] These mutations are recessive, loss of function mutations in the *AR*. The prostatic phenotypes caused by the AIS and Tfm mutations include the absence of epithelial budding and branching morphogenesis, and the absence of organ-specific cellular differentiation.

The phenotypes caused by inactivating mutations in the *steroid 5-alpha reductase 2 (Srd5a2)* gene further clarify the roles of specific androgens. Mutations in this gene include the pseudovaginal perineoscrotal hypospadias (PPSH) mutations in humans[23] and a mouse knockout allele created by homologous recombination in mouse embryonic stem (ES) cells.[24] *Srd5a2* encodes the enzyme Δ^4-3-ketosteroid-5α-reductase (5α-reductase) type 2 that converts testosterone to the more potent androgen receptor agonist dihydrotestosterone.[25] During prostatic development, testosterone is the major androgen in circulation. *Srd5a2* is expressed in the mesenchyme of the developing prostate. Thus, testosterone could activate androgen receptors in the prostate through direct binding to androgen receptors, or through local conversion of testosterone to dihydrotestosterone that could subsequently bind androgen receptors. When conversion of testosterone to dihydrotestosterone is blocked in the prostate by mutations in *Srd5a2,* the urogenital sinus is specified as prostate but prostatic growth and

Table 1. *Mutations affecting prostatic branching morphogenesis*

Gene	Symbol	Protein Type	Expression[1]	Alleles[1]	Phenotypes[1]	References
Androgen receptor	Ar	nuclear receptor	mes + epi	human AIS, mouse Tfm	no branching	(Brown et al, 1988; He et al, 1991)
Bone morphogenetic protein 4	Bmp4	secreted	mes	mouse knockout (heterozygote)	increased branching in the AP and VP	(Lamm et al, 2001a)
Fibroblast growth factor 10	Fgf10	secreted	mes	mouse knockout (homozygote)	no branching	(Don acour et al, 2003)
Growth hormone receptor	Ghr	transmembrane receptor	mes + epi	transgenic antagonist	small prostate with reduced branching	(Ruan et al, 1999)
Homeobox A10	Hoxa10	transcription factor	mes + epi	mouse knockout (homozygote)	reduced AP branching, partial AP to DLP transformation	(Podlasek et al, 1999d)
Homeobox A13	Hoxa13	transcription factor	mes + epi	mouse Hd (heterozygote)	reduced size and reduced branching in the VP and DLP	(Podlasek et al, 1999b)
Homeobox D13	Hoxd13	transcription factor	mes + epi	mouse knockout (homozygote)	small VP and DLP, reduced DLP branching	(Podlasek et al, 1997a)
Insulin-like growth factor 1	Igf1	secreted		mouse knockout (homozygote)	small prostate with reduced branching	(Baker et al, 1996; Ruan et al, 1999)
NK-3 transcription factor, locus 1	Nkx3.1	transcription factor	epi	mouse knockout (homozygote)	reduced branching	(Bhatia-Gaur et al, 1999; Schneider et al, 2000; Tanaka et al, 2000a)
Tumor protein p63	p63	nuclear	epi	mouse knockout (homozygote)	no branching	(Signoretti et al,2000)
Prolactin	Prl	hormone		mouse knockout (homozygote)	small VP	(Steger et al, 1998)
Steroid 5-alpha reductase 2	Srd5a2	enzyme	mes	mouse knockout (homozygote), human PPSH (homozygote)	small prostate	(Andersson et al, 1991; Mahendroo et al, 2001)

1. Abbreviations: mesenchyme (mes), epithelium (epi), androgen insensitivity syndrome (AIS), pseudovaginal perineoscrotal hypospadias (PPSH), Testicular feminization (Tfm), Hypodactyly (Hd), anterior prostate (AP), dorsolateral prostate (DLP), ventral prostate (VP). The tissue layer expressing the gene is indicated for the prostate only.

branching morphogenesis are greatly reduced.[23,24] These phenotypes demonstrate that dihydrotestosterone is not required for prostatic branching morphogenesis to occur, but dihydrotestosterone is required to achieve the normal extent of growth and morphogenesis. Further studies have confirmed that nonreducible analogues of testosterone will stimulate prostatic growth and branching.[6]

Genetic experiments in mice have implicated two additional circulating hormones in prostatic development. Expression of an antagonist for the growth hormone receptor (GHR) in transgenic mice resulted in a small prostate with reduced branching.[27] In mice transgenic for the growth hormone receptor antagonist, reduced prostatic morphogenesis occurred in the context of an overall dwarfism syndrome caused by the transgene. GHR antagonist transgenic mice were about 50% the size of wild type control mice. These phenotypes demonstrated that growth hormone (GH) and the GHR act as positive factors that stimulate prostatic branching morphogenesis. However, the effects of GH and GHR to stimulate growth and morphogenesis are not specific to the prostate. Inactivation of *prolactin* in a mouse knockout model resulted in a smaller ventral prostate.[28] Steger et al did not measure the dorsolateral prostate or quantify prostatic branching morphogenesis so the potential effects on these aspects of prostatic development are not clear.

Mutations That Affect Homeobox-Containing Transcription Factors

Several homeobox-containing transcription factors have been implicated in prostatic branching morphogenesis by genetic data. These include three members of the Hox gene family: *Hoxa10, Hoxa13,* and *Hoxd13.* Mice homozygous for a knockout allele of *Hoxa10* have reduced branching in the anterior prostate and a partial anterior prostate to dorsolateral prostate transformation.[29] Mice heterozygous for a deletion in *Hoxa13* [the mouse Hypodactyly (Hd) mutation] have reduced size and reduced branching in the anterior prostate and dorsolateral prostate.[30] In the case of the Hd mutation, phenotypes in other organ systems result in lethality in late gestation for homozygous mutants. Prostatic development primarily occurs after birth. Consequently, the phenotypic consequences of removing both copies of *Hoxa13* for prostatic development cannot be determined using the Hd mutation. Mice homozygous for a knockout allele of *Hoxd13* have smaller anterior prostates and dorsolateral prostate and reduced dorsolateral prostate branching.[31] These phenotypes demonstrate that each of these Hox genes acts to promote branching morphogenesis in one or more prostatic lobes.

In mice and humans, the Hox genes are organized into 4 paralogous clusters (Hoxa, Hoxb, Hoxd, and Hoxd clusters). In many tissues, a comparison of paralogous Hox genes between clusters has identified similar gene expression patterns and partially redundant gene functions. This is true for paralogous genes *Hoxa13* and *Hoxd13* that act in a partially redundant fashion in the prostate. This is shown by a more severe failure of prostatic development in compound *Hoxa13 Hd/+: Hoxd13 -/-* mutant mice than the individual mutants.[32] The most dramatic example of this is for the AP. Mice homozygous for the *Hoxd13* mutation have an AP of wild type size and wild type branching.[31] Mice heterozygous for the Hd mutation have only a modest decrease in the extent of AP morphogenesis that was not statistically significant in the study by Podlasek et al.[30] In contrast, compound *Hoxa13 Hd/+: Hoxd13 -/-* mutant mice lack the AP entirely.[32]

The three Hox gene mutations that alter prostatic morphogenesis also cause developmental defects in multiple unrelated organ systems. Mutations in a fourth homeobox-containing transcription factor, *Nkx3.1,* have developmental phenotypes that are limited to a small number of glandular organs including the prostate.[33-35] Mice homozygous for a knockout allele of *Nkx3.1* have reduced branching in all prostatic lobes without defects in prostatic size. These phenotypes demonstrate that *Nkx3.1* acts to promote prostatic branching morphogenesis independent of prostatic growth. The anatomical sites of other developmental defects in *Nkx3.1* mutant mice are the bulbourethral glands and salivary glands (see Chapter 9).

Mutations That Affect Secreted Signalling Molecules

Branching morphogenesis during prostatic development is dependent upon mesenchymal-epithelial interactions such that morphogenesis and differentiation of both the epithelium and mesenchyme are abortive if the epithelium and mesenchyme are grown separately. Not surprisingly, the genes implicated in prostatic branching morphogenesis by genetic data include three genes encoding secreted signalling molecules, *Bmp4, Igf1,* and *Fgf10,* that can mediate intercellular communication. Mice heterozygous for a knockout allele of *Bmp4* have increased branching in the anterior prostate and ventral prostate.[36] Mice homozygous for a knockout allele of *Bmp4* die during embryogenesis prior to prostatic development so the affects of the complete absence of *Bmp4* on prostatic morphogenesis cannot be determined using the allele studied by Lamm et al. These data are consistent with a role for BMP4 as a factor that limits prostatic branching morphogenesis.

Mice homozygous for a knockout allele of *Igf1* have reduced size and reduced branching for all prostatic lobes.[27,37] These phenotypes are observed in the context of a severe dwarfism syndrome. In the study by Ruan et al only 10% of homozygous mutants survived to the adult stage when prostatic phenotypes were assessed. At that age, mutant males were approximately 20% the size of wild type controls. These phenotypes demonstrated that *Igf1* acts to stimulate prostatic branching morphogenesis. However, the effects of *Igf1* to stimulate growth and morphogenesis are not specific to the prostate. In addition, *Igf1* mutant mice have decreased circulating androgen levels.[37] Consequently, the prostatic phenotypes in *Igf1* mutant mice may be indirect.

Mice homozygous for a knockout allele of *Fgf10* lack nearly all prostatic buds at birth.[9] In wild type mice, many prostatic buds and some early ductal branches have formed by birth. Homozygous *Fgf10* mutant mice die at birth due to defects in other organ systems. Donjacour et al performed rescue and grafting experiments of prostatic rudiments into immuno-deficient male host mice to determine the affects of *Fgf10* deficiency on postnatal prostatic branching morphogenesis and cellular differentiation. In these experiments, *Fgf10* mutant prostatic rudiments exhibited little growth and failed to undergo branching morphogenesis, but they did express mature differentiation markers for prostatic epithelial cells. Under the same experimental conditions wild type rudiments exhibited extensive growth and branching morphogenesis. These phenotypes demonstrate that *Fgf10* is required for prostatic growth and branching morphogenesis.

Other Mutations

The *p63* gene encodes a nuclear protein that is expressed in many epithelial tissues. Mice homozygous for a knockout allele of *p63* lack any sign of prostatic budding or branching morphogenesis at birth.[38] In wild type mice, many prostatic buds and some early ductal branches have formed by birth. *p63* homozygous mutant mice die at birth due to defects in multiple other epithelia throughout the body. These data are consistent with a requirement for *p63* in prostatic morphogenesis. However, since *p63* mutant mice die shortly after prostatic morphogenesis begins in wild type mice, it is not possible to exclude the possibility that prostatic morphogenesis is merely delayed in the absence of *p63* based on the study by Signoretti et al.

Prostatic Growth and Branching Morphogenesis Are Under Separate Genetic Control

Several pathways have been identified that simultaneously affect prostatic growth and prostatic branching morphogenesis. However, the phenotypes caused by some mutations demonstrate that these are two distinct processes under separate genetic control. For example, the ventral prostate has dramatically reduced branching in *Nkx3.1* deficient mice, but the ventral prostate grows to its normal size in these mice.[3,28,35] Similarly, the ventral prostate has increased branching morphogenesis in heterozygous *Bmp4* mutant mice without an increase in prostatic size.[16] These types of phenotypes demonstrate that genetic mechanisms are present within the prostate to regulate prostatic branching morphogenesis independently of prostatic growth.

Conclusions

Branching morphogenesis in the prostate is regulated by several factors though we are far from a comprehensive knowledge of their identity. Since the prostate is a target of steroid hormone action there is additional complexity regarding the role of androgens in growth versus branching, and much needs to be done in separating the pathways involved in these two processes. At present, it appears that many mechanisms identified in other organ systems are involved in regulating prostatic branching morphogenesis and it seems unlikely that there are prostate-specific mechanisms of branching. However, since the organ is dependent upon androgens for its development there may be specific pathways involved in the response to androgens and it will be interesting to determine how these endocrine pathways interact with those involved in growth and branching.

References

1. Takeda H, Mizuno T, Lasnitzki I. Autoradiographic studies of androgen-binding sites in the rat urogenital sinus and postnatal prostate. J Endocrinol 1985; 104:87-92.
2. Cunha GR, Chung LW. Stromal-epithelial interactions—I. Induction of prostatic phenotype in urothelium of testicular feminized (Tfm/y) mice. J Steroid Biochem 1981; 14:1317-1324.
3. Gao J, Arnold JT, Isaacs JT. Conversion from a paracrine to an autocrine mechanism of androgen-stimulated growth during malignant transformation of prostatic epithelial cells. Cancer Research 2001; 61:5038-5044.
4. Timms BG, Mohs TJ, Didio LJ. Ductal budding and branching patterns in the developing prostate. J Urol 1994; 151:1427-1432.
5. McNeal JE. Normal histology of the prostate. Am J Surg Pathol 1988; 12:619-633.
6. Sugimura Y, Cunha GR, Donjacour AA. Morphogenesis of ductal networks in the mouse prostate. Biol Reprod 1986; 34:961-971.
7. Hayashi N, Cunha GR, Parker M. Permissive and instructive induction of adult rodent prostatic epithelium by heterotypic urogenital sinus mesenchyme. Epithelial Cell Biol, 1993 2:66-78.
8. Timms BG et al. Instructive induction of prostate growth and differentiation by a defined urogenital sinus mesenchyme. Microsc Res Tech 1995; 30:319-332.
9. Donjacour AA, Thomson AA, Cunha GR. FGF-10 plays an essential role in the growth of the fetal prostate. Dev Biol 2003; 261:39-54.
10. Podlasek CA et al. Prostate development requires Sonic hedgehog expressed by the urogenital sinus epithelium. Dev Biol 1999; 209:28-39.
11. Freestone SH et al. Sonic hedgehog regulates prostatic growth and epithelial differentiation. Dev Biol 2003. In press.
12. Thomson AA et al. The role of smooth muscle in regulating prostatic induction. Development 2002; 129:1905-1912.
13. Hayward SW et al. Stromal development in the ventral prostate, anterior prostate and seminal vesicle of the rat. Acta Anat 1996; 155:94-103.
14. Tollet J, Everett AW, Sparrow MP. Spatial and temporal distribution of nerves, ganglia, and smooth muscle during the early pseudoglandular stage of fetal mouse lung development. Dev Dyn 2001; 221:48-60.
15. Tollet J, Everett AW, Sparrow MP. Development of neural tissue and airway smooth muscle in fetal mouse lung explants: a role for glial-derived neurotrophic factor in lung innervation. Am J Respir Cell Mol Biol 2002; 26:420-429.
16. Sugimura Y et al. Keratinocyte growth factor (KGF) can replace testosterone in the ductal branching morphogenesis of the rat ventral prostate. International Journal of Developmental Biology 1996; 40:941-951.
17. Thomson AA, Cunha GR. Prostatic growth and development are regulated by FGF10. Development 1999; 126:3693-3701.
18. Itoh N et al. Developmental and hormonal regulation of transforming growth factor-beta1 (TGFbeta1), -2, and -3 gene expression in isolated prostatic epithelial and stromal cells: Epidermal growth factor and TGF-beta interactions. Endocrinology 1998; 139:1378-1388.

19. Cancilla B et al. Regulation of prostate branching morphogenesis by activin A and follistatin. Dev Biol 2001; 237:145-158.
20. Marker PC et al. Hormonal, cellular, and molecular control of prostatic development. Dev Biol 2003; 253:165-174.
21. Brown TR et al. Deletion of the steroid-binding domain of the human androgen receptor gene in one family with complete androgen insensitivity syndrome: evidence for further genetic heterogeneity in this syndrome. Proc Natl Acad Sci USA 1988; 85:8151-8155.
22. He WW, Kumar MV, Tindall DJ. A frameshift mutation in the androgen receptor gene causes complete androgen insensitivity in the testicular-feminized mouse. Nucleic Acids Res 1991; 19:2373-2378.
23. Andersson S et al. Deletion of steroid 5 alpha-reductase 2 gene in male pseudohermaphroditism. Nature 1991; 354:159-161.
24. Mahendroo M.S et al. Unexpected virilization in male mice lacking steroid 5 alpha-reductase enzymes. Endocrinology 2001; 142:4652-4662.
25. Russell DW, Wilson JD. Steroid 5 alpha-reductase: two genes/two enzymes. Annu Rev Biochem 1994; 63:25-61.
26. Foster BA, Cunha GR. Efficacy of various natural and synthetic androgens to induce ductal branching morphogenesis in the developing anterior rat prostate. Endocrinology 1999; 140:318-328.
27. Ruan W et al. Evidence that insulin-like growth factor I and growth hormone are required for prostate gland development. Endocrinology 1999; 140:1984-1989.
28. Steger RW et al. Neuroendocrine and reproductive functions in male mice with targeted disruption of the prolactin gene. Endocrinology 1998; 139:3691-3695.
29. Podlasek CA et al. Hoxa-10 deficient male mice exhibit abnormal development of the accessory sex organs. Dev Dyn 1999; 214:1-12.
30. Podlasek CA, Clemens JQ, Bushman W. Hoxa-13 gene mutation results in abnormal seminal vesicle and prostate development. J Urol 1999; 161:1655-1661.
31. Podlasek CA, Duboule D, Bushman W. Male accessory sex organ morphogenesis is altered by loss of function of Hoxd-13. Dev Dyn 1997; 208:454-465.
32. Warot X et al. Gene dosage-dependent effects of the Hoxa-13 and Hoxd-13 mutations on morphogenesis of the terminal parts of the digestive and urogenital tracts. Development 1997; 124:4781-4791.
33. Bhatia-Gaur R et al. Roles for Nkx3.1 in prostate development and cancer. Genes Dev 1999; 13:966-977.
34. Schneider A et al. Targeted disruption of the Nkx3.1 gene in mice results in morphogenetic defects of minor salivary glands: Parallels to glandular duct morphogenesis in prostate. Mech Dev 2000; 95:163-174.
35. Tanaka M et al. Nkx3.1, a murine homolog of Ddrosophila bagpipe, regulates epithelial ductal branching and proliferation of the prostate and palatine glands. Dev Dyn 2000; 219:248-260.
36. Lamm ML et al. Mesenchymal factor bone morphogenetic protein 4 restricts ductal budding and branching morphogenesis in the developing prostate. Dev Biol 2001; 232:301-314.
37. Baker J et al. Effects of an Igf1 gene null mutation on mouse reproduction. Mol Endocrinol 1996; 10:903-918.
38. Signoretti S et al. p63 is a prostate basal cell marker and is required for prostate development. Am J Pathol 2000; 157:1769-1775.

CHAPTER 11

Uterine Glands

Thomas E. Spencer, Karen D. Carpenter, Kanako Hayashi and Jianbo Hu

Abstract

This chapter focuses on the comparative development and mechanisms regulating branching morphogenesis of endometrial glands in the mammalian uterus. All uteri contain endometrial glands that secrete substances, termed histotroph, essential for conceptus (embryo/fetus and associated extraembryonic membranes) survival, implantation, development and growth. Uterine adenogenesis is the process whereby endometrial glands develop and is primarily a postnatal event in domestic livestock, rodents and humans. Adenogenesis involves initial differentiation and budding of glandular epithelium from luminal epithelium, followed by invagination and extensive tubular coiling and branching morphogenesis through the endometrial stroma to the myometrium. The endocrine, cellular and molecular mechanisms regulating uterine adenogenesis are not well defined in any species. In neonatal rodents and pigs, uterine adenogenesis is ovary- and steroid- independent. Estrogen receptor alpha regulates the initial stage of endometrial gland budding and differentiation in the pig, but not rodent or sheep. However, the ovary and uterine estrogen receptor alpha regulate later stages of endometrial gland coiling and branching morphogenesis in sheep. In rodents, pigs and sheep, uterine adenogenesis is a critical period that is sensitive to the detrimental effects of steroid-based endocrine disruptors, which cause functional defects in the adult. In neonatal sheep, pituitary prolactin acts on receptors, expressed exclusively in the nascent and proliferating endometrial glands, to regulate their coiling and branching morphogenetic differentiation. Following postnatal adenogenesis, humans and menstruating primates undergo recurrent endometrial growth and uterine gland branching morphogenesis during the proliferative phase after menstruation. This regrowth is regulated by ovarian estrogen and intrinsic growth factors. In domestic animals, extensive endometrial gland hyperplasia and hypertrophy occur during gestation to provide increasing histotrophic nutrition for fetoplacental growth. In sheep, the sequential actions of ovarian steroid hormones (estrogen and progesterone), the pregnancy recognition signal (interferon tau), and somatolactogenic hormones from the placenta (placental lactogen and growth hormone) constitute a servomechanism that directly regulates endometrial gland morphogenesis and terminal differentiated function. An increased understanding of the mechanisms regulating uterine gland branching morphogenesis is important, because aberrations in this important process can cause infertility, fetal growth retardation, and disease.

Introduction and Historical Background

Uterine endometrial glands are present in all mammalian uteri. The uterine glands selectively transport or synthesize and secrete substances, termed histotroph, into the uterine lumen.[1,2] The precise components of histotroph is not known in any animal, but is complex and contains numerous binding and nutrient transport proteins, ions, mitogens, cytokines, lymphokines, glucose, enzymes, hormones, growth factors, protease inhibitors, and other

---

Branching Morphogenesis, edited by Jamie A. Davies.
©2005 Eurekah.com and Springer Science+Business Media.

substances.[3] The idea that uterine secretions nourish the developing conceptus (embryo/fetus and associated extraembryonic placental membranes) was discussed first by Aristotle in the third century BC and then by William Harvey in the 17th century. In 1882, Bonnett concluded that secretions of uterine glands were important for fetal well being in ruminants.[4]

Evidence accumulated from primate and subprimate species during the last century overwhelmingly supports a role for uterine gland secretions in establishment of endometrial receptivity to the embryo and conceptus survival, growth and development.[1-13] In marsupials, carnivores and roe deer, changes in endometrial secretory activity are proposed to regulate delayed implantation.[14,15] In rodents, two genes, leukemia inhibitory factor and calcitonin, are expressed exclusively in the uterine endometrial glands and are required for establishment of uterine receptivity and blastocyst implantation.[10,11] Similarly, uterine gland knockout (UGKO) ewes that lack endometrial glands are infertile and exhibit a defect in peri-implantation conceptus survival and development.[16,17] Pregnancy rates are also reduced in cows that exhibit partial to complete UGKO phenotypes due to exposure to a combination of progesterone and estrogen immediately after birth.[18]

Histotroph appears to be particularly important for conceptus survival and development in domestic animals, given the prolonged nature of the implantation and placentation. In sheep,[19,20] cattle,[21] pigs,[22,23] and horses,[24-26] endometrial glands undergo extensive hyperplasia and hypertrophy during pregnancy, presumably in response to increasing demands of the developing conceptus for uterine histotroph.[2,19,26] Indeed, the epitheliochorial placentae of domestic animals develop unique placental structures, termed areolae, over the mouth of each uterine gland as specialized areas for absorption and transport of uterine histotroph by fluid-phase pinocytosis.[27] In humans, uterine glands and their secretions may provide much of the nutrient supply during the first trimester when there is little maternal blood flow to the placenta and oxygen tension within the fetoplacental unit is low.[13,28] Given that the success of uterine gland morphogenesis dictates, in part, the embryotrophic and functional potential of the adult uterus, it is important that mechanisms regulating endometrial gland morphogenesis be defined and understood.

Comparative Developmental Biology of Uterine Glands

Comparative developmental biology of uterine glands in domestic animals, rodents and humans has been reviewed.[12,18] In all mammals, developmental specialization of the paramesonephric or Müllerian ducts gives rise to infundibula, oviducts, uterus, cervix and anterior vagina in the fetus.[29,30] In the adult, the mature uterine wall is comprised of two functional compartments, the endometrium and myometrium (see Fig. 1). The endometrium is the inner mucosal lining of the uterus, derived from the inner layer of ductal mesenchyme. Histologically, the endometrium consists of two epithelial cell types, luminal epithelium (LE) and glandular epithelium (GE), two stratified stromal compartments including a densely organized stromal zone (stratum compactum), and a more loosely organized stromal zone (stratum spongiosum), blood vessels and immune cells. The myometrium is the smooth muscle component of the uterine wall that includes an inner circular layer, derived from the intermediate layer of ductal mesenchymal cells, and an outer longitudinal layer, derived from subperimetrial mesenchyme.

Morphogenetic events common to development of all mammalian uteri include: (1) organization and stratification of endometrial stroma; (2) differentiation and growth of the myometrium; and (3) differentiation and morphogenesis of the endometrial glands termed adenogenesis.[12,18,31] Uterine adenogenesis is primarily a postnatal event in all species that leads to initial histological maturity. Figure 2 summarizes the timing of initial uterine adenogenesis in a number of mammals. The progressive development of endometrial GE from the LE to the inner circular layer of myometrium is a coordinated event that involves bud formation, tubulogenesis, and coiling and branching morphogenesis (see Fig. 3). However, the rodent uterus lacks the large numbers of coiled and branched uterine glands characteristic of uteri in

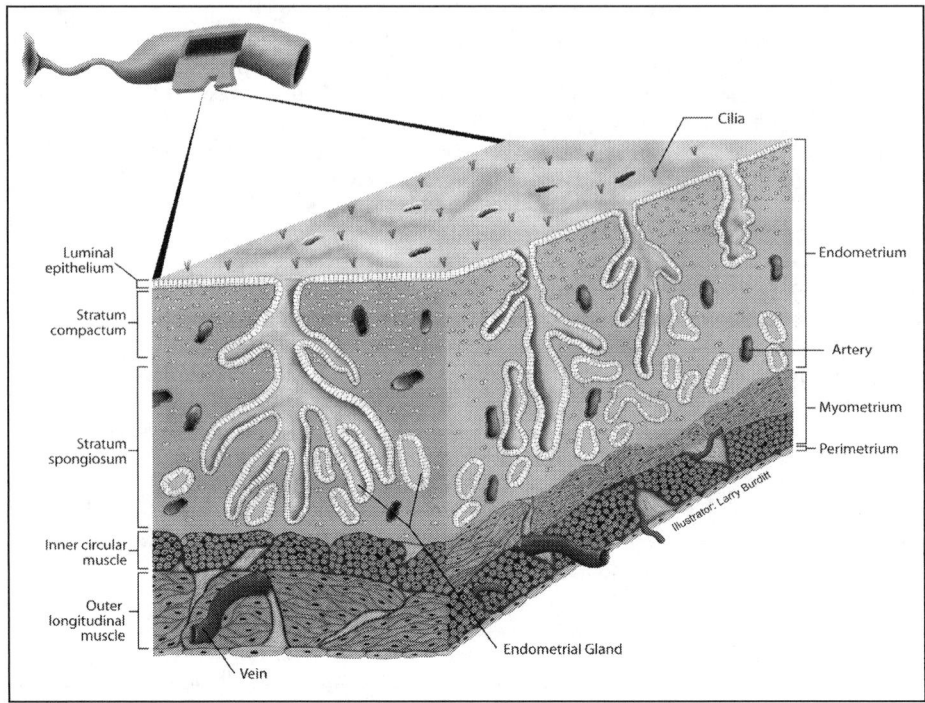

Figure 1. Schematic illustration of the uterine wall of domestic animals. Uterine glands develop from the endometrial luminal epithelium. Coiled and branched uterine glands are present throughout the stroma to the inner circular layer of myometrium. The opening of each ductal portion of the uterine glands is present on the mucosal surface, allowing for uterine gland histotroph to be secreted into the uterine lumen. (Graphic courtesy of Rodney Geisert and Larry Burdett, Oklahoma State University, Stillwater, USA).

most other mammals. Consequently, neonatal ungulates (e.g., sheep, cattle, and pigs) provide attractive models for the study of mechanisms regulating these processes in women.

Ruminants

The endometrium in adult ruminants (sheep, cattle and goats) consists of a large number of raised aglandular caruncles and glandular intercaruncular areas.[19,32] These ruminants have a synepitheliochorial type of placentation in which placental cotyledons fuse with endometrial caruncles to form placentomes that serve a primary role in fetal-maternal exchange of gases and micronutrients across the placenta.[30] The dichotomous nature of the adult ruminant endometrium, consisting of both aglandular caruncular areas and glandular intercaruncular areas, makes it an excellent model for the study of mechanisms underlying establishment of divergent structural and functional areas within a single, mesodermally derived organ.[33] Uterine adenogenesis has been described in sheep[29,33-38] and, to a lesser extent, in cattle.[32,39]

In sheep (Figs. 3, 4), endometrial gland genesis is initiated between birth (postnatal day or PND 0) and PND 7, when shallow epithelial invaginations appear along the LE in presumptive intercaruncular areas.[33,35,37] Between PND 7 and 14, nascent, budding glands proliferate and invaginate into the stroma, forming tubular structures that coil and branch by PND 21. After PND 21, the majority of glandular morphogenetic activity involves coiling and branching morphogenesis of tubular and coiled endometrial glands to form terminal end bud-like structures in deeper stroma. However, GE bud differentiation and tubulogenesis is continuous

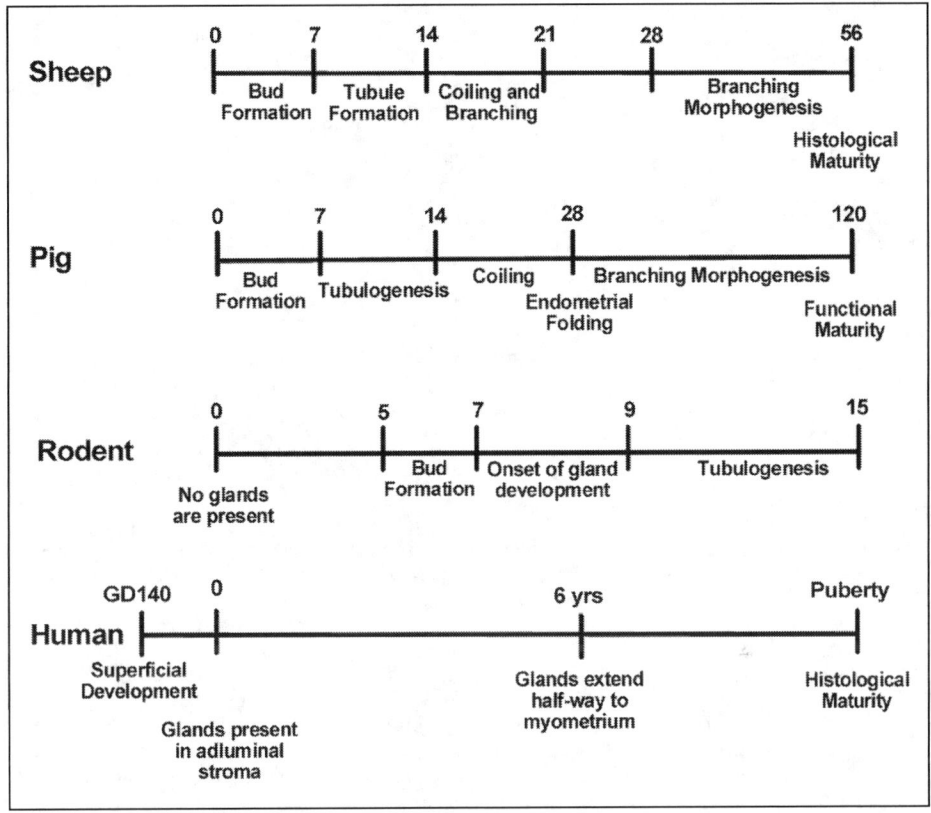

Figure 2. Comparative time line of the seminal events during endometrial gland morphogenesis in the sheep, pig, rodent and human. Postnatal day (PND) 0 is the day of birth. GD, gestational day.

between PND 7 and PND 56. By PND 56, the caruncular and intercaruncular endometrial areas are histoarchitecturally similar to those of the adult uterus. Branching development of the GE appears to direct or permit differentiation of uterine stroma into subluminal stratum compactum and lower stratum spongiosum in intercaruncular areas of the endometrium.[40]

Although the ovine uterine wall is histoarchitecturally mature by eight weeks after birth, final maturation and growth may not occur until puberty,[34] or even the first pregnancy.[19,20] As illustrated in Figure 4, extensive endometrial gland hyperplasia and hypertrophy occurs during each pregnancy.[19,20] In sheep, superficial implantation and placentation begins on Days 15-16, but is not completed until Days 50-60 of pregnancy.[41] During this period, the uterus grows and remodels substantially in order to accommodate rapid conceptus growth in the latter one-half to one-third of pregnancy. In addition to placentomal development in the caruncular areas of the endometrium and changes in uterine vascularity, the intercaruncular endometrial glands grow substantially in length (4-fold) and width (10-fold) and establish additional side-branchings during pregnancy.[19] During gestation, endometrial gland hyperplasia occurs between Days 15 and 50 followed by hypertrophy to increase surface area that allows for maximal production of histotroph after Day 60.[20] Indeed, the histotrophic capacity of the endometrial glands is well correlated with fetal growth, indicating the importance of histotrophic nutrition and placental areolae. The life cycle of endometrial glands in the uterus of domestic animals is strikingly similar to changes in the mammary glands during mammogenesis, lactogenesis and involution (see Chapter 7).

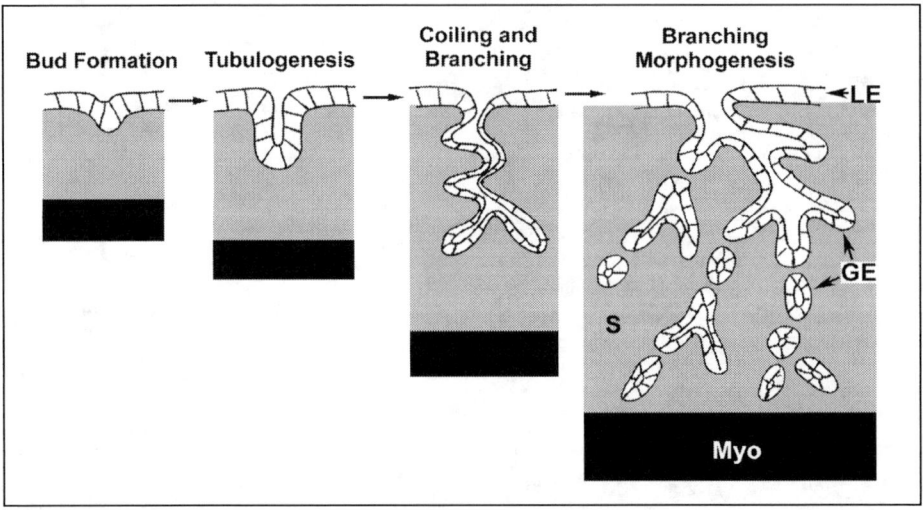

Figure 3. Diagrammatic representation of endometrial adenogenesis in the uterine wall of most species. In many cases where invagination of an epithelium is required for a developmental process, the first step is bud formation that involves conversion of a cuboidal cell to the geometry of a frustrum, a three-dimensional cone with the top removed parallel to the base. This creates a bulge in a basal portion of the epithelial cells at the point where the invagination is to occur. The uterine glands originate as shallow glandular epithelial (GE) buds from the lumenal epithelium (LE). The next step is tubulogenesis, which involves subsequent cell divisions to convert the frustrum into a tube. The simple, tubular glands then begin to coil as they proliferate in the stroma (S) towards the myometrium (Myo). The final stage of adenogenesis is the coiling and branching morphogenesis of the tubular, coiled glands throughout the stroma until the tips reach the inner circular layer of the myometrium. The overall process of endometrial adenogenesis is similar to adenogenesis that occurs during development of other epitheliomesenchymal organs, such as the mammary gland, salivary gland, and lung. Legend: GE, glandular epithelium; LE, luminal epithelium; Myo, myometrium; S, stroma.

Pigs

Transformation of the porcine uterine wall from histoarchitectural infancy to maturity occurs within 120 days of birth.[42-45] Overall patterns of endometrial morphogenesis in the postnatal pig and sheep are similar (Fig. 3). At birth, the porcine uterus consists of a simple, slightly corrugated columnar epithelium supported by unorganized stromal mesenchyme, encircled by a rudimentary myometrium.[44,45] Shallow, epithelial depressions are observed on PND 0 which appear to be GE buds.[44] The budding GE develops into simple epithelial tubes by PND 7 and 14 that extend radially from the luminal surface into the endometrial stroma. Between PND 14 and 56, tubular glands undergo coiling and branching within the stroma until they reach the adluminal border of the myometrium.[43,44] In adult pigs, uterine glands appear as simple, coiled, tubular structures with a narrow lumen.[22,23] The simple columnar GE includes both ciliated and secretory cells. At mid-pregnancy, endometrial glands are highly dilated and filled with substantial amounts of histotroph.[22,23,46] Glandular secretory activity remains high in the last third of pregnancy and also is highly correlated with the growth of the fetus.[2,5,46] Indeed, the number of areolae in the placenta and, by inference, the number of uterine glands is directly related to birth weight of the fetus in the pig.[46,47]

Rodents

At birth, uteri of mice and rats lack endometrial glands, and the uterus consists of a simple epithelium supported by undifferentiated mesenchyme.[48] On PND 5, epithelial invaginations

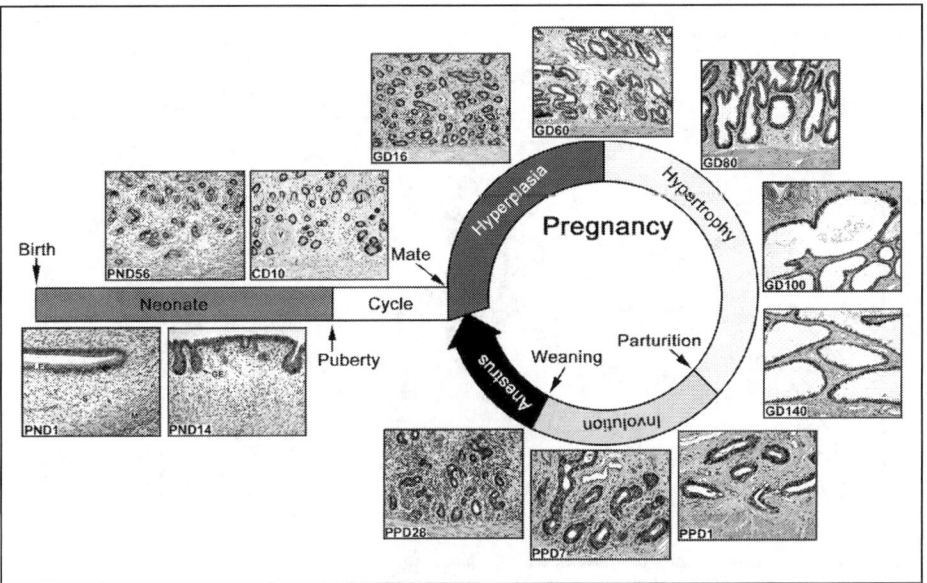

Figure 4. Life cycle of uterine glands in sheep. At birth or postnatal day (PND) 0, the uterus is devoid of endometrial glandular epithelium (GE) and consists of endometrial lumenal epithelium (LE) and stroma (S) surrounded by the myometrium (M). Between PND 0 and PND 7, the endometrial glands differentiate and bud from the lumenal epithelium. The gland buds form tubules by PND 14 and then coil and branch as they proliferate from the lumen to the myometrium. By PND 56, the uterus is histoarchitecturally mature as compared to that of adult cyclic ewes. During gestation, the endometrial glands undergo hyperplasia from gestational day (GD) 20 to 50 and then hypertrophy from GD 50 to GD 60. Maximal differentiated function of the endometrial glands occurs between GD 80 and GD 120. After parturition, the endometrial glands regress between postpartum day 1 (PPD 1) and PPD 28 during uterine involution. All photomicrographs are shown at the same magnification (20X). Legend: CD, cyclic day; GE, glandular epithelium; GD, gestation day; LE, lumenal epithelium; M, myometrium; PND, postnatal day; PPD, postpartum day.

represent formation of GE buds. Genesis of endometrial glands is not observed until PND 7 and 9 in mice and rats, respectively.[49] In the rat uterus, adenogenesis proceeds from PND 9 through PND 15,[49] resulting in development of simple, tubular and slightly coiled glands that, unlike ungulate endometrial glands, are neither tightly coiled nor extensively branched. If a successful pregnancy occurs, endometrial glands at the implantation site are ablated during stromal decidualization in response to conceptus invasion.

Humans and Menstruating Primates

As illustrated in Figure 5, the adult human and menstruating primates endometrium is stratified into the stratum functionalis and the stratum basalis as well as four zones.[50-52] Zone I is lost almost entirely during menses and consists of LE and subadjacent stroma. The endometrial functionalis, which is lost during menses, is subdivided into Zone I, consisting of LE and subadjacent stroma, and Zone II, consisting of stroma surrounding the straight, tubular or ductal portions of endometrial glands. The endometrial basalis is subdivided into Zones III and IV. Zone III consists of stroma surrounding the tortuous branched endometrial glands. Zone IV is a dynamic, but structurally stable, compartment of the primate uterus that is not eroded during menstruation or at the end of gestation and where endometrial glands terminate.[53] Although there may still be some controversy, Zone I and parts or all of Zone II are primarily shed during menses. As such, the transient functionalis is considered to be Zones I

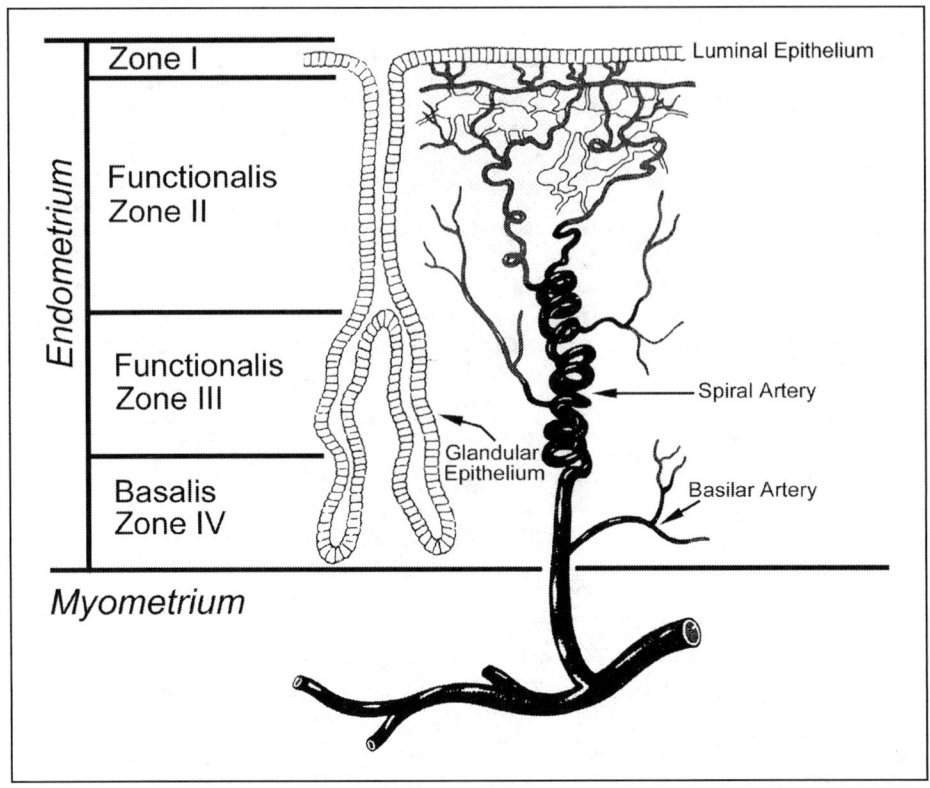

Figure 5. Schematic illustrating endometrial zonation of the rhesus monkey (adapted from Bartelmez et al[50]). The primate endometrium consists of four structurally distinct zones. The germinal basilis consists of two zones, basalis III and IV. The functionalis is composed of two zones, functionalis I and functionalis II. Endometrial regeneration after menstruation normally comes from functionalis II, but can proceed from basalis III. Estrogen-dependent glandular mitosis only occurs in functionalis Zone II and Zone I.

and II with the remaining two zones forming the "germinal" basalis from which the functionalis regenerates within each cycle or after gestation.[51]

As in other animals, initial development of uterine glands occurs from the LE downward to the myometrium. The simple columnar LE of the undifferentiated uterine body gives rise to numerous invaginations that appeared to represent primordial GE buds.[54,55] By 20 to 22 weeks of gestation in the fetus, the myometrium is well defined, but endometrial gland development is very superficial.[56] At birth, endometrial GE is sparse and limited to the adluminal stroma.[54,57] Between birth and onset of puberty, uterine glands develop slowly. By six years of age, endometrial glands extend from one-third to one-half of the distance to the myometrium, and at puberty a mature uterine histoarchitecture is observed with endometrial glands extending to the inner circular layer of the myometrium.[57]

The premenopausal endometrium in adult humans and menstruating primates undergoes programmed, phasic changes that include menses, a proliferative phase during the period of ovarian follicular development, and a secretory phase during the luteal phase of the menstrual cycle.[52,58,59] With each menstrual cycle, Zone I and parts of Zone II of the functionalis is eroded and sloughed during menstruation and regenerates from the basalis during the proliferative phase.[50] This process appears to be regulated primarily by ovarian estrogen and progesterone[60] in a cell- and endometrial zone-specific manner.[52,61-63] In uteri of women and menstruating primates, endometrial glands undergo extensive branching morphogenesis as the functionalis is reconstructed during the proliferative phase.[52,61-63] The early proliferative endometrium is thin, contains regenerating superficial epithelium of the functionalis, and endometrial glands that are narrow, short, straight and partially collapsed. During the mid-proliferative phase, the glands elongate and become more tortuous. Proliferation is primarily observed in GE on Day 3 after onset of menses, but by Day 5 proliferation of the endometrium includes GE of the functionalis and surrounding stroma. By the late proliferative phase, the glands are clearly branched. Tortuosity and branching of endometrial glands reach a maximum during the secretory phase by Day 8 post-ovulation. During the secretory phase, the glands increase in diameter, a process termed sacculation, but do not proliferate. In the macaque, the basalis proliferates during the luteal secretory phase.[52,63] If pregnancy does not occur, endometrial glands regress late in the secretory phase, comensurate with luteolysis, and deteriorate during menses.

Mechanisms Regulating Endometrial Adenogenesis

Postnatal uterine morphogenesis and endometrial adenogenesis is governed by a variety of endocrine, cellular and molecular mechanisms, for which details remain relatively undefined as compared to other epitheliomesenchymal organs.[12]

Transcription Factors

HOX genes are a family of regulatory genes that encode transcription factors and are essential during embryonic development. This system is conserved in the reproductive tract, where HOX genes are involved in Müllerian duct development and continue to be expressed in the adult uterus.[64,65] In mice, Hoxa10, Hoxa11 and Hoxa13 genes are essential both for appropriate reproductive tract development and for adult uterine function. In addition, Wnt7a signaling from the epithelium maintains the molecular and morphological boundaries of distinct cellular populations along the anterior-posterior and radial axes of the female reproductive tract.[66,67] The uteri of Wnt7a null mice have disorganized myometrium, no endometrial glands and lack expression of both Hoxa10 and Hoxa11. The nature and roles of other specific transcription factors in postnatal uterine adenogenesis remain unknown.

Epithelial-Mesenchymal Interactions

Uterine development and function depends upon epithelial-mesenchymal interactions[68-70] for local control and coordination of morphogenetically important cell behaviors including movement, adhesion, differentiation and proliferation.[71] Tissue recombination studies in rodents clearly indicate that uterine mesenchyme directs and specifies patterns of epithelial development, whereas epithelium is required to support organization of endometrial stroma and myometrial differentiation.[68-70] Studies of neonatal ovine,[38,40] porcine[44] and rodent[72] uteri, as well as of the regenerating post-menstrual primate uterus[61-63] revealed that uterine gland morphogenesis is supported and regulated through interactions between epithelium and stroma. It is through such interactions that developmentally critical tissue microenvironments, necessary to support and maintain spatially focused changes in cell behaviors associated with gland genesis, such as cell proliferation, are thought to evolve.[18,31] The initial differentiation and budding of endometrial GE from LE does not appear to require cell proliferation in the pig and sheep.[37,40,44]

Epithelial-mesenchymal interactions are mediated, in part, by changes in the composition and distribution of extracellular matric (ECM) components. Glycosaminoglycans (GAGs), oligosaccharide components of the ECM, can affect cell function directly and indirectly, by mediating access of growth factors and other molecules to their receptors or target cells. During adenogenesis in many tissues, including salivary glands, prostate, and uterus, sulfated GAGs, including chondroitins and heparans, become localized to morphogenetically inactive sites, such as the necks of glands, while nonsulfated GAGs, such as hyaluronic acid, accumulate in morphogenetically active sites, such as the tips of proliferating glands.[36,44,73] Although not investigated in the developing uterus, metalloproteinases and other factors that alter the biochemical nature of the basal lamina, affect both physical and chemical interactions between epithelium and underlying stroma in human and menstruating primate uteri during the menstrual cycle.[74] Indeed, mice lacking tissue inhibitor of metalloproteinase one (TIMP-1) gene have an increased number of endometrial glands in the uterus.[75]

Ovary and Ovarian Steroids

Jost[76] established the concept that prenatal urogenital tract development in female mammals is an ovary-independent process. Early postnatal events in rodent uterine development and endometrial adenogenesis are both ovary-independent[77] and adrenal-independent.[72,78] In the neonatal pig, ovariectomy at birth does not affect genesis of uterine glands or related endometrial morphogenetic events prior to PND 120, but does inhibit uterine weight after PND 60.[45] Similarly, ovariectomy of ewe lambs at birth does not affect patterns of uterine gland genesis on PND 14,[35] but reduces uterine weight after PND 28.[34] However, the ovary and, hence, ovarian factor(s), do regulate branching morphogenesis in the neonatal ovine uterus.[79]

Ovaries of Spring-born ewes contain significant numbers of growing and antral ovarian follicles at birth (~455 and 935 per ovary, respectively) that increase in number by PND 28 (~683 and 1100 per ovary) and then decline in number by PND 84 (~100 and 287 per ovary).[34] However, there is no evidence that these ovarian follicles secrete appreciable amounts of estrogens between birth and puberty.[34,79] Indeed, ovariectomy of ewes on on PND 7 does not affect circulating estrogen levels.[79] Moreover, administration of a nonsteroidal aromatase inhibitor, CGS 20267, from birth to PND 56 in neonatal ewes did not affect circulating levels of estrogens, uterine growth or endometrial adenogenesis.[80]

Although uterine development and adenogenesis is ovary-independent in rodents and pigs, ovariectomy of ewes on PND 7 reduced uterine growth and the number of coiled and branched endometrial glands on PND 56.[79] Thus, the ovary and an ovarian-derived factor(s) influence coiling and branching morphogenesis of uterine glands after PND 14 in ewes. The ovarian factor(s) would presumably be secreted from the abundant growing and antral follicles in the ovary. Candidate ovarian factors include follistatin, activins and inhibin as well as insulin-like growth factor (IGF-I) and IGF-II.[79,81] Interestingly, the activin-follistatin system is present in both the neonatal ovine uterus and ovary and is proposed to regulate uterine adenogenesis.[81]

In neonatal and prepubertal girls, uterine development and adenogenesis is also likely to be an ovary- and steroid-independent process.[82] However in menstruating primates, the only ovarian factors required to elicit complete maturational and functional responses of the endometrium are the ovarian steroids, estrogen and progesterone.[52,60] Endometrial morphogenesis during the proliferative phase of the menstrual cycle is regulated by ovarian estrogen.[61-63] Estrogen-dependent glandular mitogenesis only occurs in the upper functionalis zones I and II, although the cells in all zones express estrogen receptors. Estromedins, such as IGF-I, IGF-II and epidermal growth factor (EGF), in the endometrium are proposed to regulate endometrial gland branching morphogenesis during the menstrual cycle.[83,84] In the macaque, the basalis endometrium proliferates during the luteal secretory phase under progesterone influence and then regresses partially by apoptosis after progesterone falls during menses.[52] This unusual luteal phase growth of the basalis endometrium does not occur in women.

Estrogen Receptors

Endometrial adenogenesis in neonatal porcine,[85] rodent[86-88] and ovine[37] uteri involves coordinated changes in epithelial phenotype that are marked by estrogen receptor alpha (ERα) expression in nascent and proliferating endometrial GE as well as in stroma and myometrium. Homozygous ERα null mice (αERKO) have hypoplastic uteri that contain all characteristic cell types in reduced proportions.[89] Thus, ERα expression is not essential for fetal murine uterine organogenesis and endometrial adenogenesis, but is essential for normal postnatal uterine growth and development in the mouse.[89] However, administration of the antiestrogen ICI 182,780, a potent ERα antagonist, to neonatal pigs from birth inhibited endometrial adenogenesis on PND 14.[85] In sheep treated with the antiestrogen EM-800, a pure ERα antagonist, from birth to PND 56, uterine growth was not affected, but the intercaruncular endometrium contained fewer ductal gland invaginations and endometrial glands that were less coiled and branched on PND 56.[80] In the human and primate uterus, endometrial gland morphogenesis during the proliferative phase is regulated by ovarian estrogen acting through ERα present in the stroma and epithelium.[52] Therefore, the regulatory role of ERα in uterine development and endometrial adenogenesis is species- and developmental stage-specific.

Growth Factors

Stromal-derived growth factors play important roles in epithelial proliferation, differentiation and branching morphogenesis in many developing epitheliomesenchymal organs, including the uterus.[68-70] Interactions between growth factors and their receptors can involve elements of the ECM, which not only affect patterns of growth factor presentation to target cells, but may also participate as elements of cell surface receptor complexes. Although many studies have promoted the concept that local growth factors regulate organ morphogenesis and differentiated function, recent evidence indicates that systemic growth factors, such as IGF-I, are also important.[90] Thus, uterine development is likely regulated by a carefully orchestrated network of growth factors and hormones from both local and systemic origins.

Fibroblast Growth Factors (FGFs) and Hepatocyte Growth Factor (HGF)

FGF-7 is an established paracrine growth factor that stimulates epithelial cell proliferation and differentiation[91] and FGF-10, isolated originally from rat lung mesenchyme, is essential for patterning of early events in branching morphogenesis.[92] HGF functions as a paracrine mediator of mesenchymal-epithelial interactions that govern mitogenic, motogenic and morphogenic behaviors of epithelia in developing liver, lung and mammary tissues.[93] In the developing neonatal ovine uterus, FGF-7, FGF-10, HGF and their epithelial receptors (FGFR2$_{IIIb}$ and *c-met*) were identified as growth factor systems associated with endometrial morphogenesis.[38,40] Although FGF-7 mRNA was constitutively expressed in uteri from PND 1 to 56, FGF-10 and HGF mRNA levels increased markedly after PND 21, a period characterized by coiling and branching development of endometrial glands in the neonatal uterus.[38] In the human uterus, profiles of HGF and FGF-7 expression are consistent with roles in epithelial proliferation and morphogenesis during the proliferative phase of the menstrual cycle.[94-96]

Insulin-Like Growth Factors

IGF-I and IGF-II regulate cell proliferation, differentiation, and functions acting through autocrine and/or paracrine mechanisms in many organ systems including the uterus.[83,84] Null mutation of the IGF-I gene in mice demonstrated the critical role of this growth factor in normal development of the female reproductive tract,[97] as well as estrogen-induced uterine growth. In neonatal rodent and ovine uteri, the IGF system is involved in postnatal uterine morphogenesis and growth.[38,97,98] IGF-I mRNA expression in the neonatal rat uterus is confined to stroma and myometrium and increases during uterine gland genesis.[98] In the neonatal ovine uterus, IGF-I and IGF-II are expressed predominantly in the stroma surrounding the developing endometrial glands that express the IGF-I receptor.[38] In the human endometrium, IGFs mediate proliferative growth responses to ovarian estradiol.[83]

Somatolactogenic Hormones

Prolactin (PRL), placental lactogen (PL) and growth hormone (GH) are members of a unique hormone family based on genetic, structural, binding, receptor signal transduction and function studies.[99] These hormones regulate growth and differentiation of a number of epitheliomesenchymal organs, including the uterus and mammary gland.[99-102]

Prolactin

Available evidence strongly indicates that pituitary PRL is a primary regulator of endometrial gland branching morphogenesis in the neonatal ovine uterus. Indeed, PRL and PRL receptor (PRLR) are also required for development and differentiation of the lobuloalveolar portion of epithelium in the mouse mammary gland.[103-104] In neonatal ewes, circulating levels of PRL are relatively high on PND 1, reach a maximum on PND 14, and then decline slightly to PND 56.[37,105] Expression of mRNAs for both short and long PRLR is restricted to nascent GE buds on PND 7 and proliferating and differentiating GE from PNDs 14 to 56.[37] Hyperprolactinemia, induced in neonatal ewes by treatment with recombinant ovine PRL from birth to PND 56, resulted in uteri with over 60% more coiled and branched endometrial glands.[105] On the other hand, hypoprolactinemia, induced in neonatal ewes by treatment with bromocryptine, a PRL secretion inhibitor, from birth to PND 56, reduced endometrial glands by 35%.[105] The PRLR gene is also expressed in endometrial glands of humans and primates,[106,107] and PRL regulates endometrial gland function during early pregnancy.[108] Circulating PRL levels change during the menstrual cycle, but the role of PRL in humans and primates, which is estrogen responsive, in endometrial gland morphogenesis is not known.

Placental Lactogen and Growth Hormone

The placentae of a number of species, including rodents, humans, nonhuman primates and sheep, secrete hormones structurally related to pituitary GH and PRL that are termed PLs.[99,100] Ovine PL is produced by binucleate cells of the conceptus trophectoderm beginning on Day 16 of pregnancy, which is concomitant with the initiation of expression of uterine milk proteins (UTMP) by endometrial GE.[20] UTMP are members of the serpin family of serine protease inhibitors and, along with osteopontin (OPN), an ECM protein, serve as excellent markers for endometrial gland differentiation and overall uterine secretory capacity during pregnancy in ewes.[20,109,110] In maternal serum, oPL can be detected as early as Day 50 and peaks between Days 120 to 130 of gestation.[100] A homodimer of the PRLR as well as a heterodimer of PRLR and growth hormone receptor (GHR) transduces signals by ovine PL.[99] The short and long forms of the PRLR are exclusively expressed in the endometrial GE during pregnancy[20] and PL binding sites are specific to GE expressing PRLR.[111] Temporal changes in circulating levels of oPL and expression of PRLR in endometrial GE are correlated with endometrial gland hyperplasia and hypertrophy and increased production of UTMP and OPN during pregnancy.[20,111] The ovine placenta also expresses GH between Days 35 and 70 of gestation,[112] which is correlated with onset of GE hypertrophy and increases in UTMP and OPN production by GE.

Sequential exposure of the pregnant ovine endometrium to estrogen, progesterone, IFNτ, PL and placental GH is proposed to constitute a "servomechanism" that activates and maintains endometrial remodeling, secretory function and uterine growth during gestation.[102,109,111] Chronic treatment of ovariectomized ewes with progesterone induces expression of UTMP and OPN by GE.[109,111] During early pregnancy, expression of the progesterone receptor (PR) declines to undetectable levels in uterine LE by Day 11 and in GE by Day 13 in response to progesterone.[102] Down-regulation of epithelial PR is requisite for progesterone induction of GE secretory gene expression, e.g., OPN and UTMP.[109] Indeed, the combination of estrogen with progesterone increases ERα and PR expression in the GE that, in turn, markedly inhibits expression of both OPN and UTMP. Similarly, PR down-regulation is required for an increase in glycodelin expression and secretion by endometrial glands in the baboon.[113] These results indicate that a critical chronic effect of progesterone is down-regulation of epithelial PR expression to allow expression of UTMPs and OPN. Indeed, endometrial GE lack detectable PR gene expression after Day 16 of gestation (T.E. Spencer, unpublished results). Additional studies revealed that intrauterine infusions of recombinant ovine PL or GH increased UTMP and OPN expression by uterine GE of progesterone-treated ewes, but only when the ewes first received intra-uterine infusions of roIFNτ between Days 11 and 21 and then either PL or GH from Days 16 to 29 after onset of estrus.[109] The increase in UTMP expression by endometrial GE was partly attributed to effects of PL and GH to increase the number of endometrial glands. Subsequently, intrauterine infusion of PL and GH into ewes, treated with progesterone and IFNτ was found to also increase endometrial gland hypertrophy;[112] an effect not observed in ewes infused with either PL or GH alone.[109,112] These studies suggest that a developmentally programmed sequence of events, mediated by specific paracrine-acting factors at the conceptus-endometrial interface, stimulates both intercaruncular endometrial remodeling and differentiated function to increase production of histotroph for fetal-placental development and growth during gestation. Development of therapeutic strategies to enhance uterine adenogenesis and gland morphogenesis in the neonate as well as adult may be useful to increase uterine capacity in domestic animals, thereby enhancing reproductive success in terms of neonatal survivability and health.

Summary

Although a functional role for endometrial glands has been established in most mammalian species, mechanisms regulating endometrial gland branching morphogenesis are not well understood. Genetic potential for uterine function during pregnancy is defined at conception, but the success of developmental events regulating endometrial adenogenesis ultimately determines the functional capacity and embryotrophic potential of the adult uterus.[18] Studies of conceptus development in the ovine UGKO model,[16,17] in which endometrial gland development was prevented by strategic endocrine disruption of early postnatal uterine organizational events, provides definitive evidence that peri-implantation conceptus survival and growth can be related directly to the presence and state of development of endometrial glands in the adult uterus. Consistent with observations in other species, endometrial glands and their secretions appear to be critical for embryo development in humans, particularly during the first trimester.[13,28] Indeed, decreased expression of cell-surface and secretory proteins within the human uterine environment is correlated with abnormal uterine gland morphology that adversely affects uterine receptivity during the peri-implantation and early stages of placentation.[114-116] These observations reinforce the idea that endometrial organizational mechanisms are critical determinants of functional uterine capacity and, therefore, must be defined.

Mechanisms regulating endometrial development and the genesis, development and function of endometrial glands are complex, and can be altered permanently by transient exposure of tissues to endocrine disrupting compounds during critical developmental periods, which are species-specific.[80,117-119] Such periods, and the uterine organizational events that are subject to disruption under specific endocrine conditions or circumstances of xenobiotic exposure, must be defined if effective guidelines for optimizing reproductive development and uterine function are to be developed.[18,119] Thus, the high and unexplained rates of peri-implantation embryonic losses in humans and domestic animals may reflect, in part, unrecognized defects in endometrial adenogenesis or function induced during critical organizational periods. In women and menstruating primates, the long pre and peri-pubertal period during which endometrial adenogenesis occurs coupled with the cyclical nature of adult endometrial regeneration, provide significant and repeated opportunities for endometrial dysgenesis and development of pathological lesions that may contribute to infertility. Such organizationally induced alterations in human endometrial gland formation and function may lead to infertility and early pregnancy loss. An enhanced understanding of normal development will allow for the discovery of mechanisms that contribute to endometrial gland dysgenesis and dysfunction leading to infertility as well as pathologies including adenomyosis, endometriosis and uterine cancer.

References

1. Amoroso EC. Placentation. In: Parkes AS, ed. Marshall's Physiology of Reproduction. 3rd ed. London: Longmans Green, 1952:2:127-311.
2. Bazer FW. Uterine protein secretions: Relationship to development of the conceptus. J Anim Sci 1975; 41:1376-1382.
3. Kane MT, Morgan PM, Coonan C. Peptide growth factors and preimplantation development. Hum Reprod Update 1997; 3:137-157.
4. Bonnett R. Die uterinmilch und ihre bedeutung fár die fruct. Beträge zur Biologie als Fetgabe dem Anatomen und Physiologen. Th von Bischoff Stuttgart 1882; 221-263.
5. Bazer FW, Roberts RM, Thatcher WW. Actions of hormones on the uterus and effect of conceptus development. J Anim Sci 1979; 49(Suppl 2):35-45.
6. Bell SC. Secretory endometrial/decidual proteins and their function in early pregnancy. J Reprod Fertil 1988; 36(Suppl):109-125.
7. Roberts RM, Bazer FW. The function of uterine secretions. J Reprod Fertil 1988; 82:875-892.
8. Bell SC, Drife JO. Secretory proteins of the endometrium-potential markers for endometrial dysfunction. Baillieres Clin Obstet Gynaecol 1989; 3:271-291.
9. Fazleabas AT, Donnelly KM, Hild-Petito S et al. Secretory proteins of the baboon (Papio anubis) endometrium: Regulation during the menstrual cycle and early pregnancy. Hum Reprod Update 1997; 3:553-559.
10. Carson DD, Bagchi I, Dey SK et al. Embryo implantation. Dev Biol 2000; 223:217-237.
11. Bagchi IC, Li Q, Cheon YP. Role of steroid hormone-regulated genes in implantation. NY. Ann Acad Sci 2001; 943:68-76.
12. Gray CA, Bartol FF, Tarleton BJ et al. Developmental biology of uterine glands. Biol Reprod 2001; 65:1311-1323.
13. Burton GJ, Watson AL, Hempstock J et al. Uterine glands provide histiotrophic nutrition for the human fetus during the first trimester of pregnancy. J Clin Endocrinol Metab 2002; 87:2954-2959.
14. Given RL, Enders AC. The endometrium of delayed and early implantation. In: Wynn RM, Jollie WP, eds. Biology of the Uterus. 2nd ed. New York: Plenum Medical Book Company, 1989:175-231.
15. Renfree MB. Diapause, pregnancy, and parturition in Australian marsupials. J Exp Zool 1993; 266:450-462.
16. Gray CA, Taylor KM, Ramsey WS et al. Endometrial glands are required for preimplantation conceptus elongation and survival. Biol Reprod 2001; 64:1608-1613.
17. Gray CA, Burghardt RC, Johnson GA et al. Evidence that an absence of endometrial gland secretions in uterine gland knockout (UGKO) ewes compromises conceptus survival and elongation. Reproduction 2002; 124:289-300.
18. Bartol FF, Wiley AA, Floyd JG et al. Uterine differentiation as a foundation for subsequent fertility. J Reprod Fertil Suppl 1999; 53:284-300.
19. Wimsatt WA. New histological observations on the placenta of the sheep. Am J Anat 1950; 87:391-436.
20. Stewart MD, Johnson GA, Gray CA et al. Prolactin receptor and UTMP expression in the ovine endometrium during the estrous cycle and pregnancy. Biol Reprod 2000; 62:1779-1789.
21. King GJ, Atkinson BA, Robertson HA. Development of the intercaruncular areas during early gestation and establishment of the bovine placenta. J Reprod Fertil 1981; 61:469-474.
22. Perry JS, Crombie PR. Ultrastructure of the uterine glands of the pig. J Anat 1982; 134:339-350.
23. Sinowatz F, Friess AE. Uterine glands of the pig during pregnancy. An ultrastructural and cytochemical study. Anat Embryol (Berl) 1983; 166:121-134.
24. van Niekerk CH, Allen WR. Early embryonic development in the horse. J Reprod Fertil Suppl 1975; 23:495-498.
25. Gerstenberg C, Allen WR. Development of the equine endometrial glands from fetal life to ovarian cycling. J Reprod Fert 1999; 56(Suppl):317-326.
26. Samuel CA, Allen WR, Steven DH. Studies on the equine placenta III. Ultrastructure of the uterine glands and the overlying trophoblast. J Reprod Fert 1977; 51:433-437.
27. Dantzer V. Scanning electron microscopy of the exposed surfaces of the porcine placenta. Acta Anat 1984; 188:96-106.
28. Burton GJ, Jaunaiux E. Maternal vascularization of the human placenta: Does the embryo develop in a hypoxic environment. Gynecol Obstet Fertil 2001; 29:503-508.
29. Davies J. Comparative embryology. In: Wynn RM, ed. Cellular Biology of the Uterus. New York: Appleton-Century Crofts, 1967:13-32.
30. Mossman HA. Vertebrate fetal membranes. New Brunswick, NJ: Rutgers University Press, 1987.
31. Bartol FF, Wiley AA, Spencer TE et al. Early uterine development in pigs. Reprod Fert Suppl 1993; 48:99-116,100.

32. Atkinson BA, King GJ, Amoroso EC. Development of the caruncular and intercaruncular regions in the bovine endometrium. Biol Reprod 1984; 30:763-764.
33. Wiley AA, Bartol FF, Barron DH. Histogenesis of the ovine uterus. J Anim Sci 1987; 64:1262-1269.
34. Kennedy JP, Worthington CA, Cole ER. The post-natal development of the ovary and uterus of the merino lamb. J Reprod Fertil 1974; 36:275-282.
35. Bartol FF, Wiley AA, Goodlett DR. Ovine uterine morphogenesis: Histochemical aspects of endometrial development in the fetus and neonate. J Anim Sci 1988; 66:1303-1313.
36. Bartol FF, Wiley AA, Coleman DA et al. Ovine uterine morphogenesis: Effects of age and progestin administration and withdrawal on neonatal endometrial development and DNA synthesis. J Anim Sci 1988; 66:3000-3009.
37. Taylor KM, Gray CA, Joyce MM et al. Neonatal ovine uterine development involves alterations in expression of receptors for estrogen, progesterone, and prolactin. Biol Reprod 2000; 63:1192-1204.
38. Taylor KM, Chen C, Gray CA et al. Expression of mRNAs for fibroblast growth factors 7 and 10, hepatocyte growth factor and insulin-like growth factors and their receptors in the neonatal ovine uterus. Biol Reprod 2001; 64:1236-1246.
39. Marion GB, Gier HT. Ovarian and uterine embryogenesis and morphology of the nonpregnant female mammal. J Anim Sci 1971; 32(Suppl 1):24-47.
40. Gray CA, Taylor KM, Bazer FW et al. Mechanisms regulating norgestomet inhibition of endometrial gland morphogenesis in the neonatal ovine uterus. Mol Reprod Dev 2000; 57:67-78.
41. Guillomot M. Cellular interactions during implantation in domestic ruminants. J Reprod Fertil 1995; 49:39-51.
42. Hadek R, Getty R. The changing morphology in the uterus of the growing pig. Amer J Vet Res 1959; 20:573-577.
43. Bal HS, Getty R. Postnatal growth of the swine uterus from birth to six months. Growth 1970; 34:15-30.
44. Spencer TE, Bartol FF, Wiley AA et al. Neonatal porcine endometrial development involves coordinated changes in DNA synthesis, glycosaminoglycan distribution, and 3H-glucosamine labeling. Biol Reprod 1993; 46:729-740.
45. Tarleton BJ, Wiley AA, Spencer TE et al. Ovary-independent estrogen receptor expression in neonatal porcine endometrium. Biol Reprod 1998; 58:1009-1019.
46. Knight JW, Bazer FW, Thatcher WW et al. Conceptus development in intact and unilaterally hysterectomized-ovariectomized gilts: Interrelations among hormonal status, placental development, fetal fluids and fetal growth. J Anim Sci 1977; 44:620-637.
47. van Rens BT, van der Lende T. Piglet and placental traits at term in relation to the estrogen receptor genotype in gilts. Theriogenology 2002; 57:1651-1667.
48. Brody JR, Cunha GR. Histologic, morphometric, and immunocytochemical analysis of myometrial development in rats and mice: I. Normal development. Am J Anat 1989; 186:1-20.
49. Branham WS, Sheehan DM, Zehr Dr et al. The postnatal ontogeny of rat uterine glands and age-related effects of 17_-estradiol. Endocrinology 1985; 117:2229-2237.
50. Bartelmez GW, Corner GW, Hartman CG. Cyclic changes in the endometrium of the rhesus monkey (Macaca mulatta). Contrib Embryol. Carnegie Institut 1951; 34:99-146.
51. Padykula HA. Regeneration in the primate uterus: The role of stem cells. Biology of the uterus. In: Wynn RM, Jollie WP, eds. New York: Plenum Medical Book Company, 1989:279-287.
52. Brenner RM, Slayden OD. Cyclic changes in the primate oviduct and endometrium. In: Knobil E, Neill JD, eds. The Physiology of Reproduction. New York: Raven Press, 1994:541-569.
53. Bensley CM. Cyclic fluctuations in the rate of epithelial mitosis in the endometrium of the rhesus monkey. Contrib Embryol Carnegie Instit 1951; 34:89-98.
54. O'Rahilly R. The embryology and anatomy of the uterus. In: Norris HJ, Hertig AT, Abell MR, eds. The Uterus. Baltimore, MD: Williams and Wilkins Company, 1973:17-39.
55. O'Rahilly R. Prenatal human development. In: Wynn RM, Jollie WP, eds. Biology of the uterus. New York and London: Plenum Medical Book Company, 1989:35-56.
56. Song J. The human uterus: Morphogenesis and embryological basis for cancer. Springfield, IL: Charles C Thomas, 1964.
57. Valdes-Dapena MA. The development of the uterus in late fetal life, infancy, and childhood. In: Norris HJ, Hertig AT, Abell MR, eds. The Uterus. Baltimore: Williams and Wilkins Company, 1973:40-67.
58. Noyes RW, Hertig AT, Rock J. Dating the endometrial biopsy. Fertil Steril 1950; 1:3-25.
59. Wynn RM. The human endometrium: Cyclic and gestational changes. Biology of the Uterus. In: Wynn RM, Jollie WP, eds. New York: Plenum Medical Book Company, 1989:289-332.
60. Hisaw FL, Hisaw Jr FL. Action of estrogen and progesterone on the reproductive tract of lower primates. In: Young WC, ed. Sex and internal secretions. Baltimore: Williams and Wilkins, 1961:556-589.

61. Padykula HA, Coles LG, Okulicz WC et al. The basalis of the primate endometrium: A bifunctional germinal compartment. Biol Reprod 1989; 40:681-690.

62. Padykula HA, Coles LG, McCracken JA et al. A zonal pattern of cell proliferation and differentiation in the rhesus endometrium during the estrogen surge. Biol Reprod 1984; 31:1103-1118.

63. Okulicz WC, Ace CI, Scarrell R. Zonal changes in proliferation in the rhesus endometrium during the late secretory phase and menses. Proc Soc Exp Biol Med 1997; 214:132-138.

64. Ma L, Yao M, Maas RL. Genetic control of uterine receptivity during implantation. Semin Reprod Endocrinol 1999; 17:205-216.

65. Taylor HS. The role of HOX genes in the development and function of the female reproductive tract. Semin Reprod Med 2000; 18:81-89.

66. Miller C, Sassoon DA. Wnt-7a maintains appropriate uterine patterning during the development of the mouse female reproductive tract. Development 1998; 125:3201-3211.

67. Parr BA, McMahon AP. Sexually dimorphic development of the mammalian reproductive tract requires Wnt-7a. Nature 1998; 395:707-710.

68. Cunha GR. Stromal induction and specification of morphogenesis and cytodifferentiation of the epithelia of the Mullerian ducts and urogenital sinus during development of the uterus and vagina in mice. J Exp Zool 1976; 196:361-370.

69. Cunha GR, Chung LWK, Shannon JM et al. Hormone-induced morphogenesis and growth: Role of mesenchymal-epithelial interactions. Recent Prog Horm Res 1983; 39:559-595.

70. Cunha GR, Young P, Brody JR. Role of uterine epithelium in the development of myometrial smooth muscle cells. Biol Reprod 1989; 40:861-871.

71. Sharpe PM, Ferguson MW. Mesenchymal influences on epithelial differentiation in developing systems. J Cell Sci Suppl 1988; 10:195-230.

72. Ogasawara Y, Okamoto S, Kitamura Y et al. Proliferative pattern of uterine cells from birth to adulthood in intact, neonatally castrated and/or adrenalectomized mice, assayed by incorporation of [125I] iododeoxyruridine. Endocrinology 1983; 113:582-587.

73. Cunha GR, Lung B. The importance of stroma in morphogenesis and functional activity of urogenital epithelium. In Vitro 1979; 15:50-71.

74. Curry Jr TE, Osteen KG. Cyclic changes in the matrix metalloproteinase system in the ovary and uterus. Biol Reprod 2001; 64:1285-1296.

75. Nothnick WB. Disruption of the tissue inhibitor of metalloproteinase-1 gene results in altered reproductive cyclicity and uterine morphology in reproductive-age female mice. Biol Reprod 2000; 63:905-912.

76. Jost A. Studies on sex differentiation in mammals. Rec Prog Horm Res 1973; 29:1-41.

77. Clark JH, Gorski J. Ontogeny of the estrogen receptor during early uterine development. Science 1970; 169:76-78.

78. Branham WS, Sheehan DM. Ovarian and adrenal contributions to postnatal growth and differentiation of the rat uterus. Biol Reprod 1995; 53:863-872.

79. Carpenter KD, Hayashi K, Spencer TE. Ovarian regulation of endometrial gland morphogenesis and activin-follistatin system in the neonatal ovine uterus. Biol Reprod 2003; 69:843-850.

80. Carpenter KD, Gray CA, Bryan TM et al. Estrogen and anti-estrogen effects on neonatal ovine uterine development. Biol Reprod 2003; 69:708-717.

81. Hayashi K, Carpenter KD, Gray CA et al. The activin-follistatin system in the neonatal ovine uterus. Biol Reprod 2003; 69:851-860.

82. Apter D. Serum steroids and pituitary hormones in female puberty: A partly longitudinal study. Clin Endocrinol (Oxf) 1980; 12:107-120.

83. Giudice LC, Chandrasekher YA, Cataldo NA. The potential roles of intraovarian peptides in normal and abnormal mechanisms of reproductive physiology. Curr Opin Obstet Gynecol 1993; 5:350-359.

84. Giudice LC, Saleh W. Growth factors in reproduction. Trends Endocrinol Metab 1995; 6:60-69.

85. Tarleton BJ, Wiley AA, Bartol FF. Endometrial development and adenogenesis in the neonatal pig: Effects of estradiol valerate and the antiestrogen ICI 182,780. Biol Reprod 1999; 61:253-263.

86. Korach KS, Horigome T, Tomooka Y et al. Immunodetection of estrogen receptor in epithelial and stromal tissues of neonatal mouse uterus. Proc Natl Acad Sci USA 1988; 85:3334-3337.

87. Yamashita S, Newbold RR, McLachlan JA et al. Developmental pattern of estrogen receptor expression in female mouse genital tracts. Endocrinology 1989; 125:2888-2896.

88. Fishman RB, Branham WS, Streck RD et al. Ontogeny of estrogen receptor messenger ribonucleic acid expression in the postnatal rat uterus. Biol Reprod 1996; 55:1221-1230.

89. Lubahn DB, Moyer JS, Golding TS et al. Alteration of reproductive function but not prenatal sexual development after insertional disruption of the mouse estrogen receptor gene. Proc Natl Acad Sci USA 1993; 90:11162-11166.

90. Sato T, Wang G, Hardy MP et al. Role of systemic and local IGF-I in the effects of estrogen on growth and epithelial proliferation of mouse uterus. Endocrinology 2002; 143:2673-2679.

91. Rubin JS, Bottaro DP, Chedid M et al. Keratinocyte growth factor. Cell Biol Int 1995; 19:399-411.
92. Bellusci S, Grindley J, Emoto H et al. Fibroblast growth factor 10 (FGF10) and branching morphogenesis in the embryonic mouse lung. Development 1997; 124:4867-4878.
93. Weidner KM, Hartmann G, Sachs M et al. Properties and functions of scatter factor/hepatocyte growth factor and its receptor c-Met. Am J Respir Cell Mol Biol 1993; 8:229-237.
94. Koji T, Chedid M, Rubin JS et al. Progesterone-dependent expression of keratinocyte growth factor mRNA in stromal cells of the primate endometrium: Keratinocyte growth factor as a progestomedin. J Cell Biol 1994; 125:393-401.
95. Siegfried S, Pekonen F, Nyman T et al. Expression of mRNA for keratinocyte growth factor and its receptor in human endometrium. Acta Obstet Gynecol Scand 1995; 74:410-414.
96. Sugawara J, Fukaya T, Murakami T et al. Hepatocyte growth factor stimulate proliferation, migration, and lumen formation of human endometrial epithelial cells in vitro. Biol Reprod 1997; 57:936-942.
97. Baker J, Hardy MP, Zhou J et al. Effects of an Igf1 gene null mutation on mouse reproduction. Mol Endocrinol 1996; 10:903-918.
98. Gu Y, Branham WS, Sheehan DM et al. Tissue-specific expression of messenger ribonucleic acids for insulin-like growth factors and insulin-like growth factor-binding proteins during perinatal development of the rat uterus. Biol Reprod 1999; 60:1172-1182.
99. Gertler A, Djiane J. Mechanism of ruminant placental lactogen action: Molecular and in vivo studies. Mol Genet Metab 2002; 75:189-201.
100. Anthony RV, Limesand SW, Fanning MD et al. Placental lactogen and growth hormone: Regulation and action. In: Bazer FW, ed. The Endocrinology of Pregnancy. New Jersey: Humana Press, 1998:78:61-490.
101. Freeman ME, Kanyicska B, Lerant A et al. Prolactin: Structure, function, and regulation of secretion. Physiol Rev 2000; 80:1523-1631.
102. Spencer TE, Bazer FW. Biology of progesterone action during pregnancy recognition and maintenance of pregnancy. Frontiers in Bioscience 2002; 7:d1879-1898.
103. Horseman ND, Zhao W, Montecino-Rodriguez E et al. Defective mammopoiesis, but normal hematopoiesis, in mice with a targeted disruption of the prolactin gene. EMBO J 1997; 16:6926-6935.
104. Brisken C, Kaur S, Chavarria T et al. Prolactin controls mammary gland development via direct and indirect mechanisms. Dev Biol 1999; 210:96-106.
105. Carpenter KD, Gray CA, Noel S et al. Prolactin regulation of neonatal ovine uterine gland morphogenesis. Endocrinology 2003; 144:110-120.
106. Jones RL, Critchley HO, Brooks J et al. Localization and temporal expression of prolactin receptor in human endometrium. J Clin Endocrinol Metab 1998; 83:258-262.
107. Frasor J, Gaspar CA, Donnelly KM et al. Expression of prolactin and its receptor in the baboon uterus during the menstrual cycle and pregnancy. J Clin Endocrinol Metab 1999; 84:3344-3350.
108. Jabbour HN, Critchley HO. Potential roles of decidual prolactin in early pregnancy. Reproduction 2001; 121:197-205.
109. Spencer TE, Gray CA, Johnson GA et al. Effects of recombinant ovine interferon tau, placental lactogen, and growth hormone on the ovine uterus. Biol Reprod 1999; 61:1409-1418.
110. Johnson GA, Burghardt RC, Joyce MM et al. Osteopontin is synthesized by uterine glands and a 45-kDa cleavage fragment is localized at the uterine-placental interface throughout ovine pregnancy. Biol Reprod 2003; 69:92-98.
111. Noel S, Herman A, Johnson GA et al. Ovine placental lactogen specifically binds to endometrial glands of the ovine uterus. Biol Reprod 2003; 68:772-780.
112. Lacroix MC, Devinoy E, Servely JL et al. Expression of the growth hormone gene in ovine placenta: Detection and cellular localization of the protein. Endocrinology 1996; 137:4886-4892.
113. Srisuparp S, Strakova Z, Fazleabas AT. The role of chorionic gonadotropin (CG) in blastocyst implantation. Arch Med Res 2001; 32:627-634.
114. Klentzeris LD, Bulmer JN, Seppala M et al. Placental protein 14 in cycles with normal and retarded endometrial differentiation. Hum Reprod 1994; 9:394-398.
115. Dockery P, Pritchard K, Warren MA et al. Changes in nuclear morphology in the human endometrial glandular epithelium in women with unexplained infertility. Hum Reprod 1996; 11:2251-2256.
116. Lessey BA, Damjanovich L, Coutifaris C et al. Integrin adhesion molecules in the human endometrium: Correlation with the normal and abnormal menstrual cycle. J Clin Invest 1992; 90:188-195.
117. Iguchi T, Sato T. Endocrine disruption and developmental abnormalities of female reproduction. Am Zool 2000; 40:402-411.
118. Halling A, Forsberg JG. Acute and permanent growth effects in the mouse uterus after neonatal treatment with estrogens. Reprod Toxicol 1993; 7:137-153.
119. Tarleton BJ, Braden TD, Wiley AA et al. Estrogen-induced disruption of neonatal porcine uterine development alters adult uterine function. Biol Reprod 2003; 68:1387-1393.

Physical Mechanisms of Branching Morphogenesis in Animals:
From Viscous Fingering to Cartilage Rings

Vincent Fleury, Tomoko Watanabe, Thi-Hanh Nguyen,
Mathieu Unbekandt, David Warburton, Marcus Dejmek,
Minh Binh Nguyen, Anke Lindner and Laurent Schwartz

Introduction*

From a physicist's point of view, and regardless of the genetic controls, the branching mechanisms of many organs and glands look similar. Most generally, an epithelium forms a pouch-like sheet which elongates and branches repeatedly. During the final steps of organogenesis, the mesenchyme is vascularized in a pattern greatly influenced by the branched epithelium so that main vessels go down (arteries) and up (veins) the main ducts towards distal branches where exchange with capillaries is performed over a very large total surface area. This principle of construction can produce a secretory or filtering or breathing organ and most glands and organs are built in this way. There is either a common phylogeny to all branching organs (see Chapter 1), or there is some simple building principle which implies easy construction and hence straightforward evolutionary convergence (Fig. 1).

In this chapter, we shall not consider how the vasculature connects to the branches; during the early stages of branching morphogenesis a full vasculature is absent in any case, although capillaries may exist. Vascular development is described in Chapter 6. We shall instead concentrate on epithelial ducts and will describe of how an organ develops in terms of a moving boundary between the pouch-like epithelium and the mesenchyme. In the introduction, we shall first present general principles of moving boundary problems and show why this description applies to branching morphogenesis. Then, after describing the simplest system of all, the penetration of one fluid into another (a phenomenon called viscous fingering), we shall go towards more complex systems including the invasion of one non-Newtonian fluid by another, the self-similar extension of membranes and the elasto-plastic growth of ducts and T-shaped ampullae. We will stress the fact that a peculiar region exists at the tips of the ducts, called the apical region, and discuss how it may form and how it finds its way in a physical or chemical "field". In a last part we will discuss briefly features which are linked to the directional nature of biological tissue. At the end of the chapter, we will reconcile physics and biology by listing all the genetic elements which a physical model requires, before being able to predict anything. At each step, we shall try to remain as close as possible to specific experimental examples or situations.

There are 23 generations of branching in a human lung. Attempts have been made to reconstruct the geometry of organs by directly measuring parameters such as branching angles,

---

* A glossary for this chapter begins on page 230.

Branching Morphogenesis, edited by Jamie A. Davies.
©2005 Eurekah.com and Springer Science+Business Media.

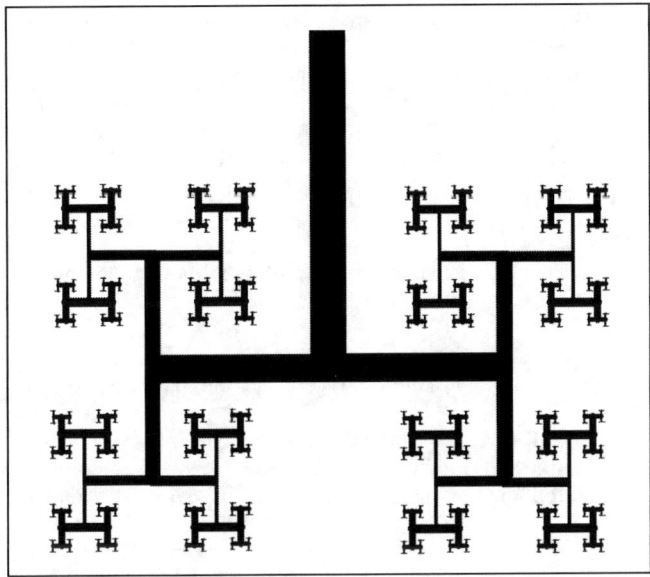

Figure 1. A fractal interative model of 'lung', inspired by a model by Mandelbrot.[2]

length and diameter ratios, etc. Although such numbers can be statistically significant, when one tries to actually re-build an organ by implementing the same rules iteratively, one fails to form a plausible organ, except for visual impressions. This is because, although a few branching events may be performed by simplistic iterative means, inspiring (or inspired by) fractal geometry, the reiteration of the rules generally leads to aberrant crossing of tubes and anatomically impossible geometries.

On another hand even if one does implement successfully a set of iterative rules to form animage of a plausible organ with many iterations, the rule cannot accomodate any anatomical modification such as the presence of a big vessel, of a bone, or of the heart, or the gentle compression by the organ capsule, by surrounding fascia or even the neighbouring organs. In order to be able to construct a plausible organ, on has to invoke many additional ad hoc rules (nine in this case).

Actually, the formation of branching organs, such as the kidney (see Chapter 8), shows that the branching events are very plastic, depending "on the context". By this it is meant that a growing tubule may experience dichotomous growth, budding, stopping, left or right turns, depending on the exact geometrical and physical situation which surrounds it; there is not a single outcome to the forward push of an epithelium. Branching may be spontaneous, or be caused by collision with the capsule, or by head-on collision of branches against each other, etc. The possibility of local adaptations is essential to the generation of such complex structures as organs, but at the scale of one actively growing duct, these adaptations are not trivial. Although an incredible number of branches form, the entire organ (e.g., kidney) keeps a bean-like shape gently surrounded by its capsule, and almost uniformly filled with branches. Therefore, although genetics may provide a set of tools for making branches iteratively, the precise positioning of branches is very dependent on the entire history and geometry of the growth, and on the spatial distribution of "fields" (diffusion, elastic, hydrodynamical...) around a given growing duct. The first few divisions may well seem very stereotypic, but reproducibility is lost after a few branching events, such that distal parts are more and more different between individuals, including identical twins, a classical observation, common to all branching organs and to the vasculature. In effect, the entire geometry and physical field around the branching pattern provides an epigenetic context which will influence the exact pattern. Still, organs are produced with some constants: general shape of kidneys, lungs or livers, regular septa formation, sometimes perfect angular divisions of tubes, and even some stereotypy such as frequent 90° rotation of subsequent divisions, as in the kidney or lung. Later on, the vascularization process shows again enough plasticity to accomodate all the micro-events which have lead to one given organ growth. How all this is physically possible is the object of this chapter, which can be regarded as a discussion of what epigenetic cues actually are.

Moving Boundary Problems

The growth of a branched organ is characterized by a soft epithelium/basement membrane/mesenchyme interface which expands and forms a tree. This is most easily seen in GFP mice (Fig. 2).[3]

For a physicist, there exists a medium A (inside the lumen), which pushes a medium B (the mesenchyme), and the interface I between the two (epithelium+basement membrane) moves (Fig. 3).

There are many physical displacements and deformations during this process. The process by which medium A pushes towards medium B, with an interface whose growth speed depends on the physical and chemical content inside B, is known in physics as a 'moving boundary

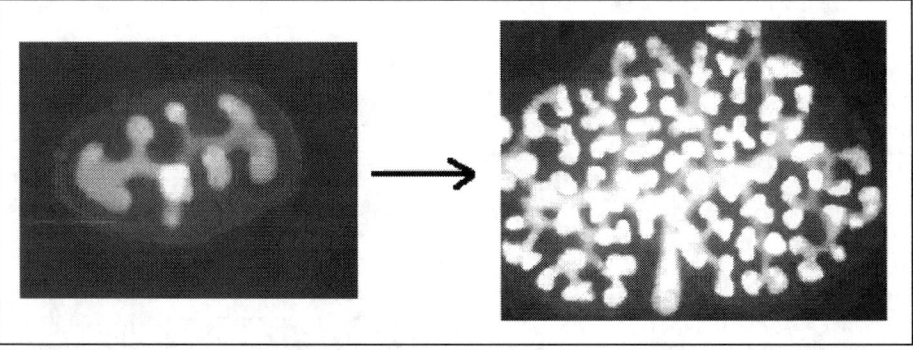

Figure 2. Optical microscope observation of a kidney growth of a GFP mouse, showing the epithelium develoment during the first two days.

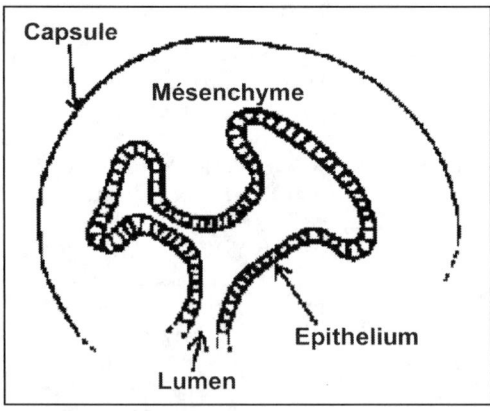

Figure 3. A simplistic view of the geometry of the problem.

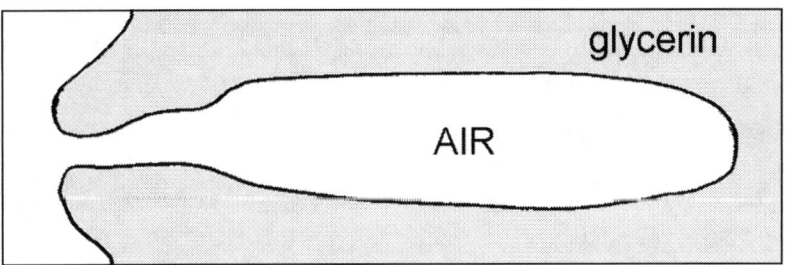

Figure 4. Drawing of a viscous finger in a channel. The white "bubble" is an area of air, pushing a viscous fluid (glycerin, in this experiment).[5]

problem', or 'free boundary problem'.[4] The simplest of all moving boundary problems is known as viscous fingering, and it shares several similarities with organ growth. Viscous fingering is the process by which a given fluid penetrates into a <u>more viscous</u> one.[5,6] It has been known for about three hundred years that, in a thin-cell apparatus (2 dimensions), the interface between such two liquids is unstable. In a small channel, such an interface will take the shape of a pouch or bubble, which is called a "viscous finger" (Fig. 4).

This shape is very reminiscent of a lung or kidney bud. In a large open medium (which is rarely the case in biology, all organs coming with a surrounding capsule), viscous fingers are unstable, and branch repeatedly to form very complex, self-organized trees. The interface may take the form of a self-similar tree (Fig. 5).

The equations of this process were derived by Saffman and Taylor,[5] who discovered the fingering solutions which occur in a channel in a thin-cell apparatus. Many similar duct-like solutions were found after them.[7-11]

The equations are:

$$\partial^2 P/\partial^2 x + \partial^2 P/\partial^2 y = 0 \tag{1}$$

The Laplace equation shown above (where x and y are spatial coordinates) is just the conservation of flux of a non-compressible viscous fluid in a 2D thin cell. In addition there are boundary conditions.

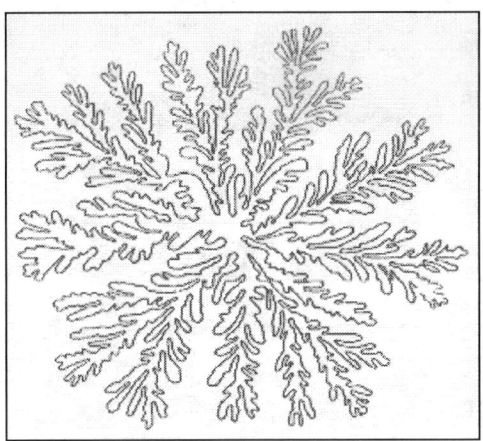

Figure 5. In an infinite isotropic medium, a bubble between air and oil spontaneously develops itself into a very complex branching pattern (courtesy of Yves Couder).

P = P$_1$ inside (some high pressure inside the lumen)
P = P$_2$ on the capsule or in the surrounding medium (some low pressure far away)

$$P_{exterior} - P_{interior} = \gamma/R \text{ at the interface} \qquad (2)$$

(γ is the surface tension, R the radius of curvature)
zero flux on the boundaries (walls of a channel, for example)

And an additional equation for how the interface moves, generally called 'the kinetics' of the interface, is

$$v = -kgradP \qquad (3)$$

(speed proportional to pressure gradient along the interface).

More recently, it has been shown that repeated viscous fingering and branching indeed leads to the structures known as "dendrites", as observed by Scheuchzer[12,13] three hundred years ago (Fig. 6).

The origin of the fingering and branching processes is rooted in the Mullins-Sekerka instability. This instability is described as follows. First, we have a pressure field outside the pattern, in a viscous fluid such as oil, that satisfies the Laplace equation because of fluid incompressibility and the thin-cell geometry.[5] Then, there is the other fluid, such as air, pushing. The kinetics of

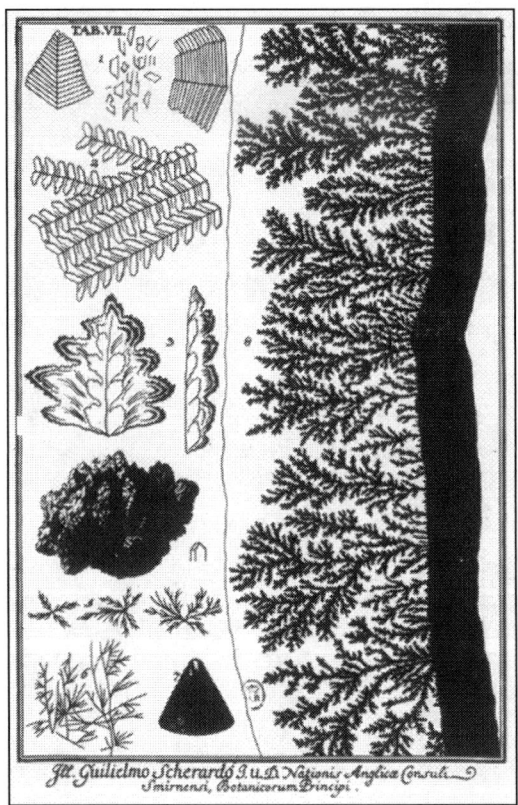

Figure 6. The branching of viscous fingers which leads to self-similar dendrites was already observed and described three hundred years ago.

the interface are very simple: by definition the interface is the last fluid surface, before the air; and it is well known that it moves (in a thin cell) with a speed just proportional to the pressure gradient (this is identical to the flow of water in pipes: the speed of a cylinder of water passing in a pipe is directly proportional to the pressure difference between the pressures at either ends of the water). The mathematics of the Laplace equation have the following consequence: as soon as a protrusion bulges forward, this protrusion encounters a sharper gradient of pressure, so it tends to grow faster and to increase in size. Similar instabilities arise with gradients of diffusible molecules, instead of a pressure gradient,[14,15] and also in electric fields. This is known in physics as the point effect, and it has the same origin as the attraction of lightning towards lightning rods (sharp protrusions tend to generate higher fluxes). The same point effect is observed in the growth of bacteria colonies in Petri dishes: since they grow quicker in the directions of high fluxes of nutrients, and since they use up the nutrient, they self-organize in a branching pattern which has elongated "pseudopods", "processes" or simply "dendrites".[16]

This sort of growth has been identified in many areas of research: metallurgy, electrochemistry, bacteria growth, fungi growth, clay or plaster erosion, vessel formation, etc.[14] It is generally called Diffusion Limited growth, after the seminal work of Witten and Sander.[18]

Intuitively, the reason why the interface is unstable is the following: in order to grow forward, the interface needs to push the medium B facing it. However, the more of B it has to push, the slower the speed, as for water flowing through pipes. When a bulge starts forming, it has less B to push, and it leaves more of B to its sides. Then, the bulge can grow faster, while its sides grow more slowly, and the instability generates a bulge exponentially. However, capillarity tends to limit the curvature of the interface. Capillary forces are tensions which are developed in a fluid interface as a direct consequence of **making** such a piece of interface. The "interfacial" energy per element of surface area dA of the interface is γdA; this interfacial energy is just the binding energy of the "molecules" or "elements" of the interface. In a simple 2D picture, it is exactly equal to the work of the capillary force or tension γ per unit length ($\gamma dx\, dy = \gamma dA$) along the contour. By definition, the tension exerts itself tangentially in the interface, just as if the surface were a piece of elastic. The equilibrium of the pressure inside, the pressure outside, and the capillary force along a curved contour, gives a relationship between the pressure drop across the interface, the radius of curvature and the interfacial energy:

$$P_{ext} - P_{int} = \gamma/R$$

Consider a flat or slightly curved interface. In order to bulge out, an interface not only has to push the material ahead, but it has to bend the interface more. Suppose a pressure P_{int} (with respect to the external pressure far away) is used to push **and** bend the interface. Part of this pressure is dedicated to bending the interface, and only the remaining fraction of this pressure is left to serve the purpose of pushing the material ahead. The pressure drop across an interface is all the bigger as it is more curved. So, bending the interface requires sacrificing a fraction of the available pressure, and this sacrificed fraction is all the bigger when the growing bud is sharper. So, an interface which "wants" to bulge out has to make a compromise between making a sharp bud, and being able to push ahead the rest of the material. The compromise between the two gives a width criterion for the size of the "fingers" of A penetrating into B: in an **infinite** thin cell, where the interface would be completely free to develop, the typical width of "a duct" would be

$$L = 1/12[b^2\gamma/\mu V]^{1/2}$$

where b is the cell thickness, γ the surface tension, V the speed of growth, and μ the viscosity. This width is obtained by what is known as a linear stability analysis, a technical tool very much used in the field of moving boundary problems.[4] It corresponds to a ratio between viscous forces and capillary forces. This is the size of a bulge strong enough to push the viscous fluid: smaller bulges cannot push at all, and larger bulges rapidly break off into smaller bulges which have this size.

So the shape of the growing "duct" is basically related to the growth speed, the surface tension and material parameters of the fluids. There is no actual "sensing" of the size: the size is only the consequence of some physical dynamics, with given parameters, which themselves are genetically determined. It is, of course, extremely difficult to imagine a way of measuring such parameters in vivo, in a developping human embryo, and to our knowledge, such measurements have never been performed.

If the thin-cell apparatus is lined by **walls** (and we may think of the capsule of an organ as forming such a wall), but infinite in the forward direction, the finger occupies **one half** of the channel. This comes from a subtle stabilizing effect of the walls,[19] the duct is lead naturally towards the middle between the two walls, where it grows stably.

To return to organ growth: **IF** the push of the epithelium were a simple pressure driven push towards the mesenchyme, **IF** the mesenchyme and the fluid were simple viscous fluids with a simple surface tension at the interface and **IF** the capsule were very flat, and infinite and narrow, then a single lung or kidney duct would look like Figure 4, and be described by the model given above (Equs. 1-3). In this spirit, the growth of a duct is almost entirely epigenetic, and quite easy: all pouches which grow by osmotic push should be able to generate finger-like structures.

Although a single field does the job, we can conceptually divide it into two contributions (this distinction will be needed in what follows): first, the duct has a specialized "apical" region where growth is promoted—or made easier—by the presence of a higher field. This apical region is not maintained by a specific cellular bio-chemistry but is the result of the self-organization of the field. Second, the tip of the duct either bifurcates or is driven gently towards the center of a channel, not by cellular bio-chemistry but, again, by the fields around the duct, which self-organize in such a way as to generate either stable ducts, or branching patterns, depending on boundary conditions. So this is the simplest example of a completely physical process for branching morphogenesis, it may have existed in very primitive organisms, and it is observed in some simple biological contexts.

However, the reality is not so simple, in that not only the epithelium but the mesenchyme also grows. The capsule is not infinite and itself expands. The tissue does not behave like a simple fluid, it has an elastic component and it adapts by growth and cell reorganization on a longer time scale, in a plastic way. Also, the problem is not a simple 2D problem in a thin cell, it is more of a 3D problem. Finally, the living tissue is made of several cell types, and they tend to form an orientational order. This is especially true of the fibroblasts in the mesenchyme, which are most generally described as elongated cells which lay down collagen in an orderly fashion.[20-22]

This is apparent in the final shape of lungs, for example, which exhibit a very conspicuous orientational structure, known as cartilage rings, which obviously play a role in the mechanical equilibrium of the tubes.[22]

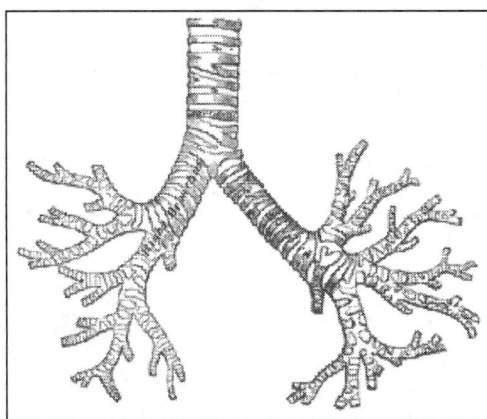

Figure 7. Classical picture of cartilage rings in a lung.

No model is at present able to incorporate in a comprehensive way all the features mentioned in this list. However, for each of these facts, a partial description can be reached.

3D Aspect

In three dimensions, an interface between fluids is much less unstable, because the viscous force acting on the sides of a bulge and preventing the rest of the interface catching up is smaller. In simple terms, the drag on a 3D cylinder is much less than the drag between two plates, as in Figure 4. When, say, air is pushed into a viscous material in a 3D cylinder, a finger of air propagates leaving part of the viscous material on the wall. The air bubble, pouch or finger occupies approximately 90% of the width. In other words, the apical region is wider in 3D than in 2D, if this active region is determined by self-organization of fluid streamlines. (Note that in many cases, the capsule of organs has some sort of a lenticular shape, which is somewhere in between a 2D and a 3D geometry, an additional difficulty). An experimental demonstration is shown in Figure 8.

It resembles the penetration of epithelium into mesenchyme. It is interesting to note that, eventually, there remains only 10% of supporting material around such a "duct". This figure is similar to the one observed, for example, in a lung, whose supporting material occupies approximately 10% of the volume.[25]

However, in 3D the interface between fluids is much more stable and, although some kind of "finger" forms, it tends not to tip-split dichotomously spontaneously. So, it is unlikely that branching proceeds via a Mullins-Sekerka-type of instability, in 3D, unless this instability is driven by a field of **diffusible molecules**, instead of a pressure field. (The deep root of this 3D instability comes from the fact that the diffusion equation is the same in 2D and 3D, and

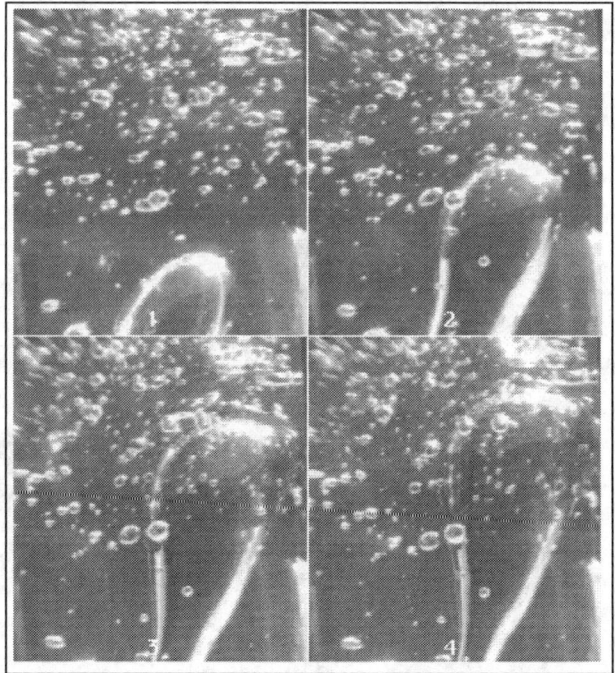

Figure 8. A viscous finger, in 3D. The experiment is much more difficult to perform, because of gravity, and of bubbles which get trapped in the gel during preparation.

generates a point effect in all geometries. The viscous fingering equations are not the same in 2D and in 3D; they amount to a **diffusion equation**—Equ. 1—only in 2D; i.e. in 3D the point effect is much weaker than in 2D).

Finite Capsule

The capsule is a finite bag surrounding the mesenchyme forming the boundary of the organ. It is not always present in in vitro experiments, but in some of these instances the surrounding gel or the mesenchyme-medium interface may act as a virtual boundary, with similar conclusions as explained here. As the kidney duct (for example) extends, it forms very conspicuous T-shaped patterns, upon collision with the limit of the tissue (Fig. 9). But, as stated before, it may aswell simply make a right turn (Fig. 10).

It would be daring to suggest a very complex position information in the form of diffusible molecules, which would lead to such a T-shaped dichotomy, or to just a turn. The T-shape is very common in physics of fluids, including fingering experiments, when an air bubble or pouch collides with a "wall". This is a mechanism of branching which is different from the Mullins-Sekerka instability. Figure 11 shows a typical numerical simulation of this kind of event.[26] It describes a bubble rising through a sink and "hitting" the ceiling.

It suggests that the T-shape dichotomous events are driven by the mechanics of the collision of the epithelium-basement membrane interface against the capsule, or the surrounding culture medium. That the epithelium should not succeed in reaching the boundary itself is a straightforward consequence of the visco-elastic nature of the mesenchyme. The important issue in this last sentence is that the distance of "collision" is linked to a material property of the tissue, and not to a distance fixed by some threshold in a gradient of molecules. In this view, the cells shape the ducts by participating in the dynamical process which generates them and by modifying the set of physical parameters. If there are specific genetically-regulated pathways for a shape, it is a matter of how the physical parameters of the problem are modified by biochemistry, in addition to possible inhibition/activation pathways.

In many instances, the T-shaped dichotomous event is the result of a collision between ducts that grow towards each other, instead of towards the limit or boundary surrounding the tissue (Fig. 12). The concomitant deformation of the ducts is straightforward in this sort of models (symmetry of boundary conditions).

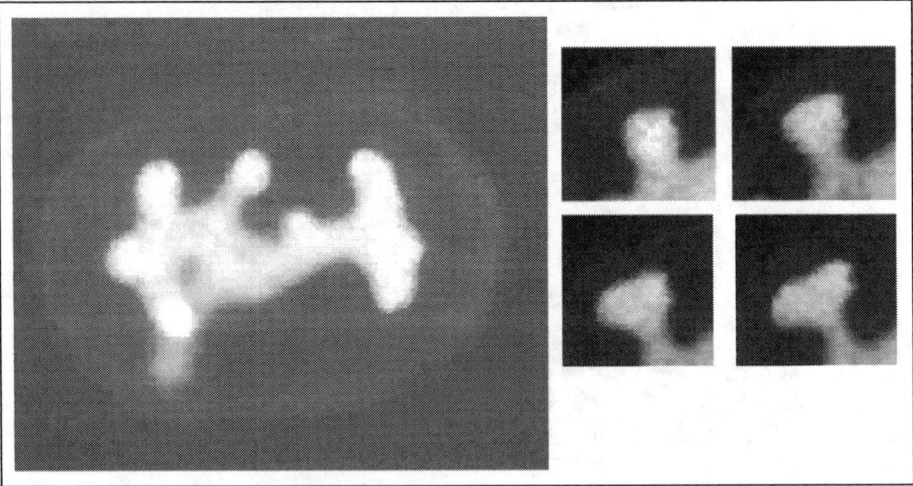

Figure 9. "Collision" of the ducts against the capsule induces a T-shape dichotomy.

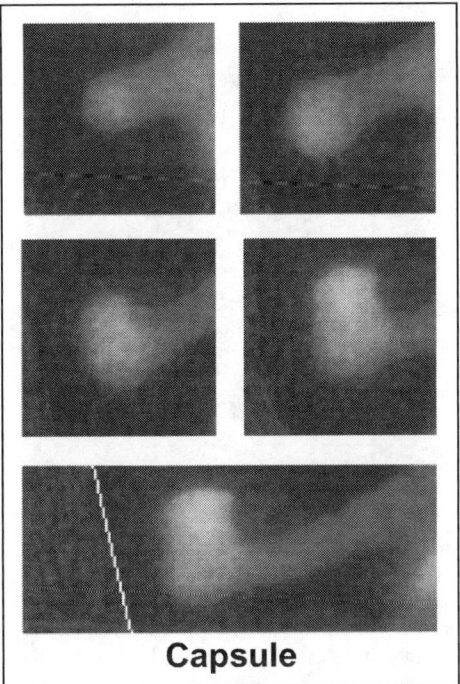

Capsule

Figure 10. Close view of a right turn of a duct which gets close to the boundary and makes a right turn.

There is not one specific pathway for head-on collision and one for collision on the capsule. Many seemingly different events are just the same physics with different boundary conditions.

Non-Newtonian Fluids

Viscous fingering instability, and the formation of "ducts" or "fingers" between two immiscible fluids has been described mainly for Newtonian fluids inside a channel geometry. Newtonian fluids have a uniform constant viscosity; for example, flow of Newtonian fluids in pipes is directly proportional to gradients of pressure. It is very unlikely that the mesenchyme should be such a fluid, because the dry weight of mesenchyme is 50% collagen (in the adult). It is well known that gelatin or collagen suspensions in high concentrations have shear-thinning (or "pseudo-plastic") properties: they flow more easily under shear stress, and

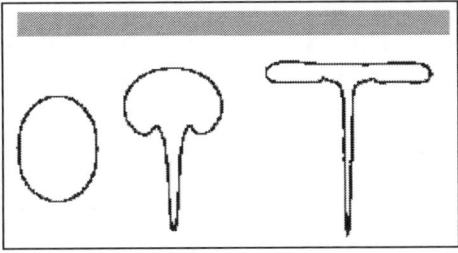

Figure 11. Collision of a rising bubble against a "ceiling" (after Pozrikidis[26]). The bubble never touches the "ceiling". Note that this image has cylindrical symmetry.

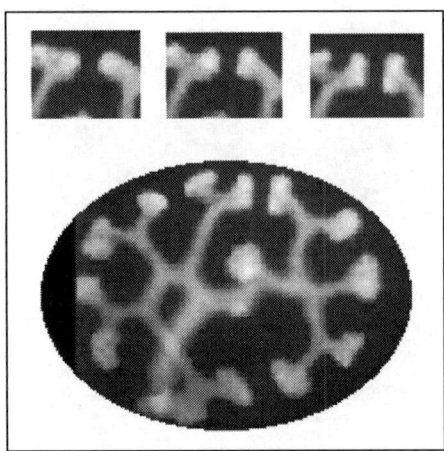

Figure 12. T-shaped dichotomies are also observed when ducts "collide" head on.

they exhibit a threshold. This means that below a certain threshold of shear, called the yield stress, the material behaves like a solid gel, although quite elastic (Fig. 13). Above that threshold, the material behaves like a fluid and flows all the more easily as it is more sheared. It has long been known that viscous fingers also form in shear-thinning materials, with finger widths much smaller than in Newtonian fluids.[27] Recent work has shown that the Mullins-Sekerka instability also exists for non-Newtonian fluids.[28,29] In this case, the width criterion is different, although it is still obtained by writing the ratio of capillary vs viscous forces :

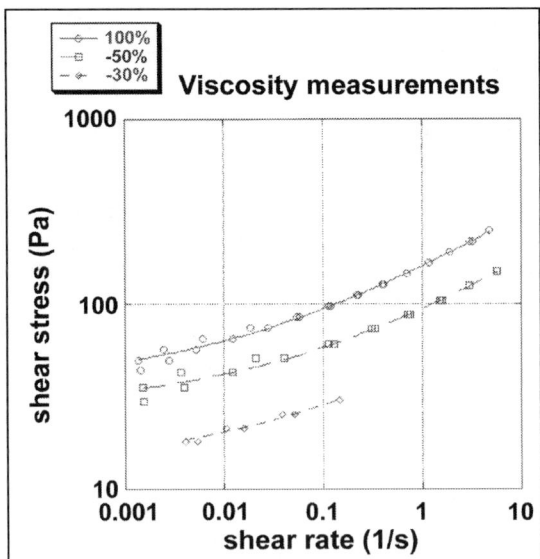

Figure 13. Top. Typical physical behaviour of gel-like materials. Below some threshold, they behave like elastic solids : the shear *rate* is zero, although there is stress. Above a threshold, they flow, and the shear rate increases with stress. As expected, the threshold (yield stress) depends on gel dilution. The more diluted the collagen, the lower the threshold.

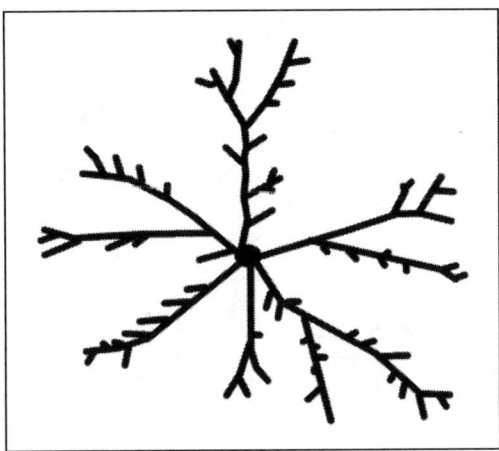

Figure 14. Drawing of radial fingering pattern with non-Newtonian fluids with shear-thinning properties : the fingers are much thinner and look like filaments.[30]

$$L = 2\pi[3\gamma b/(2\sigma)]^{1/2} \tag{4}$$

(in the Newtonian case: $L = 1/12[b^2\gamma/\mu V]^{1/2}$)

where σ is the stress, and γ the shear rate. γ is now a function of shear **rate** such that $\sigma = \sigma_{yield} + \alpha\gamma^n$. This last identity describes the non-Newtonian behaviour of the material: below the yield stress σ_{yield}, the deformation is static, there is no shear **rate**. Above the yield stress, there is a rate of shear, as for a liquid: the material moves. Again, in vivo the value of the yield stress is completely determined by the chemical content of the tissue (especially, collagen density and bridging), therefore giving a leverage for morphogenesis of elongated filaments.

The important issue to remember is that the stress is more important at the tip, and hence the material is more fluid, and likely to be pushed ahead, in the region of the tip, than outside the region of tip (see Fig. 14). The consequence is the formation of more elongated duct-like stuctures.

So, we may say that there exists a spontaneous "apical region" around ducts although this does not require any specific metabolic activity, because living material tends naturally to be softer at tips. In this view, the apical region is special in two respects: the pushing field is higher there (as for viscous fingering processes), but in addition the effect of the push is enhanced by the fact that, in presence of a high field, the material properties are changed (it is softer, more fluid). In 3D, the situation is analogous to the 2D situation, and in fact, such images as Figure 7, were taken with a non-Newtonian fluid (hair gel, which contains a lot of collagen). Again, the Mullins-Sekerka instability with non-Newtonian fluids is much weaker in 3D. It is clear that playing with the material properties, as cells do, will greatly modify the morphogenesis.

Elastic Properties

As mentioned above, it is uncertain whether the living material of an organ behaves like a fluid. It might behave like an elastic material with a threshold, since (1) it contains a lot of collagen (2) the shear regimes are very small (the growth is slow), so possibly below the thresholds for fluid flow. Therefore, the penetration of the epithelium might be treated with the laws of mechanics instead of fluids. From a fundamental point of view, the laws of mechanics are different in fluids and in solids. When a force is applied to a fluid, it moves. When a force is applied to a solid, it deforms. In the simplest hypothesis—the linear elasticity and small deformation—a supposed small push of the epithelium simply generates a small deformation, which

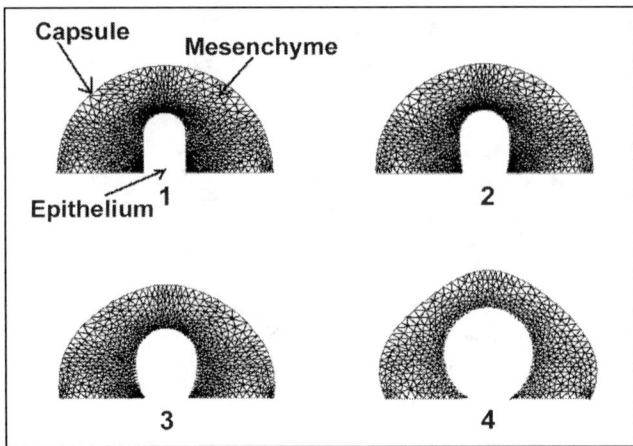

Figure 15. Simple elasto-plastic growth of a "bud" which generates a round pouch.

is static. If the force is removed, the shape resumes its original reference configuration, unless there is some plasticity. If we wish to develop a "finger" or duct into a solid, we need to consider a plastic material, which adapts to the new stress and deformed shape, in such a way that the deformation is absorbed, and the energy (energy = stress.deformation) is dissipated. If we remove the pushing force, the bud-like shape remains. Therefore, we may, in the simplest view, consider a small forward motion of a duct into the mesenchyme at a very small "flow" rate, as an elastic step, followed by a plastic step: the mesenchyme is first deformed, the deformation-stress relationship implies a high stress in the cells, then cell rearrangements, divisions and metabolism diminish the individual cell deformation so that the stress is reduced. The reduction of stress is both linked to morphology changes and to material changes (number of cell layers, shape and volume of cells etc.). This is described by Figure 15. In this picture, the deformation rate is proportional to the elastic deformation, with a proportionality constant which fixes the time scale for long time permanent adaptation. The growth and deformation continues, until some stress set point is reached. In this view, the mechanical state of the cell is a very important component of its mitotic rate and activity.

By so doing, we can, perform step-by-step, an elasto-plastic growth of a bubble inside a medium which itself expands. The simplest 2D simulation shows that it simply swells the entire pattern and does not develop a duct (Fig. 15).

This is also the case in 3D. In order for a tip to grow and form a tube, in this sort of elastic problem, one needs a more localized zone of deformation in the region of the tip. There could also exist mechanical instabilities, like buckling. It is a fact, however, that many human diseases are associated to excess of "ballooning" of the organs (polycystic kidney, emphysema, etc.).

The introduction of a special localized apical zone leads to what is known as "tip growth". In this class of models, the special zone at the tip is not self-organized by the field, but genetically predetermined. A recent elegant self-similar solution of this problem was proposed by Goriely and Tabor.[31] It is assumed that the elastic properties of the material depend, for "genetic" reasons, on the distance to the tip, so that the tip is softer than the shaft between a distance L from the apex and the said apex. This increase in softness may be produced, for example, by membrane degradation enzymes. In this case a tubular membrane or duct forming a "tip" growing at a constant speed, with a constant profile can be generated. This solution was derived in order to describe the uniform growth of an algal tip, but we may think that it applies to branching morphogenesis in a scenario where the mesenchyme is totally absent. Also, there exist many branching patterns, especially in invertebrates, which may be described by similar models.

The profile is known as a self-similar profile, i.e. although the tip elongates, the profile does not change. This problem is different from viscous fingering, in that the active apical region is due to a localized variation of material properties (while in viscous fingering the deformation localizes itself spontaneously, by self-organization of a diffusive field, either chemicals or pressure). In these models, something emanates from the tips and diffuses down the duct to a region where the round cap joins the trunk. So there is an activated tip, which is limited to a narrow region, and which pushes itself forward. The shape and growth speed is determined by the mechanical push (essentially the osmotic balance across the membrane), the rates of material production to accomodate the elongation, and the dimensions of the apical zone. This problem has a flavour of viscous fingering with a non-Newtonian fluid, in that a softer region is localized at the tip.

Reaction-Diffusion

The softer region which induces buds may have a more complex shape than just a spot at the tip. In this spirit, essential ingredients of morphogenesis of tubular structures may be brought into play by inhibition/activation genetic pathways. It has been suggested recently by molecular biological approaches that a complex interplay of activation-inhibition reactions gives rise to distribution of spots of mitotic activity, corresponding to the regions where ducts elongate. This is known as a "prepattern" model. This word means that although the model does not actually treat the true problem of how the boundary extends, it claims that the buds extend in specific regions on the surface, in some spatio-temporal regularity, which is determined by a couple of chemical reactions, in the spirit of Alan Turing's "leopard spots" models.[32] These models have also been put forward in models of plant growth.[33] In the case of lung growth, Warburton et al.[34] have reviewed the data concerning the complex interplay of biomolecules (peptides, FGFs, sprouty etc.) around a duct. Many genes are expressed in the lung, belonging to different pathways (shh, bmp, fgf, ...). It is known that several of these molecules are activators or inhibitors of each other. In the first place, FGFs act as chemoattractants of epithelium. A ridge of FGF10 exists close to the buds, which seems to provide the biological incentive for growth and elongation. This comprises certainly, albeit indirectly, the mechanical incentive. Disrupting fgf10 expression causes profound abnormalities. Coupling of this global incentive for growth with inhibitors, such as sprouty, may be the cause for regular and stereotypic branching.

In effect, localized "spots" of sprouty exist and split concomitantly in the region of formation of the new tips. This model may explain the origin of splitting and of budding, somewhere outside the apical region, down the duct, and mathematical models are in progress to incorporate as much molecular detail as necessary. However, at this stage, it is unclear whether gene expression is the cause or a consequence of physical forces, or even whether this question even has any meaning. Although spots of active areas have indeed been identified, they are generally found in regions which are morphologically singular: convexities, edges, corners, centers of symmetry, apices etc.

In general terms, there are problems with Turing models. First of all, they require a complex interplay of molecules, and the generation of "spots" can be achieved only if activators and inhibitors have very different diffusion constants. This implies that they have probably very different sizes, and hence correspond to very different biochemical kinetics. The second problem is that in order to maintain the structure, molecules must be produced continuously. Although the pattern seems static and stable, it corresponds to a constant flux of molecules, which is not very favorable in terms of metabolism. Immunochemistry visualizes concentrations, and not fluxes. Thirdly, in Turing's models, when something like a spot is seen to be displaced, say, from left to right, by ten microns, it can only do so by being dissolved at the left, and recreated at the right, which is not coherent with what is generally observed: tips or branching points actually move. In addition, in Turing models, when a tissue of type A (say mesenchyme) eventually occupies a region formally of type B (say epithelium), it can do so only by transformation of B cells into A cells. It is likely that the description of the problem is more in terms of a moving boundary, which should incorporate inhibited or activated regions along the surface, which modify locally the material constants (softness/hardness) and the distribution of forces (push-pull). For example: if shear and bending induce a mitotic activity of different cell types, gene expression will be naturally localized in different specific regions, giving a possible impression of one gene inhibiting the other. The next paragraph addresses the question of how active regions may move.

Filament Navigation in a Field

Be it duct growth by elastic deformation of an apical zone, or by viscous fingering,, these phenomena also share a similarity with a problem studied long ago by Meinhardt.[35] Meinhardt describes the growth of filaments which have an active localized tip and which "chemoattract" themselves forward.

In this model, there is an autocatalytic production of a substance, say A (for activator) there is a field S of morphogen, and a field Y which describes the cell states. This is how patterning proceeds. In a medium of finite concentration of morphogen S, a small excess of A is able to trigger autocatalytically a local peak of A. This peak propagates in the shape of a travelling wave. If, in the back of the wave, the concentration of S returns to zero, then the wave of A also returns to zero, because in the absence of S, the concentration of A goes down to zero exponentially. Y is a two-state switch of cellular activity, which turns cells from a state 1 to a state 2 irreversibly. In the state 2, S is consumed, so, there is a diffusion field of S between a low level close to the pattern and sources of S "far away". Since the exponential increase of A depends on the magnitude of S, we find a moving front of A which navigates inside a landscape of S, with a speed depending locally on gradients of S, close to the wave. Since this landscape corresponds to the diffusion of S, it has a "point effect" wired in. The model generates filaments which branch randomly, in response to noise (random obstacles or random noise generated by the mitotic activity, for example).

In this class of models **diffusing molecules** are responsible for forward growth of an active tip, instead of **mechanical features**, but, to be honnest, the field of S introduced by Meinhardt could just as well be a pressure field, so similar are the equations. The model neglects, of course, the elasticity of the tissue. The equations corresponding to what has been described above contain a set of activator-inhibitor equations which maintain a localized, but moving, active region. The reader will easily identify in Equ. 5 the term cA^2S, which induces an increase of A wherever S and A are non-zero, the term $-mA$, which makes A go down to zero in the absence of S, and the term $-eYS$, which makes S be consumed in the state where $Y = 1$

$$\partial A/\partial t = cA^2S - mA + D_a\Delta A \tag{5}$$

$$\partial S/\partial t = c - cA^2S - gS - eYS + D_s\Delta S \tag{6}$$

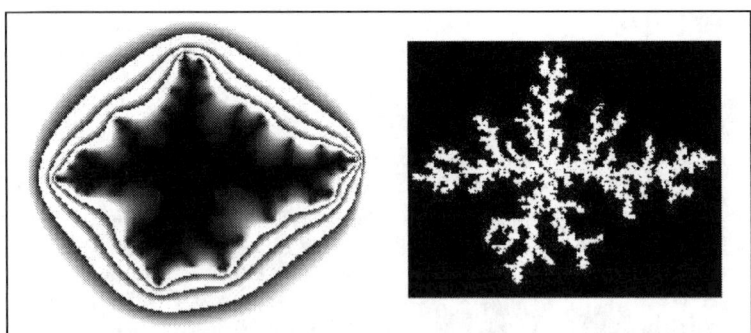

Figure 16. Navigation of filaments in a diffusible landscape of a morphogen S. This model is in its spirit very close to fractal viscous fingering. It requires noise at the moving tips of the filaments to induce dichotomous branching.

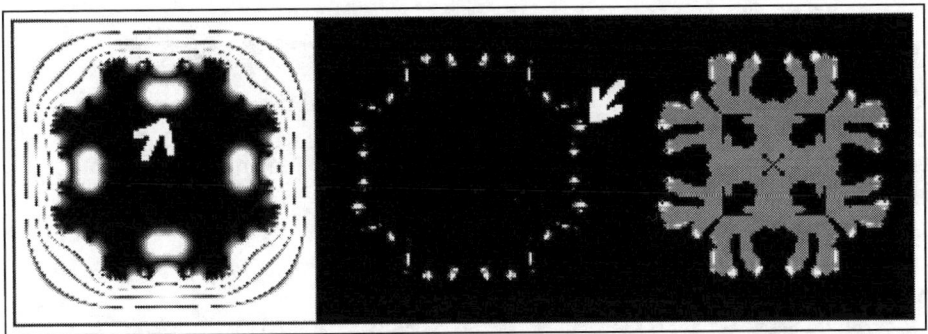

Figure 17. A self-organized growth of filaments, which avoid self-organized obstacles. The result is a set of branches surrounding obstacles, and filling space uniformly (pay attention to the regular distribution of active spots along the perimeter). The arrow in the left image points to large spots of inhibitors (there exist smaller ones at the curently branching tips) and the arrow in the middle points to moving spots of activators.

The two-state dynamical switch necessary to model the jump of cells from one state (state 0) to the other state (state 1) is given by :

$$\partial Y/\partial t = dA - eY + Y/(1 + fY^2) \tag{7}$$

Although these equations may look complex, they are conceptually very simple,[33] and we recommend to the reader to make the effort of reading some technical explanations about them.[33,35,36]

It is simple to incorporate in these models a feedback of the surrounding "tissue", which inhibits the growth when some threshold of S is reached. In such a case, one can form a more regular pattern which grows by avoiding self-organized obstacles (Fig. 17).

The image is typical of what a reaction diffusion process can give, in 2D, with active regions avoiding their inhibitors. The arrows point to the pools of inhibitor (in the left) and to the pools of activators (in the middle). These spots of chemicals form all by themselves, and the genetically programmed parameters are the parameters in the equations above. However, although such models may contain the correct molecular interplay, they are unable to provide realistic images, because they do not incorporate the mechanics of the tissue which "fingers".

Therefore, Meinhardt's model of wave propagation in a landscape by chemical reactions is weaker than the moving boundary models. It requires a different, additional set of concepts to explain why the filaments are actually tubes, how the cap is constructed, how the surrounding medium is pushed away, how the tubes slowly reorganize to make branching points at 120°, etc. Especially, the formation of ducts is generally **not** a matter of mesenchyme cells (fibroblasts) being turned into epithelial cells. There is a true displacement of all cells, which push themselves forward. This is absent in Meinhardt's models. In moving boundary models these features come all together by the laws of deformable bodies or of fluid flow.

If a given duct is constructed mechanically by a localized apical region (not self-organized by the epigenetic field), the duct is still, not growing alone, in vacuo as is for example the beautiful solution of Goriely and Tabor.[31] The duct has to find his way across a mesenchyme. In the simplest picture, a free duct moves forward in a straight fashion, while a confined duct, inside a capsule or surrounded by a stiffer external medium, either moves up the gradients of a diffusible molecule (for example, if the duct produces degradation enzymes which soften the mesenchyme, then it grows in the directions of the gradient of it), or, otherwise navigates following the stress. This is because an additional force due to the surrounding tissue, breaks the symmetry of the duct,[31] and the duct grows laterally in response to forces in the mesenchyme (see Equ. 5 of ref. 31). In the simplest case, the growth amounts to elongation of a duct that follows a localized active region and navigates inside a diffusive or mechanical field (the

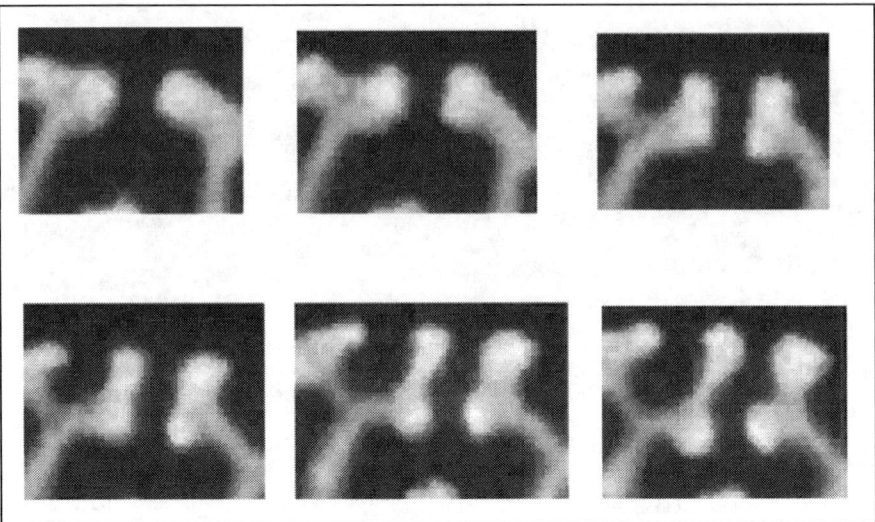

Figure 18. In situ observation of head on collision of ampullae.

pressure, if the capsule is a flat bag). Regardless of the mechanical detail, this is analogous to what Meinhardt has described.

However, although the pressure in 2D (inside a flat capsule) is a diffusible quantity, it makes a huge difference if the actual field is a mechanical field and not a chemical field, because the geometric parameters of the shape are "recorded" for good in material properties and not in levels of concentration of molecular fields produced by the cells, which are more variable quantities. In particular, the diffusion of activators across the tissue is formally replaced by the propagation of a mechanical stress in the tissue, which is much more robust.[36] In addition, many features are readily explained. For example, a mechanical field induces a subtle symmetry breaking of duct collision. As seen in Figure 18, it is observed that when ducts collide "head on" (instead of against the capsule) and branch, they tend to slowly shift sideways, in order that the branching points form an S shape, instead of remaining parallel.

This slow shift is easily accounted for by a mechanical buckling. Indeed, when a collision occurs head on, the region of contact between the two Ts is akin to a flat sheet, composed of several layers (epithelium, basement membrane, mesenchyme, basement membrane, epithelium). Such a composite material folds or buckles as a consequence of uniaxial tangent stress, a configuration with a broken symmetry (shift of the branching points) being more favorable (see below).

Active Growth of the Mesenchyme

So far, we neglected the active growth of the mesenchyme. This was taken into account by Meinhardt[35] as a mere homothetic dilation of the tissue, a good start for that time, although it is not quite realistic. It is obvious from inspection of experimental data that the epithelium/ basement-membrane interface does not grow forward everywhere. In some regions it recedes (Fig. 19). The two basic regions where it recedes are the trunk of the ducts, and the top of the T shape. The recession of the tissue in the older parts of the ducts (Fig. 19) amounts to a strangling of the ducts.

So branching morphogenesis is not just a matter of the apical region being softer, but also of the trunk being contracted. This strangling has several consequences. First, the tube diameter narrows. Second the tubes become straighter and apparently more rigid, and third, the strangling pushes forward the region of the ampulla. In the region of the ampulla, at times when it

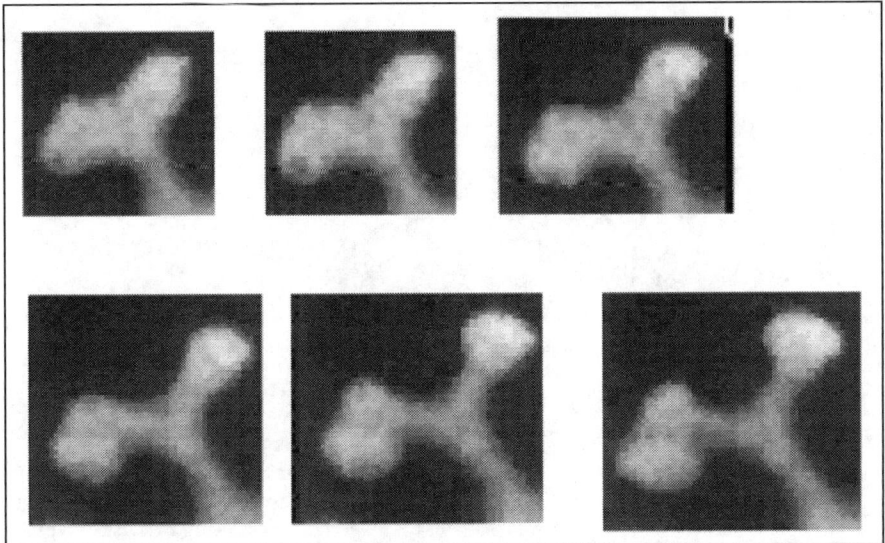

Figure 19. In situ observation of epithelium recession or strangling,

collides with the capsule and has a T shape, one observes a push of the mesenchyme which transforms the T shape ampulla into a 3-fold branching region (Fig. 20) with a saddle point.

It is clear that the mechanics of the transformation of the T into a branching point are the same as the mechanics of the contraction. They even are often concomitant. The difference lies only in the geometrical situation: either the push acts on a cylinder, or the push acts on a T-shaped ampulla. Therefore, we have to incorporate active deformation, growth and push of the mesenchyme towards the epithelium, a feature completely absent of Newtonian or non-Newtonian fluids, or of models of tip-growth. It is a well known fact that mechanical pushes of boundaries induces regular polygonal motives. This is very well known in the context

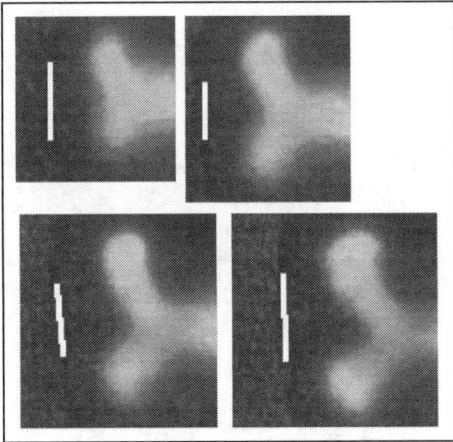

Figure 20. The formation of an interconnection of 3 tubes requires a recession of the T-shape, with respect to the capsule. Indeed, the mesenchyme between the capsule and the tip grows *against* the epithelium (the white bar is the border of the capsule).

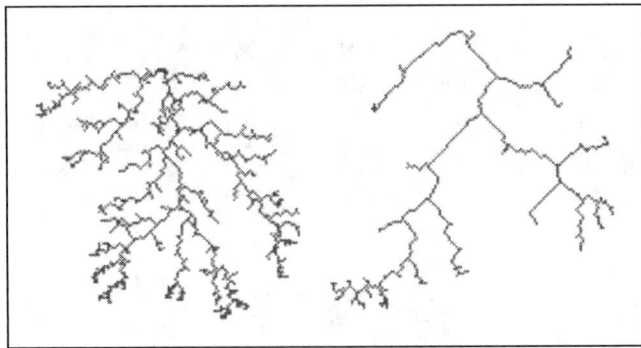

Figure 21. Transformation of a random tree into a more regular tree by push of the domains in between branches (2D).

of Dirichlet domains, and it has been suggested for long that such mechanical pushes may explain the regular ordering of spots, hairs, scales, in the animal realm. More recently, it has been shown that such mechanical pushes explain in part the regular ordering of vessels. Figure 21 explains the logic, on a dendritic tree in 2D.

A dendritic growth is performed in the regime where fractal dichotomous dendritic branching exists. But a push of the interstitial tissue is added, which acts only outside the apical region. The mechanical push is self-organized. Pressure particles are released which move randomly until they touch the tree. There, the tree is displaced by one step, in the direction opposite to where the push came from. The global result is a progressive straightening of the strands, and formation of more regular angles, which accomodate the difference between the left-right pushes on each strand. The same can be performed in 3D, it gives such results as Figure 22.

A very important point of mechanical push by the mesenchyme is the correlated deformation of neighboring tubes. Indeed, when a mass of tissue pushes, by the principle of action-and-reaction, the surrounding tubes are pushed in a correlated manner. This is to say that the shape of a tube is not something per se, the shape of all tubes surrounding a given mass of mesenchyme is under the influence of that mass, considered as one "ball". This is easily seen, again, using GFP mice, where one clearly sees the push of "balls" of mesenchyme which gener-

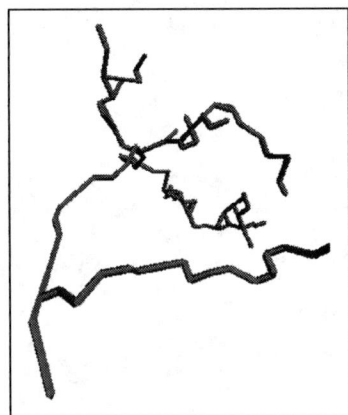

Figure 22. A random tree transformed into a more regular tree by push of the domains in between branches (3D).

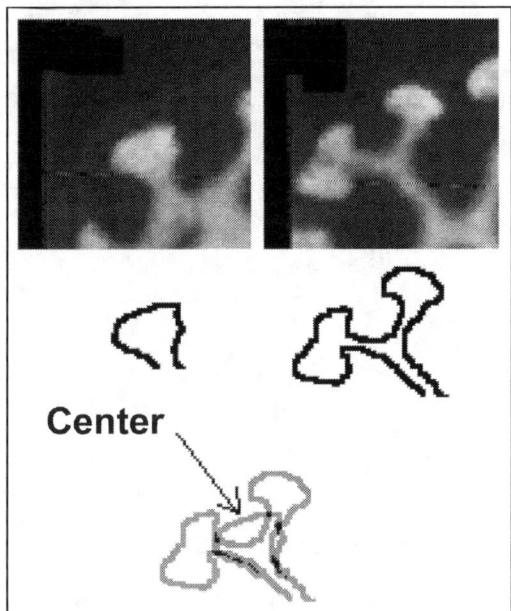

Figure 23. Image analysis of a sequence of growth. The 2 images on top show two stages of an initial T. The contours are extracted by thresholding the imlage (same threshold). Then the contours are superimposed. The result shows that the ampulla has receded, and has been pushed away from a region now occupied by mesenchyme. However, the middle of the diagonal of the gulf formed by the branches after splitting of the T, lies at the top and middle of the initial T. This shows that the position acquired by the two branches are not independent, they find themselves on either sides of a center of push, whose position can be traced back to the "ignition" point of the epithelium recession.

ate tubes eventually find themselves along the edges of polyhedrons. So, one cannot predict the position of one branch in isolation. A very pictorial evidence of the mesenchyme push is provided by the following image analysis. One starts from the film of a dichotomy event. Then the film image (grayscale) is thresholded, and the contours are extracted. Then the contours of the last and first image are superimposed on the first image. A very striking feature appears: the middle of the T of the initial T-shape is the starting point of a recession of the epithelium, as if it were pushed by the mesenchyme, but this point remains the middle of the diagonal of the "gulf" eventually formed by the two branches on either sides of the invisible ball of mesenchyme. So, there is a **correlated** motion of the epithelium on either sides, which gets away from this center at the same pace on the branch to the left, and the branch to the right, by the principle of action-reaction (it is not always the case that the situation is a symmetrical as here). In the Figure 23, the "context" of one branch is the presence of the other branch and of the tissue in between.

Now, in such images as Figures 20 and 23, the 2D projection is somewhat misleading, in that the region of the branching point where 3 tubes are well identified is not flat (contrary to what a projection shows) it has the shape of an hyperboloid (saddle) because of the topology of the interconnection of three tubes. If we state that tubes eventually follow polygonal motives along edges of mesenchyme polyhedrons, we need to explain the origin of the vertices (bifurcation points) and explain how they form, starting from a T-shape undergoing such a push by a mesenchyme. It is rather difficult to imagine a Reaction-Diffusion process à la Turing generating such hyperbolic shapes (saddle). However, mechanical push generates hyperboloids or saddle-shapes very simply. It is a general fact in mechanics that a tangential stress in a flat disk generates a saddle-shape spontaneously.[38,39] It does not matter whether the stress is centripetal (contraction) or centrifugal (dilation). This is why potato-crisps have a saddle shape (the tangential stress inside the crisp comes from dessication by frying). Therefore, if we consider the flat T in contact with the mesenchyme, a small radial growth of the mesenchyme facing the epithelium will spontaneously turn the disk surface forming the top of the T into a saddle, and hence favor the transformation of the T into a Y, by breaking the symmetry of the disk. This description completes the T-shape dichotomy of the Figure 11 above (the drop colliding against a ceiling) which cannot be complete since it has cylindrical symmetry.

Intuitively, the reason why this flat disk is unstable is not the same as the fingering instability. In this case, the instability comes from the fact that when a force is exerted tangentially in a flat object (foil, sheet, plate) the object tends to buckle hyperbolically. Why is it so? When a force is exerted tangentially, the flat object deforms, and stores elastic energy. This elastic energy is the product of the deformation by the force, summed over the entire object. One could imagine that the object deforms uniformly (it shrinks, or dilates). This is not so, because a uniform deformation, multiplied by a uniform force and integrated over the surface, gives a bigger energy than a buckled configuration in which the deformation is indeed larger, but the integral of the product (force.deformation) is smaller. During the process of buckling, the object always tries to minimize the energy, so it adopts the saddle shape. This is exactly how potato crisps buckle, and similar models have been proposed for brain convolutional development.[40,41] This mechanism of transformation of the T-shaped tip into a Y-saddle point cannot be described by a purely 2D model.

There exist more complex deformation modes, which may lead to more complex branching, and which have been invoked in the case of plant growth.[42,43] A bizarre prediction of this model is that if the organs were grown more rapidly, they would exhibit a phyllotactic order (branches budding along spirals), like plants.[44,45] Such branching patterns, to our knowledge, are not observed in the animal kingdom. Only trifurcation events are also observed in branching morphogenesis.

However, the saddle buckling mechanism gives a very simple explanation to the growth of buds at right angles, outside the apical region, down the trunk. Indeed, when the saddle buckling of the T-shaped ampulla occurs, two points lying at either ends of one axis of the saddle grow forward and become the branches emanating from the T (Fig. 24 below), while the two other points lag behind and are driven down the existing trunk (Fig. 25). It is known that the mesenchyme carries pools of Fgf 10, which favor forward growth of the epithelium.

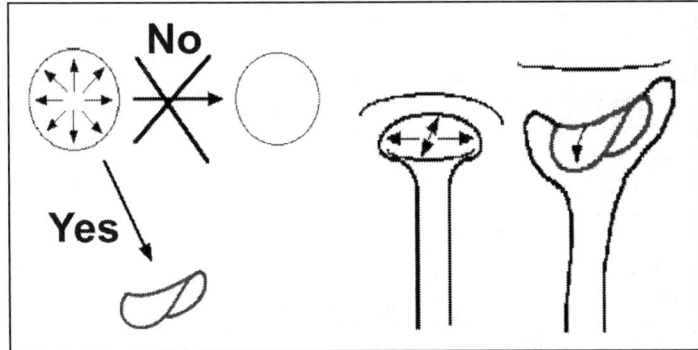

Figure 24. Left) When a disk (for example the T-shape seen from top) undergoes a uniaxial stress (the mesenchyme facing it increases or dilates radially), it is deformed spontaneously into a saddle (symmetry breaking). This explains naturally how a flat disk-like ampulla squeezed against a growing mesenchyme transforms itself into a bifurcation point. Right) Pictorial representation of buckling at a T-shaped interface. For clarity, the saddle region has been overlayed.

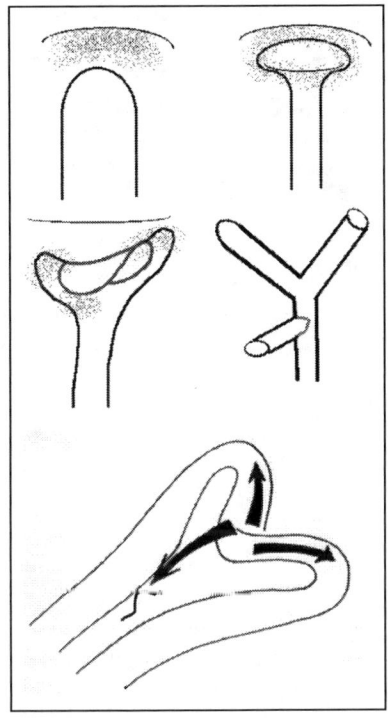

Figure 25. Saddle buckling of the epithelium and of the mesenchyme. Top left, a bud prior to be flattened; left. The T-shape. Middle left, buckling of the T, middle right: the final tube aspect. In presence of a pool of Fgf10 (shaded gray), one expects the growth promoter to be driven ahead in the regions of two siblings, but also downstream, towards "the back" of the T-shaped event, hence resuming growth at right angles, by budding. As growth resumes, the pool of Fgf10 is carried downstream by mesenchyme growth, and finds itself, eventually, somewhere down. Bottom, a tentative 3D sketch incorporating the deformation of the mesenchyme, at the moment of buckling, and just at the begining of budding. The arrows explain the physical displacements. The small tube is the internal epithelium, mesenchyme is around. Budding will resume down the trunk, with respect to the bifurcation point, in the region pointed by the arrow. This mechanism is still speculative (but see Chapter 8 for in vivo evidence).

The saddle buckling induces a breaking of the pool of Fgf10 into several pieces, which passively follow the deformation of the mesenchyme, in addition to possible inhibition-activation. As the mesenchyme is driven down the trunk in the direction of the small axis of the saddle, it will favor growth of buds in two regions, located at right angles of the main axis of the saddle, somewhere down the apical region and pointing outwards in the 3D dimension, with respect to the plane formed by the initial bud and the two axes defined by the tip-splitting event. This simple explanation of budding predicts that the well known structure of lungs with dichotomies "following each other at 90°" is in fact a misinterpretation of branching events produced somewhere in their past by dichotomous events. By this it is meant that as ducts elongate and tip-split, the dichotomy induces a new bud along the trunk from which the branches of the T have emerged. So, when an anatomist examines the branches, he or she may think that budding occured sequentially, first one branch and then another, because this is the spatial order which is seen on the final geometry, but the temporal order was actually the opposite, and so was the causality.

So much for the push in the apical region. The push of the mesenchyme along the trunk can also be implemented by finite element methods to show how a duct becomes a more ampulla-like or T-shaped pattern after strangling (Fig. 26).

This sequence of images is important, although the calculation is simplistic. First, it shows that recession of the epithelium has a stabilizing virtue. Indeed, for sport, we used a wavy edge of the growing duct, as may happen in vivo, spuriously. The mesenchyme push tends to smooth out the irregularity, and to reshape the tube to make it straighter. But there is more to this strangling: the ampulla adopts somewhat of a T-shape. If we have a more careful look at the T shape, we may notice that it has something like a smooth corner not at the top, but somewhere in the region where the trunk joins the ampulla. Such a "corner" or "parrot-beak" is actually seen in the experiments. The reader may identify such corners somewhat down the apex, in many images above, and in the more global image below (Fig. 27) in which the "parrot-beaks" are ubiquitous.

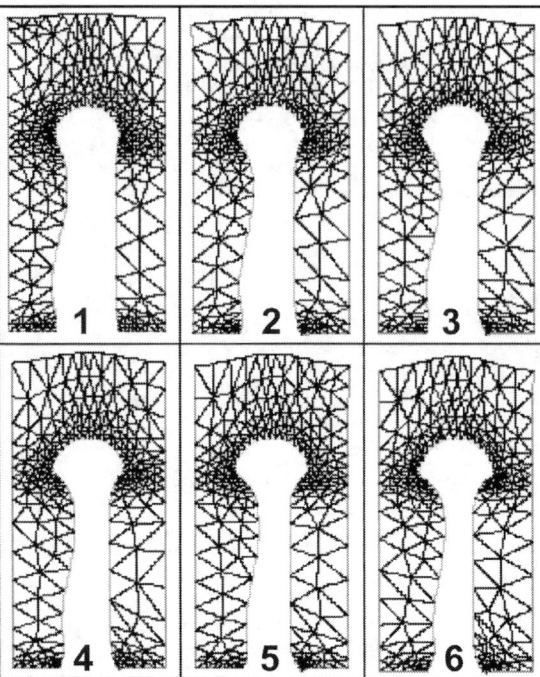

Figure 26. Ampulla and thin tube, obtained by a mechanical elasto-plastic push of the mesenchyme towards a duct, outside an apical region. The distance, outside the apical region, at which the mesenchyme push commences is chosen arbitrarily.

A very important feature of this "parrot-beak" is that it is **not** located at the junction between the region where the epithelium moves forward and the region where it recedes. Not at all. Because of the laws of elasticity, if there is a threshold below which the epithelium recedes (as is the case here) a corner bowing backwards appears, which finds itself in the region where the epithelium grows forward, neither at the tip, nor along the trunk, but on the side of the

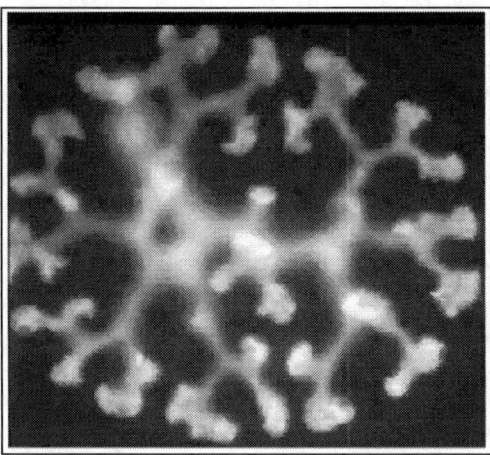

Figure 27. Typical kidney growth. The reader may identify many regions where the duct apex has a typical "parrot-beak" structure. The apex curves, or bows, with a sharp corner oriented backwards, with respect to the original growth direction. This backwards turn is due to the torque exerted on the epithelium apex.

region growing forward. As explained before, fingering is favored in sharp protruding regions. Therefore, we expect the growth to be accelerated in the region of the sharp corner, and hence to induce sibling ducts on either sides of the T. Moreover, if expression patterns are linked to curvature or stresses along the surface, the region where the contour is sharper will induce a ring of expression around the growing duct. This is indeed the pattern of expression of sprouty and of Fgfr2 during lung growth.[34]

However, in all these images, one needs to incorporate a threshold such that, outside the apical region, mesenchyme growth (hence recession of the epithelium) occurs. This threshold may well be given by some distance to the tip (decay of some diffusible molecule), by some time delay, or by some stress threshold which triggers mesenchyme development or maybe a conformatioal change of cell shape. This is an entirely open question, to our knowledge. Athough many features are known (e.g., change from cuboidal to squamous cells) it is not clear which feature is the cause and which features are mere consequences. However, it sounds awkward that one threshold should be needed to explain the existence of a soft apical region, and another threshold should be needed to introduce the growth and contraction of the mesenchyme around the growing ducts. It is more likely that there is a single explanation to these two facts. Above the threshold, the mesenchyme recedes and the duct moves forward, below the threshold, the mesenchyme moves forward and the epithelium recedes. There seems to be an imbalance rocking between epithelium and mesenchyme. The term rocking is in fact very apt: such a push-pull action around a threshold amounts to rotation around a fixed point (the boundary between the two regions). Then, the effect of the forces is to make the ducts rotate, in addition to growing forward. Movies of duct growth give this impression of systematic turns: the ducts swerve permanently. The result is the typical ampulla-like duct snaking between obstacles in the apical region, and straight and geometric tubes outside the apical region further down the tubes. The global result is an increase in size, but with a repeatedly branching structure, and isolated round balls of mesenchyme around each duct being pressed together in the capsule and forming pea-like aggregates, with septa separating different levels of the hierarchy.

One may wonder whether the threshold which is hypothesized in tip-growth (a diffusible molecule), could not itself be mechanical, in such a way that the process would, again, be entirely epigenetic instead of genetically-driven. Indeed, a mechanical stress setpoint might come into play in the following way. When a rigid medium is bent, the inner shells experience a contraction (compressive stress) and the outer shells experience a tensile stress. Although it is not very intuitive, a mechanical push inside a rigid elastic cylinder induces a compression close to the push, and a tension somewhat afar. The overall force, integrated over the entire body is tensile. However, in 3D, there exists 2 principle axis of curvature, for any surface. A cylinder has a finite radius of curvature R in one direction an infinite one in the other direction. The tissue is bent in one direction only. The internal pressure is sustained by a small deformation of the cylinder. At the tip of a duct, the surface is more spheroidal, and 2 radii of curvature, along two perpendicular axis come into play. The tensile or compressive stress is then less for a cell located at the tip of a duct, than along the cylinder. If there is a strangling of the duct, as described above, the situation is even less favorable for the tip: the tension is higher in the region of the parrot-beak. Then, if a mechanical set-point, either compressive or tensile triggers cell proliferation, it is more likely to trigger cell proliferation somewhat in the back of a duct, outside the apical region. Again, this comes from the 3D character of ducts, and it cannot be described by a 2D model. Whether the same threshold appplies to the recession in the region of the T-shape remains to be studied.

Orientational Order

So far, we have not taken into account the orientational order of the tissue. It is debated whether such an order already exists at the tip of a growing duct during growth.[22] It is at least not apparent in the epithelium, but it is very likely to exist in the mesenchyme.[20,45] Fibroblatsts are seldom represented without an orientational order (Fig. 28).

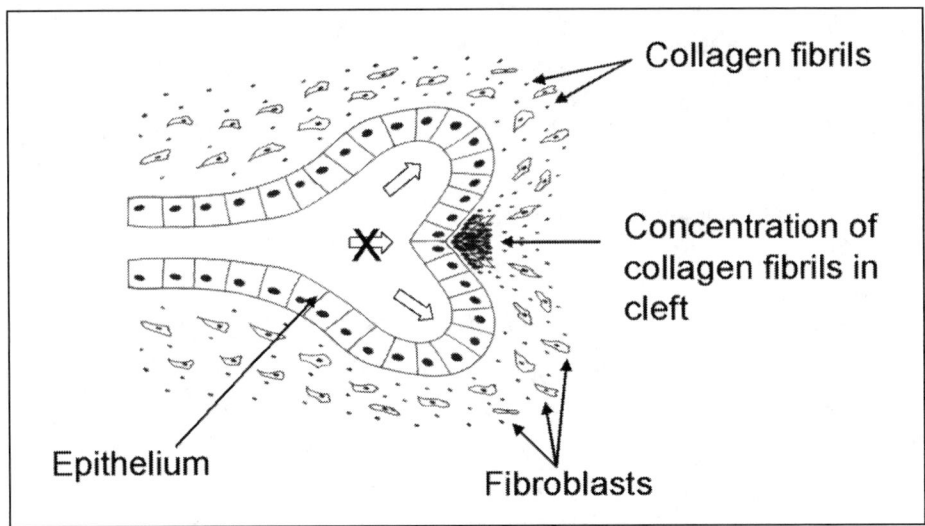

Figure 28. Typical representation of fibroblasts around a growing duct (after Nakanishi et al[46]).

This raises two questions: how does the shape and stress state of the surface influence the alignments of cells and even their own shape, and second, does the alignment of cells influence the pattern in return? This is the core of the **material** aspect of morphogenesis. If there is stereotypy, symmetry breaking, tubular structures, ampullae, push-pull actions, foldings etc., are all these features a consequence or a cause of the presence of morphogenic fields, be them concentration of chemicals or force fields?

Basically, the question arises of how the presence of an orientational order modifies the behaviour of the material, and how it couples to all the fluidics and mechanics explained above. This is truly a fomidable task. However, simple features can be explained. The orientational order induces streams of aligned cells, and hence of deposited collagen. When a surface develops in 3D, an orientational order experiences a topological problem: certain surfaces cannot be dressed with a uniform orientation in a perfect manner, there always remain topological singularities. These singularities deform the surface. This is obvious on spheres, and at branching points. One may think also of fingerprints. In the simplest picture, an orientational order on a sphere, with cylindrical symmetry around a "singularity" induces a protrusion in the region of the "singularity". If one imagines the drawing of parallels and meridians on a sphere, the singularities are simply the north and south poles. But there exist more complex solutions, such as the pattern of stitching on a tennis ball.

In order to show that the existence of singularities has an impact on the shape, we calculate the deformation of a spherical "bubble" which would be fibered like a tennis ball (Fig. 29, left). This drawing is more complex than parallels or meridians, but it is commonly observed on fingers, and also on cartilage rings. A distribution of little rods which can flip statistically, but which interact with a tendency to align themselves with each other can be modelled by an energy which is called the Frank-Oseen elastic energy.[47] In such a model, the cost of making an element of surface depends on the local drawing of the lines. If the lines are bent, or twisted, or fanned out, instead of being gently parallel, there is an additional energy cost. Making a singularity is "very expensive" in terms of energy, and spontaneously, all the fibroblasts would prefer to be all parallel to one and the same direction. However, if little elongated elements (and we may think of fibroblasts as being such elongated elements) try to align themselves on a sphere, there is no way that they can all be parallel. Hence, they

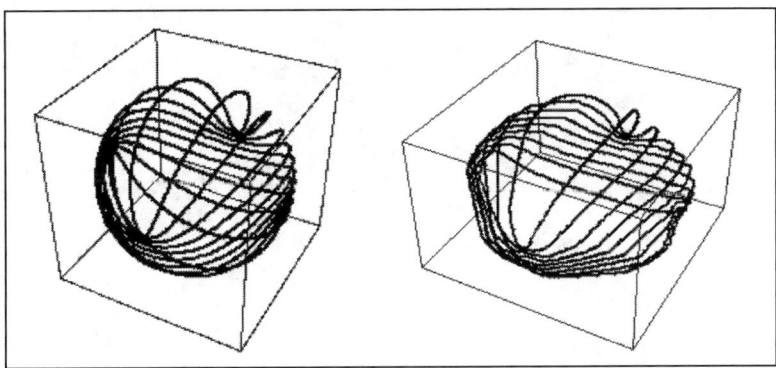

Figure 29. Calculation of the shape of a sphere dressed with fibers, in a simple Frank-elasticity model. Left the initial distribution of lines (e.g., fibroblasts), right the actual equilibrium shape. Hence, such a sphere... is not spherical. This shows that the laying down of collagen in an orderly fashion implies deformations of the surface.[49]

will make a pattern with singularities, and the surface energy associated to the drawing will locally depend on the specific drawing at each point. We have seen that there is a tension associated to the energy. If the tension is not uniform, then the "sphere" will not be spherical. One can find the surface which matches the equilibrium of the internal and external pressures and the non-uniform tension (Wulff construction).[43] This is how the equilibrium shape of crystals such as quartz or ice are mathematically constructed. In the case of the "tennis ball" it gives the following result (Fig. 29).

The region of the singularity (here the U-turn of the lines), induces, or favors, a bump in the region of the loop. However, the bump is not located right in front of the singularity, but shifted somewhat sideways. This appears in Figure 29 (right) in the fact that the region of the U-turn is not the foremost one pointing towards the reader. The part of the surface which protrudes most towards the reader is slightly below the U-turn (Fig. 29, right). The origin of this shift is in the symmetry breaking of the material properties, in the region of the singularity. This can be observed on anyone's fingers: if you find a loop on the drawing of your epidermal ridges, the center of the loop is not below the summit of the relief, but shifted more proximally, because a region of more curved lines is somewhat stiffer (think also of folding an edge of cloth and sewing the two half-edges: one finds a protrusion in the center, as for example on socks). Similar effects will occur with cartilage rings.

Therefore, in general terms, the presence of a singularity favors a forward growth in that direction. The mismatch between the maximum height of the relief and the position of the singularity is again associated to a torque (of surface tension), which bends the surface in this non-intuitive shape. The very center of the core is singular and must be accomodated by the tissue either with different cells, or with a hole, or with a uniform medium. Data about what happens in these regions are scarce.

The effect of an orientational order on a **branching region** has not, to our knowledge, ever been discussed. The problem of the branching of a uniform membrane has been adressed only recently.[50] However, a simple calculation (Fig. 30) of a cross-like membrane with vertical load and either parallel or transverse orientational order shows that, for identical pressure conditions, one has narrower necks in the region of the interconnection. Therefore, it is expected that the anisotropy of the fibroblasts and smooth muscle cells will contribute to build sharper connections between tubes (than with a uniform medium), if the cells run azimuthally, and to build rounder connections (than with a uniform medium) if the cells run longitudinally. Interestingly, this is seen in plants.

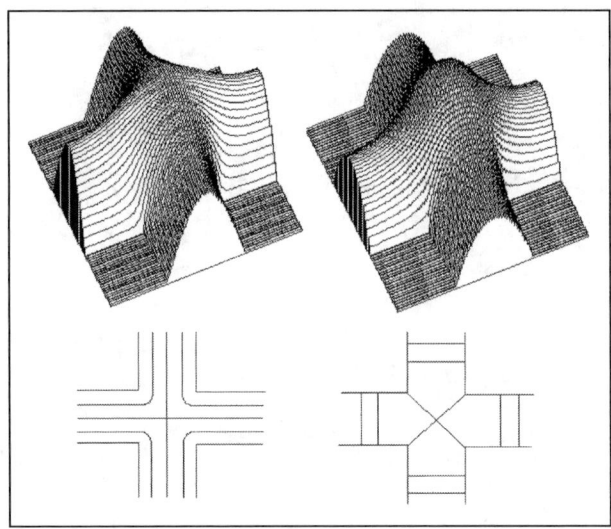

Figure 30. A membrane is stretched on a cross-like frame. The membrane is supposed to be constructed with fibers which have an anisotropic Young modulus, such that the membrane is more elastic in one direction (perpendicular to the lines) than along the lines. The membrane is flexed under a vertical load. The lines may be chosen longitudinally, or transversally to the arms of the cross. The deformation is not the same. If the stiff direction is transversal, the region of the interconnection is more curved (which amounts to sharper connections of tubes).

Conclusion

As a conclusion, a complete picture of branching morphogenesis, by no means quantitative, is probably as follows. In the presence of growth factors, especially FGFs, the tissue grows. The growth process is mediated by a mechanical push of osmotic origin. Under pressure, the mechanical (mostly tensile) setpoints of the pouch are overtaken; this tends to generate ducts with an active apex. The apex is curved and moves forward because there is an active region there, but the region is also active **because** the apex is curved. There is anyway an active region, even in the absence of any specific chemical activity, because of physical principles. This fact solves the chicken-and-egg problem raised by such a situation. The genetic actions, such as softening of membranes by degradation enzymes, "take over" and enhance physical cues.

The "pouch" is composed of two kinds of tissue, a more internal one (epithelium), and a more external one (mesenchyme). They do not experience the same mechanical fields, but still, they both grow. The ducts elongate more in the apical region by visco-elasto-plastic growth (fingering), while the mesenchyme pushes more outside the apical region (Dirichlet polygonal domains). In the apical region, a softer region, maybe even close to a fluid, forms a self-perpetuating finger-like structure, which swerves as it is contracted in its back, and collides repeatedly against the capsule, or against other fingers, finding its way in a self-organizing landscape of branches and pushing mesenchyme. When it collides, the duct is transformed into a disk, which undergoes a saddle-buckling which selects two siblings out of the disk, rarely more, and also induces buds down the trunk, depending on the context. Outside the apical region, the mesenchyme push straightens the tubes and makes them more polygonal.

The orientational order of the fibroblasts contributes to the process in several respects. As the fibroblasts adopt an orientational order around tubes, they get more susceptible to circumferential stress (as receptors of tension, they are more sensitive), but, in addition, the pattern of force which they exert on the tube is more radial and centripetal. As a consequence, they tend to select or even create well defined, stable, directions of growth for the soft apex, and sharp branching points. They probably contribute to selecting the direction of saddle-buckling of T-shaped ampullae, because the pattern of lines around the apex has a broken symmetry already (like fingerprints).[49]

It should be noted that the description given in this chapter implies that branching morphogenesis is a completely 3D process conceptually, and it cannot be described by any 2D scenario. This is because: (1) tubes are 3D entities which have 2 axis of principal curvature, a fact which disappears in 2D, (2) A T-shaped ampulla is a 3D entity which has a disk facing the mesenchyme, a fact which disappears in 2D, (3) the saddle-branching of a disk is a 3D phenomenon whose 2D equivalent produces only folds (like brain convolution), (4) an orientational order of fibroblasts is meaningless on a 2D drawing of a cross section of branches. (5)growth of a 3D organ requires symmetry breaking so that out-of-plane branching is favored.

It should also be insisted that a deterministic immutable branching rule cannot generate a confined tree. The construction of confined trees such as the organs is possible because small adjustments of the rules are always taking place at the scale of individual ducts.

It must also be said that the entire pattern implies both a push perpendicular to the surface and a push tangential to the surface. Whatever the molecular detail performing these pushes, a tangential push can only generate folds (as in brain), while a push perpendicular to the surface generates fingers, T's or pouches without branches. The perpendicular push is needed to move the ducts forward, the tangential push is needed to have them buckle. The perpendicular push is linked to osmotic balance, while the tangential push is linked to mitotic activity, with mitotic planes always perpendicular to the surface (cell anisotropy, or "polarity").[51] Hence, the mechanisms of these pushes are entirely different, and the balance between the two will generate different patterns, between super-folded structures (like *polymicrogyria* in brain), super-branched structures (like organs) or only "balloons" (like in polycystic kidney disease).[52,53]

What kind of genes are needed to perform these feats? First of all "geometrical" genes, i.e. genes which preserve the geometry of the system. These are essentially the polarity genes of the epithelium, and the genes regulating contact inhibition. Thanks to them, the epithelium remains a sheet, facing the lumen, with the basement membrane on the other side, facing the mesenchyme. Polarity genes also play a role in the shape of fibroblats and hence on the way collagen is deposited (and vice-versa). In the absence of polarity genes, a mass of epithelium could still penetrate into a mass of mesenchyme, but the filtering, breathing, etc. properties would be lost (no lumen). Such aberrant epithelia are observed frequently in squamous cell carcinoma.[51] These genes probably act also on the very shape of the epithelial cells, with an influence on growth. The original organ in the phyllogenetic order was probably already a bag or pouch, prior to evolving the branching. A second set of genes must rule the push of the epithelium and that of the mesenchyme. This set of genes acts on osmotic balances, in response to growth factors and to geometry, and on mitosis. Since one expects the epithelial push to be localized in the region of the apex, and the mesenchyme push outside that region, one is tempted to speculate that some non-linear autocatalytic reaction-diffusion system localizes the active regions. However, it is obvious from simple inspection of the growth that the system is very much influenced by mechanical cues, since it is able to turn, split or stop in the region of obstacles. Therefore it is possible that the dynamics of the growth is greatly influenced or even fully generated by a mechanical setpoint, probably a curvature or tension-related setpoint.[54,55] A third set of genes must monitor the material aspects, i.e. essentially density of fibroblasts and laying down of collagen in an orderly fashion, akin to a liquid crystal phase.[22]

Typical anatomical dimensions are certainly adjusted or fixed by material properties also, and not only by chemical gradients. It may be thought that a partial or even complete theoretical description of this branching morphogenesis, useful at a clinical level and even allowing one to design regeneration machines, is not too remote. At time this book is published, regeneration machines will probably be already under trial.

Acknowledgements

The authors thank Frank Constantini, Mathis Plapp and Jean-François Gouyet for their help. Vincent Fleury, Thi-Hanh Nguyen, Mathieu Unbekandt: Laboratoire de Physique de la Matière Condensée Ecole Polytechnique 91128 Palaiseau cedex, France. Anke Lindner: Laboratoire des Milieux Désordonnés et Hétérogènes Université Paris 6 4 place Jussieu-case 86, 75252 Paris Cedex 05, France. Laurent Schwartz: Laboratoire LPICM, Ecole Polytechnique 91128 Palaiseau Cedex, France. David Warburton: Department of Pediatric Surgery, Childrens Hospital, University of Southern California, 1975 zonal Avenue KAM 110. Los Angeles California, 90083-9023, USA.

Appendix 1

There exists to our knowledge one alternative model of epithelial morphogenesis. This model considers contraction waves propagating in the epithelium. It may contain part of the picture.[55] However, while it is easy to imagine a spontaneous contraction wave in a 1D circular epithelium, it is less so easy to imagine this scenario in 3D, because it amounts to proposing that tubes form by emission of self-organized rings of contraction waves emanating from the tip of the duct, a much less plausible structure. It is much more likely that the ring-like structures which are necessary to maintain the ducts have a material origin instead of a diffusible one. The rings which contract and maintain tubes, in this picture are true rings of cells, as classically observed in histological preparations. Although contraction waves may propagate, they need not carry themselves a morphological information. Also, it is a remarkable fact that ducts may collide against each other and form T-shaped dichotomies facing each other. If contraction waves in the epithelium were responsible for a T-shaped dichotomy event, then one would have to imagine a correlation between contraction waves in two tips facing each other, although their distance along the epithelium may be enormous, a very unlikely feature. Considering that the T-shaped dichotomy is almost the same (apart from the symmetry breaking, see text) whether tips collide against each other or against the capsule, it is tempting to state that the "information" needed to perform a T-shaped dichotomy is zero.

Glossary

Adaptation: the process by which cells, which sense mechanical forces, proliferate or change shape until they reach some mechanical setpoint which inhibits further mitotic activity.

Ampulla: the ducts are generally wider at the apex, thus having the shape of a lamp bulb (ampulla). Ampullae are easily obtained in falling or raising drops problems, they are characteristic of interfaces flowing or hanging in presence of a uniaxial force (like gravity, see nuclear mushrooms). The growth of a soft apex, which is fixed to a more polymerized trunk is somewhat analogous to the problem of hanging tethered drops.[56]

Anisotropy: a surface or physical process may admit a preferred direction of growth. For example, in crystal growth problems, the crystal lattice induces an anisotropy, which leads to beautiful six fold branching of snow-flakes, or conspicuous four-fold symmetry of pyrite. The nematoid order of cells, on a spheroidal surface, generates an anisotropy, in the direction of the center of the singularity. Anisotropy may cause (or be the result of) a symmetry breaking.

Apex, apical region: a narrow region, at the tip of a duct, where something different is going on, which is responsible for elongation and growth of the duct.

Buckling: mechanical "folding" by which a flat plate undergoing tangent stress breakes its symmetry. In 1D, buckling induces wavy folds. This type of folding may explain the folds of the hindbrain or of intestine villi. In 2D a flat sheet becomes folded in several directions, as a consequence of buckling. The simplest buckling mode of a circular or eliptical plate is the saddle.

Budding: a process by which a tip emerges on the side of a trunk. It is debated whether there is a specific molecular instruction which provokes budding, or whether it is epigenetic. The first branching events of a lung are more of a budding type than those of a kidney.

Dichotomoy: a process by which a tip splits into two (see budding). The first branches of a kidney are more dichotomous than those of a lung.

Duct: a tube of tissue (epithelium+basement membrane) with a roundish end which extends progressively, either in a culture medium, or in a supporting tissue.

Elastic deformation: reversible deformation of a body, with respect to a reference configuration when some external or internal force is applied. If the force is removed, the solid returns to the reference shape. The work of the force stores an energy into the solid which maintains the solid in the deformed state; when the force is removed, the body resumes its initial form, and gives the energy back.

Epigenetic: specified or controlled by an element of the context (temperature, pressure field, geometry, etc.) and not directly by genes. The epigenetic cue may of course be mediated by some genetic transduction. The number of fingers is genetically determined, the implantation of each hair, for example is epigenetic.

Field (biology): a domain containing cells which eventually will find themselves in the same organ or structure.

Field (physics): a non uniform quantity which is defined in space and whose value at each point is given by a mathematical function of the space coordinates, and possibly also of time. For example $T(x,y,z)$, the temperature field, $C(x,y,z)$ the field of concentration of some chemical, $\varepsilon(xy,z)$ the deformation field, etc. Fields of different physical origins may sometimes satisfy analogous if not identical mathematical equations (because of "conservation laws"). For example, the diffusion equation is identical for temperature and for concentration fields, although they carry different names (Fick's law, Fourier's law). The fields may be more complex than just one scalar quantity, for example: fluid velocity is a vector, strain and deformation in a body are tensors.

Gradient: spatial variation of a quantity, such as pressure or growth factor concentration. High gradients correspond to big slope in the pressure or concentration landscape. Most growth processes tend to navigate up or down high gradients (fluxes and flows are greater in presence of higher gradients).

Minimal surface: a minimal surface is a surface which follows a thermodynamic equilibrium, but which, in addition, feels the same pressure on both sides. The result is a surface which has zero mean curvature (one concave and one convex). A typical minimal surface is a hyperboloïd, or a catenoid. A bubble is not a minimal surface (it has internal pressure). The name "minimal" comes from the fact that it is constructed automatically with the smallest possible area allowed by a given boundary condition.

Newtonian fluid: a fluid whose viscosity is constant and uniform, not dependent on velocity (e.g., water, oil). An intuitive characteristic of a Newtonian fluid is that objects sink into a Newtonian fluid, although sometimes very slowly (in honey, for example).

Nematoid order: order of oblate or elongated objects which interact with a tendency to align themselves. The simplest example is given by crystal liquids. Fibroblasts exhibit a nematoid order, epidermal ridges, cartilage rings, etc. exhibit a nematoid order.

Plastic Deformation: irreversible deformation of the body, with respect to a reference configuration. Plastic deformation requires energy dissipation in the body. In a solid, the dissipation is mostly by defects rearrangements inside the solid (up to rupture). In a living tissue, energy dissipation is both by physical polymer-polymer disentanglement (yield), and by cellular metabolism.

Prepattern: a distribution of chemicals on a biological surface, which is obtained by a Reaction-Diffusion mechanism à la Turing. The patterns are usually in the shape of spots, or lines. The spots usually adopt a square or hexagonal lattice, although more complex patterns exist. If one of the biochemicals is a growth factor, it is expected that growth will occur exactly in the region of one of the spots, and generate a bump, a duct, a needle, etc. However, mechanics must be incorporated at some stage to complete the growth model.

Setpoint: a specific value of the surrounding epigenetic context at which a genetic switch modifies the behaviour of a cell.

Stereotypy: the laying down of branches tends to follow some regular pattern, which is somewhat of a constant. For example: dichotomies tend to repeat at constant interval, and turn each time by 90°. The stereotypy of kidney and of lung are different (there is more budding in lung than in kidney).

Surface tension (interfacial energy): if a surface is produced by a molecularly fluctuating material, and if the thermodynamical cost (interfacial energy) of an element of area is γ, then there exists a tension (force = energy per unit length) inside the surface equal to γ. The shape of the interface is the result of the equilibrium between the internal and external pressures, and the tension in the membrane. For a spherical object, this writes Pint - Pext = γ/R, where R is the radius of curvature of the sphere at equilibrium. Shapes which are not under thermodynamic fluctuations never reach equilibrium. They can have any shape, with or without tension (for example: a solid spherical shell, in the reference state, with or without residual tension, may be taylored with any diameter).

Shear-thinning fluid: a fluid which is less viscous, and hence flows more easily if it is sheared. Only big and heavy enough objects will sink in the fluid. Below some shear threshold, they behave like solids (typical of gels, wax, albumin, paints etc.)

Symmetry breaking: a mechanism by which, upon variation of some physical parameter, a structure becomes less symmetrical. For example, if elongated particles, deposited on a sphere, start interacting, then they tend to form filaments, but the orientational order of filaments on a sphere generates a symmetry-breaking, because the filaments organize themselves around singularities (for example a north and a south pole), which break the symmetry of the sphere.

Surface stiffness: the surface tension is generally supposed to be a constant. But in the true world, this is not so generally true. If the surface tension is not a constant, the relevant quantity, in order to find an equilibrium shape, is not the surface tension but the surface stiffness $\{\gamma + \partial^2\gamma/\partial^2\theta\}/R$ where $\partial^2\gamma/\partial^2\theta$ contains the effect of the spatial variation of γ, with respect to a polar direction θ. The equation to solve becomes Pint - Pext = $\{\gamma + \partial^2\gamma/\partial^2\theta\}/R$. This comes from the fact that the variation in surface tension acts as an additional torque which bends the surface. In many materials for which the elasticity of the surface interacts with the bulk, the concept of surface tension does not permit alone to find the shape.

T-shape dichotomy: when tips split they tend to have, transiently, a T shape which progressively transforms into a Y shape. This T-shape is easily obtained in moving drop problems, upon collision on a flat obstacle.

Viscous finger: elongated duct-like bubble which forms when one fluid penetrates into a more viscous one, especially in 2D. Viscous fingers are also observed in non-Newtonian fluids.

References

1. Kitaokal H, Takaki R, Suki B. A three-dimensional model of the human airway tree. J Appl Physiol 1999; 6:2207-2217.
2. Mandelbrot B. The fractal geometry of nature. San Francisco: Freeman & Co, 1983.
3. Srinivas S, Goldberg MR, Watanabe T et al. Expression of green fluorescent protein in the ureteric bud of transgenic mice : a new tool for the analysis of ureteric bud morphogenesis. Dev Genetics 1999; 24:241-251.
4. Pelcé P. Dynamics of curved fronts. London: Academic Press, 1991.
5. Saffman PG, Taylor G. The penetration of a fluid into a porous medium or Hele-Shaw cell containing a more viscous liquid. Proc R Soc London Ser A 1958; 245:312-329.
6. Homsy G. Viscous fingering in porous media. Ann Rev Fluid Mech 1987; 19:271-311.
7. Howison SD, Ockendon JR. Singularity development in moving boundary problems. J Mech Appl Math 1985; 38(3):342-360.
8. Howison SD. Cusp development in Hele-Shaw flow with a free surface. SIAM J Appli Math 1986; 46(1):20-26.
9. Howison SD. Fingering in Hele-Shaw cells. J Fluid Mech 1986; 16:439-453.

10. Bensimon D, Pelcé P. Tip-splitting solutions to a Stefan problem. Phys Rev A 1986; 33(6):44774478.
11. Mineev-Weinstein MB, Ponce-Dawson S. Class of non-singular exact solutions for Laplacian pattern formation. Phys Rev E 1994; 50(1):R24-R27.
12. Scheuchzer JJ. Herbarium Diluvianum. litt D Gesneri 23, Zürich,1709, 1711.
13. Fleury V, Arbres de Pierre. la croissance fractale de la matière. Paris: Flammarion, 1998.
14. Vicsek T. Fractal Growth Phenomena, Second Edition. Singapore: World Scientific, 1992.
15. Fleury V, Gouyet JF, Leonetti M, eds. Branching in Nature. Paris: Springer/EDP Sciences, Berlin, 2001.
16. Ben Jacob E, Shochet O, Tenenbaum A et al. Evolution of complexity during growth of bacteria colonies. In: Cladis PE, Palffy-Muhoray, eds. Spatio-temporalpPatterns in non-equilibrium complex systems, Santa Fe Institute studies in the sciences of complexity. Addison Weseley Publishing Company, 1995:619-634.
17. Marcus Dejmek, Thesis. Palaiseau: Press of the Ecole Polytechnique, 2002.
18. Witten TA, Sander LM. Diffusion Limited Aggregation as a critical phenomenon. Phys Rev lett 1981; 47:1400-1403.
19. Combescot R, Dombre T, Hakim V et al. Shape selection of Saffman-Taylor fingers, Phys Rev Lett 1986; 56(19):2036-2039.
20. Gilbert SF. Developmental Biology. Sunderland: Sinauer Associates Publishers, 1994:Chapter 18.
21. Bard J. Morphogenesis. Cambridge Cambridge: University Press 1992.
22. Fleury V, Watanabe T. How collagen and fibroblasts break the symmetry of growing biological tissue, CR Acad Sci Biologies 2002; 325: 571-583.
23. From reference 21, itself from Elsdale TR, Wasoff FL, Whilh. Roux' Arch dev Biol 1976; 180:121-47.
24. Gray H. Anatomy of the human body. Philadelphia: Lea & Febiger, 1918.
25. Weibel E, The pathway for oxygen, Structure and function in the mammalian respiratory system, Massachussets and London: Harvard University Press Cambridge, 1984.
26. Pozrikidis C. The deformation of a liquid drop moving normal to a plane wall. J Fluid Mech 1990; 215:331-363.
27. Van Damme H. Flow and interfacial instabilities in Newtonian and colloidal fluids, in The fractal approaches to heterogeneous chemistry. Avnir D, John Wiley and sons limited, 1989.
28. Lindner A, Coussot P, Meunier J. Phys Fluids 2000; 12:256.
29. Lindner A Coussot P, Bonn D. Phys Rev Lett 2000; 85:314-317.
30. From reference 27, itself from Daccord G, Nittmann J, Stanley HE. Phys Rev Lett, 1986; 56:336.
31. Goriely A, Tabor M. Self-similar tip growth in filamentary organisms. Phys Rev Lett 2003; 90(10) 108101:1-4, and references therein.
32. Turing AM, The chemical basis of morphogenesis AM. Phil Trans Roy Soc B 1952; 237:32-72.
33. Koch AJ and Meinhardt H, Biological pattern formation: from basic mechanisms to complex structures, Reviews of Modern Physics 1994; 66(4):1481-1507.
34. Warburton D, Bellusci B, Del Moral PM et al. Growth factor signaling in lung morphogenetic centers: automaticity, stereotypy and symmetry, Respir Res 2003; 4(1):5—Biomed central article http://www.pubmedcentral.nih.gov/articlerender.fcgi?artid=185249.
35. Meinhardt H. The morphogenesis of lines and nets. Differentiation 1976; 5:117.
36. Fleury V. Branching morphogenesis in a reaction diffusion model. Phys Rev E 2000; 6(4)4158-4156.
37. le Noble F, Eichmann A, Nguyen TH et al. Engineering vascular architecture, submitted.
38. Chai H. Buckling and post-buckling behavior of elliptical plates, Part I-analysis, J Appl Mech 1990; 57:981-994.
39. Zhang Y, Hobbs BE, Ord A et al. Computer simulation of single layer buckling. J Struct geol 1996; 18(5):643-655.
40. Caviness VR. Mechanical model of brain convolutional development. Science 1975; 189:18-25.
41. Fleury V. Des pieds et des mains. Flammarion, Paris 2003.
42. Green PB, Pattern formation in shoots, a likely role of minimal energy configurations of the tunica. Int J Plant Sci 1992; 153(3):S59-75.
43. Dumais J, Kwiatowska D. Analysis of surface growth in shoot apices, Plant J 2001; 31 (2):229-241.
45. Schwabe WW, Clewer AG. Phyllotaxis, a simple computer model based on the theory of a polarly translocated inhibitor. J Theor Biol 1984; 109:595-619.
45. Douady S, Couder Y. Phyllotaxis as a physical self-organized growth process. Phys Rev Lett 1992; 68:32098-2100.
46. Nakanishi Y, Sugiura F, Kishi JI et al. Scanning electron microscopy observation of mouse embryonic submandibular glands during initial branching: preferential localization of fibrillar structures at the mesenchyme ridges participating in cleft formation. J Embryol Exp Morph 1986; 96:65-77.

47. De Gennes PG, Prost A. The physics of liquid crystals. Oxford: Clarendon, 1993.
48. Godrèche C, Solids far from equilibrium, coll. Aléa Saclay, Cambridge University Press, 1992.
49. Nguyen MB, Fleury V, Gouyet JFG. Epidermal ridges: Positional information coded in an orientational field. In: Noval M, ed. Fractals and complex systems. Singapore: to be published World Scientific, 2004.
50. May S, Yardena B, Avinoam BS. Molecular theory of bending elasticity and branching of cylindrical micelles. J Phys Chem B 1997; 101:8648-8657.
51. Fleury V, Schwartz L. Numerical investigation of the influence of cell polarity on cancer morphology and invasiveness. Fractals to appear, 2003.
52. Igarashi P, Somlo S. Genetics and pathogenesis of polycystic kidney disease. J Am Soc Nephrol 2002; 13:2384-2398.
53. Lubarsky B, Krasnow M. Tube morphogenesis, making and shaping biological tubes. Cell 2003; 112:19-28.
54. Taber LA. Biomechanics of growth, remodeling and morphogenesis. Applied Rev Mech 1995; 48(8):487-545.
55. Odell GM, Oster G, Alberch P et al. The mechanical basis of morphogenesis. Dev Biol 1981; 85:446-462.
56. Nye JF, Lean HW, Wright AN. Interfaces and falling drops in a Hele-Shaw cells. Eur J Phys 1984; 5:73-80.

CHAPTER 13

Afterword

Jamie A. Davies

The subject of this book—branching morphogenesis—may seem to be very narrow, yet its chapters extend into a surprising number of aspects of modern biological science. The systems examined range from genes and signal transduction pathways, through morphogenetic apparatus within single cells to processes that take place at the scale of the whole organism. The techniques described include biochemistry, molecular genetics, cell biology, anatomy, developmental biology, biophysics and computer modelling. The applications discussed range from pure scientific discovery through biotechnology to improved clinical management of disease. It would be presumptuous for any editor to try to tell readers how they might integrate the material in their own minds but, especially for readers who have no time to read the book in its entirety, I shall use these last few pages to summarise my own thoughts on the key questions of the field.

Do We Yet Understand Branching Morphogenesis in Any System?

Frankly, no, and no one system is even close. An encouraging observation to emerge from this book, however, is that different systems have different areas of strength and weakness and, often, what is unknown for one system is understood well in another. This means that, unless different systems turn out to manage branching very differently (see below), it should be possible for researchers to apply techniques already developed by others for other systems, to gain information in the weak areas of their own system with relative ease.

Regulation of Branching

Regulation of branching, especially by paracrine factors released by other tissues, is one of the best-understood aspects of branching morphogenesis. A combination of organ culture experiments and analyses of mutant animals has revealed a large number of proteins that are required for normal morphogenesis. Some encourage branching, while others inhibit it. Most are produced by other tissues in the immediate vicinity of the branching structure, but others, such as the sex hormones discussed in Chapters 7 and 10, originate far away in the body and control development in conjunction with more local signals. Most systems studied so far depend absolutely on paracrine signals, and careful examination of these signals reveals the potential for feedback loops to control branching very tightly (see Chapter 8). Paradoxically, however, there are culture systems for several of these branching structures that can generate apparently organotypic branching, even without any obvious potential for feedback via another tissue.

Regulation by autocrine signalling also seems to exist ('autocrine' in the sense that the branching tissue is signalling to itself—it is possible that the cells producing and receiving these signals will be found to be in such different states of differentiation that the word 'autocrine' becomes inappropriate). This type of signalling is vital to the patterning of branched structures that do not develop in the context of other tissues, for example the fungi described in Chapter 4 and the hydroids described in Chapter 5. It may also account for patterning in the metazoan cell

Branching Morphogenesis, edited by Jamie A. Davies.
©2005 Eurekah.com and Springer Science+Business Media.

culture systems mentioned above, and is certainly important in some metazoan organs such as the tracheae of *Drosophila*.

Within tissues, regulation comes not only from small, diffusible molecules but also from the extracellular matrix, and in organs such as the mammary gland (Chapter 7) regulation by the matrix is of key importance to development and also, probably, to cancer. The feedback loops involved here, between a branching epithelium activating matrix remodelling and the matrix controlling the epithelium, are complex but worth understanding from the points of view of both biology and of medicine. Regulation of blood vessel growth (Chapter 6) is also of great medical importance, offering a potential way to control the growth of a neoplastic tumour by restricting its access to nutrients, oxygen and waste disposal.

Morphogenetic Mechanisms

Generally, the morphogenetic mechanisms actually responsible for creating new branches are the least well-understood aspects of branching morphogenesis. In some systems, such as neural extensions, a proper understanding of mechanism does seem to be in reach thanks to recent advances in areas of cytoskeletal and membrane cell biology. In most multicellular systems, methods of culture, assay and intervention have only just reached the stage at which it is reasonable to study the dynamics of the cytoskeleton, of the extracellular matrix, of cell division and of locomotion in living tissues as they branch. Biophysical approaches, such as those described in Chapter 12, offer a valuable complement to molecular genetics and illustrate how, given appropriate physical parameters, branching may be the 'natural' way of a fluid tissue to grow. Certainly, these ideas offer a new perspective on studies of matrix biology and stress the importance of physical aspects of matrix components as well as their importance as ligands for receptors involved in cell signalling. What would be very useful, now, is the use of these models to generate testable hypotheses about the effects of specific experimental modification to the matrix; performing such modifications will be a valuable indication of the validity of the biophysical models.

Constructing Models

The two chapters in this book that make substantial use of mathematical modelling make a refreshing change from many traditional attempts at modelling biology, in that their authors are very concerned with biological realism and are fully acquainted with the results of recent cell biological research. Indeed, most of their authors have contributed directly to this research as well as to modelling. The chapters address very different systems, and also different aspects of biology, the model in Chapter 4 being concerned mainly with modelling a developmental programme controlled by diffusible factors, and Chapter 12 being concerned more with purely physical forces operating on a tissue. They both illustrate, however, the feasibility of constructing a model based on cell-biological/ biochemical observations and of using such a model to generate hypotheses. The interplay between modelling and experiment promises to be one of the most powerful processes we have to increase our understanding of branching.

How Conserved Is Branching?

The key idea behind this book is that different branching systems have enough in common that researchers working on branching in one system can learn from those working in another. It is clear that, within a single animal where conservation would be expected to be at its highest, there is good evidence that branching systems do have components in common. In epithelial tubules, for example those of the mammary, prostate, salivary and uterine glands (Chapters 7, 9-11), the main activator of branching is usually FGF7 or FGF10. Some epithelia, such as those of the kidney (Chapter 8) use a different primary morphogen, but one that probably acts through the same intracellular pathways. These organs also share inhibitors of branching, notably TGF β and BMPs. Endothelial branching within animals (Chapter 6) uses similar intracellular pathways, although they are driven by different external activators. Endothelia also show intussusceptive branching as well as 'normal' arborization, and it is not yet clear how similar is the regulation of the two forms.

At the other extreme, it would probably be surprising if the cell biology of branching fungi, with their different habitat and walled cells, were very similar to that of animal epithelia. Nevertheless, the use of diffusible molecules to space out and direct branch points is highly relevant and may follow very similar dynamics. If so, what is learned from branching of simple organisms such as fungi and hydroids can be of tremendous use to those studying the complex anatomies of vertebrates and insects as well.

It is still too early to answer whether branching morphogenesis uses general mechanisms across phyla and tissues, but it is no longer too early to ask. Those of us who have been involved have gained much inspiration from writing this book—we hope that you may also have gained some from reading it.

Index